Robert Walters • Leon Tr

Data Protection Law

A Comparative Analysis of Asia-Pacific and European Approaches

European Union
Singapore
Australia
India
Indonesia
Malaysia
Thailand
Japan

Robert Walters
Victoria University
Melbourne, VIC, Australia

Bruno Zeller
University of Western Australia
Crawley, WA, Australia

Leon Trakman
Faculty of Law
University of New South Wales
Kensington, NSW, Australia

ISBN 978-981-13-8112-6 ISBN 978-981-13-8110-2 (eBook)
https://doi.org/10.1007/978-981-13-8110-2

© Springer Nature Singapore Pte Ltd. 2019
This work is subject to copyright. All rights are reserved by the Publisher, whether the whole or part of the material is concerned, specifically the rights of translation, reprinting, reuse of illustrations, recitation, broadcasting, reproduction on microfilms or in any other physical way, and transmission or information storage and retrieval, electronic adaptation, computer software, or by similar or dissimilar methodology now known or hereafter developed.
The use of general descriptive names, registered names, trademarks, service marks, etc. in this publication does not imply, even in the absence of a specific statement, that such names are exempt from the relevant protective laws and regulations and therefore free for general use.
The publisher, the authors, and the editors are safe to assume that the advice and information in this book are believed to be true and accurate at the date of publication. Neither the publisher nor the authors or the editors give a warranty, express or implied, with respect to the material contained herein or for any errors or omissions that may have been made. The publisher remains neutral with regard to jurisdictional claims in published maps and institutional affiliations.

This Springer imprint is published by the registered company Springer Nature Singapore Pte Ltd.
The registered company address is: 152 Beach Road, #21-01/04 Gateway East, Singapore 189721, Singapore

Foreword

This book provides a comparison and practical guide for academics, students and business communities grappling with the current day data protection laws across the Asia Pacific (Australia, Singapore, India, Indonesia, Japan, Malaysia and Thailand) and the European Union. During the course of researching this book, the original proponent of the World Wide Web, Sir Tim Berners-Lee, published a letter raising serious concerns over the development and use of the Internet today.[1] Below is an excerpt from his letter of 12 March 2017.

Today marks 28 years since I submitted my original proposal for the World Wide Web. I imagined the web as an open platform that would allow everyone, everywhere to share information, access opportunities and collaborate across geographic and cultural boundaries. In many ways, the web has lived up to this vision, though it has been a recurring battle to keep it open. But over the past 12 months, I've become increasingly worried about new trends, which I believe we must tackle in order for the web to fulfill its true potential as a tool which serves all of humanity. "We've lost control of our personal data".

The current business model for many websites offers free content in exchange for personal data. Many of us agree to this – albeit often by accepting long and confusing terms and conditions in documents – but fundamentally we do not mind some information being collected in exchange for free services. But, we're missing a trick. As our data is then held in proprietary silos, out of sight to us, we lose out on the benefits we could realise if we had direct control over this data and chose when and with whom to share it. What's more, we often do not have any way of feeding back to companies what data we'd rather *not* share – especially with third parties. This widespread data collection by companies also has other impacts. Through collaboration with – or coercion of – companies are also increasingly watching our every move online, and passing extreme laws that trample on our rights to privacy. It creates a chilling effect on free speech and stops the web from being used as a space to explore important topics, like sensitive health issues, sexuality or religion.

[1] World Wide Web Foundation, https://webfoundation.org/2017/03/web-turns-28-letter, accessed 17 December 2017

This is a complex problem, and the solutions will not be simple. We must work together with web companies to strike a balance that puts a fair level of data control back in the hands of people, including the development of new technology like personal "data pods" if needed and exploring alternative revenue models like subscriptions and micropayments. We must push back against misinformation by encouraging gatekeepers to continue their efforts to combat the problem, while avoiding the creation of any central bodies to decide what is "true" or not. We need more algorithmic transparency to understand how important decisions that affect our lives are being made, and perhaps a set of common principles to be followed.

Sir Tim Berners-Lee's letter cannot be underestimated. The point about transparency and a common set of principles has to function across the entire data chain. Doing so creates and establishes a level of trust and certainty in technology and a law protecting people's personal data. That data chain needs to be framed in light of a definition of personal data that is both legally defensible and economically sustainable. That data chain also needs to encompass such diverse elements as the collection, consent, use, analysis, disclosure, retention and limitation of data, along with the backend systems that collect and store the data and to acknowledge the importance of government and industry regulation of data. It needs to go along with definitions of personal data, including inevitable tensions between them. These issues are no different to those faced by any other industry. Transparency is about knowing the unknown, because technology is no different to polluting the ocean, and, therefore, providing a level of trust in the systems. What is under the ocean is out of sight and thus out of mind. The authors, in writing this book, seek to stimulate the reader and respond to Sir Tim Berners-Lee's concerns. There are fundamental questions that need to be considered. What is possibly the solution to promote greater legal convergence and harmonisation in data protection law and policy? Does the current co-regulatory approach adopted by governments and industry work effectively? Will it safeguard personal data into the future? What is the best regulatory model? Is there yet a different way forward? This book will examine competing approaches towards data protection, based on the three models that have been identified. The models identified in this book include the European, Singaporean and Australian approaches to data protection and privacy over the Internet. This book calls for a different approach to redress their deficiencies and build on their strengths through an international model. It also highlights other areas of the law where personal data is being considered such as intellectual property, anti-trust, transnational contracts and cybersecurity.

Preface

The free flow of personal data as a tradable commodity is becoming an important part of the global economy. Personal data is transcending human rights, antitrust law, intellectual property, transnational contracts, cybercrime and criminology, among many other areas of law.

Largely emerging out of developments in the European Union (EU), data protection law is considered as new area of law in the contemporary digital economy. With the recent implementation of the EU's General Data Protection Regulation (GDPR), the EU is arguably influencing the development of data protection and privacy law across the world. Therefore, this book is timely, coming 1 year after the implementation of the GDPR.

The book is the first of its kind, comparing the data protection laws of the EU to the divergent laws across Asia and the Pacific. The time is also right for the governments around the world to consider how to formulate policies and to develop laws that embrace the new digital economy, while protecting individuals from the potential harmful use of their personal data. The national and regional responses to this data revolution have been diffused. Some jurisdictions, such as the EU and Australia, have had data protection and privacy laws in place since the 1980s. For other countries in Central, Southeast and East Asia, these laws are a recent phenomenon. For example, Singapore, Malaysia and Japan have all established data protection laws that, while differing from one another, are no more than a decade old. India and Indonesia, in turn, have adopted a sectorial approach and are in the process of developing specific data protection laws.

Despite the attempts by some jurisdictions and international organisations to establish a baseline of concepts and principles that can be found in most data protection laws, the approach to data protection remains fragmented and inconsistent in both law and policy. The challenges for the government are not easily addressed as technology changes at a rapid rate and legal systems are often slow to respond. This slow legal and policy response is epitomised in the perpetuation of a general focus on regulating data use, such as regulating data controllers and processors. However, there is little to no government regulation of the actual Internet systems, platforms, servers and infrastructure. With the expansion of big data, practice-based data analytics, blockchain and development of quantum technology, the collection of personal data will only increase. At issue, the security framework to protect this data is far from fully understood. Moreover, the regulatory and policy framework underpinning the management of data is fragmented, requiring greater vigilance by nation states to ensure the appropriate policy and legal response are developed for the future. Thus, there is the potential for criminal activity arising from the misuse of personal data, which is likely to result in privacy breaches increasing.

Equally important is the need for more pervasive international responses to improve data protection law(s), including the need to redress tensions across regional and national institutions, and bodies responsible for regulating the use of personal data. With the internationalisation of the Internet and technology, there is also a growing need for both resources and information to fill the gaps in the legal and policy responses to these deficiencies. Complicating these international aspirations is the absence of an agreed best model or combination of models that adequately provide a balance between the many competing and conflicting areas of law and policy pertaining to data. In other words, as people become more aware of how organisations are using their personal data for monetary or some other gains, they may call for more regulation or less regulation. At this point in time, the current models that have emerged see the EU version taking a greater focus on human rights, while Singapore has implemented a business-friendly model. Australia, on the other hand, sits somewhere between the two. Therefore, this book calls for greater legal and policy convergence and harmonisation, at an international level, in data protection law and in the many other areas of law that pervades. It must be noted that it is out of scope if this book to examine other possible models for data protection, such as, North America, China or the Middle East.

Thanks to the Springer publisher team.

Melbourne, VIC, Australia Robert Walters
Kensington, NSW, Australia Leon Trakman
Crawley, WA, Australia Bruno Zeller

Acknowledgements

A special acknowledgement and great thanks go to Associate Professor Dr. Sinta Dewi Rosadi, Cyber Law Center, Faculty of Law, Padjadjaran University, Indonesia, who has coauthored journals with the authors of this book, for providing a specialist advice on Indonesian data protection law. The work continues today in this complex but important area of law.

The authors would also like to thank Professor Marc Bungenberg, Director of the Europa-Institut Saarbrücken, European Law and Public International Law at Saarland University in Germany, for his advice on technical matters within the GDPR.

Further acknowledgement and thanks go out to Dr. A. Nagarathna, Associate Professor and Coordinator, Advanced Centre on Research, Development and Training in Cyber Law and Forensics, National Law School of India University, for providing valuable comments on the draft Indian chapter and input into the Indian component in relation to cybercrime/security.

Thanks to Dr. Sonny Zulhuda, Assistant Professor, International Islamic University, Malaysia, for his review of the Malaysia chapter. Without his wise counsel, the Malaysian chapter would not have been completed.

A special thanks also goes out to Dr. Jompon Pitaksantayothin (PhD), Lecturer, Department of Society and Health, Faculty of Social Sciences and Humanities, Mahidol University, Thailand, for providing valuable input and assistance in relation to the current status of data protection law in Thailand.

The authors would also like to acknowledge the work of Professor Dr. Graham Greenleaf and Professor Dr. Simon Chesterman in the area of data protection and privacy throughout Asia and Australia.

The book would not have come to publish without the assistance of many individuals and organisations from the various countries.

Chia Swee Yik – Chia, Lee & Associates, Advocates & Solicitors, Kuala Lumpur, Malaysia
Dr. Nilubol Lertnuwat, Assistant Professor of Law, Thammasat University, Bangkok, Thailand
Amber Sinha, Senior Programme Manager, The Centre for Internet and Society and Practicing Lawyer, India
Linklaters – DLA Piper – International Comparative Legal Guide

Contents

About the Authors

Robert Walters Dr. Robert Walters is a Lecturer of Law, Victoria University, Melbourne, Australia; is an Adjunct Professor with the European Faculty of Law, The New University, Slovenia, Europe. Dr. Walters is a Member of the ASEAN Law Association, Singapore, and of the United Nations Commission on International Trade Law Coordination Committee for Australia.

A very special thanks and enormous gratitude go to the coauthors of this book, Professor Dr. Bruno Zeller and Professor Dr. Leon Trakman, for allowing Dr. Walters the fortunate opportunity to work with them in this complex but, ever increasingly, important area of law and public policy. Without their combined wise counsel, this book would not have been completed.

This book is a dedication to his late father, Mr. Gordon Walters, and mother, Pam Walters. Special thanks to his wife, Catherine Tan-Walters, who provided him the space and support to complete this work.

Since completing his PhD, Dr. Walters legal research areas now include data protection, artificial intelligence, technology, cybercrime/security law, and international trade, finance and investment law with a particular focus on Australia, Asia Pacific (APEC and ASEAN countries), including Europe. Dr. Walters is admitted as an Australian Lawyer.

Leon Trakman Professor Trakman is former Dean (2002–2007) and currently Professor of Law, Faculty of Law, University of New South Wales, and Director, Masters in Dispute Resolution. He is also Academic Disability Advisor at the UNSW. His appointments include the following: Distinguished Visiting Professor, University of California, Davis (1999–2000); Professor of Law, Schulich School of Law at Dalhousie University (1975–1999); Visiting Professor, University of Wisconsin Law School (1992–1993); Visiting Professor of Law, University of Cape Town (1990); Bora Laskin National Fellow in Human Rights, Canada (1997–1998); Killam Professor, Killam Foundation (1986); Visiting Professor, Tulane University School of Law (1983), and Bolton Visiting Professor, McGill University Faculty of Law (1982).

Professor Trakman specialises, inter alia, in contracts, international commercial arbitration, trade and investment law. He is an author of 8 books and over 100 articles in international recognised journals in his areas of specialty and has been Lead Chief Investigator on a Discovery Grant from the Australian Research Council (2014–2018). In addition, he has received significant fellowships including a Harvard Doctoral Fellowship, a Bora Laskin National Fellowship (one awarded annually across all disciplines in Canada), a Killam Senior Fellowship and various grants from the Canada Council and the Social Sciences and Humanities Research Council of Canada.

Trained as an International Commercial Arbitrator and Mediator, he has served as Presiding Arbitrator or Arbitrator in more than 100 international disputes and as mediated in over 30 disputes. These have included disputes in the fields of contracts, sales, construction, IP, sales, franchise, insurance law and executive remuneration, among others. He also chairs or has chaired and serves on various panels, boards and associations devoted to arbitration and mediation on four continents.

Bruno Zeller Professor Zeller, who has finished his PhD, BCom and BEd at the University of Melbourne and Master of International Trade Law at Deakin University, is a Professor of Transnational Commercial Law at the University of Western Australia; an Adjunct Professor at Murdoch University, Perth; a Fellow of the Australian Institute for Commercial Arbitration, Panel of Arbitrators, MLAANZ; and a Visiting Professor of the Institut fur Anwaltsrecht, Humboldt University, Berlin, and Stetson University College of Law, Florida. His areas of expertise are international trade law, international arbitration, conflict of laws and maritime law. He has published extensively on the CISG, arbitration law, harmonisation of contract law and carbon trading.

List of Diagram

Part I

Chapter 1
Problem Definition, Structure and Methodology

Abstract This Chapter begins by outlining the problem in defining and understanding the interrelationship between privacy and data protection law in Australia, India, Indonesia, Japan, Malaysia, Singapore, Thailand and the European Union. This Chapter will demonstrate and discuss how the concept of privacy is considered an important feature of the modern era. In other words, it is argued that there has been wide acceptance and a convergence of privacy that now transcends, government, countries, cultures religion over the Internet. This convergence of the concept of privacy, has resulted in nation states adopting to varying degrees, data protection and privacy laws. However, it will be highlighted that the current day approach needs further development and greater convergence and harmonization of data protection law and policy at the international level. This will be important as the trade in personal data continues to grow.

It will be argued in this Chapter that the privacy and data protection law of these jurisdictions is far from settled. It is further argued that data protection and privacy law has two dimensions. First, is to protect personal data and information of individuals, as a human right. Second, is balancing the protection of personal data with current and future economic activity (trade) of personal data. Moreover, data protection and privacy cannot be restricted to a single country or region of the world. It is international, and has been underpinned by Internet technology and infrastructure that knows no [national] borders. Thus, these laws, while being developed by nation states for their own particular sovereign needs, the internationalization of the Internet poses significant challenges to the future law and policy in this area. They are likely to continue to be challenged and require reviewing and updating, as technology continues to change. Being a recent addition to the law, data protection is also challenging and is arguably in conflict with other areas of the law, such as intellectual property, competition, transnational commercial contract law, and cybercrime-security law. This Chapter also highlights the structure of the overall book in recognizing and responding to these differences in data protection and privacy law. It argues that data protection is a tool of Internet privacy. At the recent June 2019 meeting of the G20, the leaders' declaration called for respect of national and international regulation of data and technology. The importance of this declaration highlights the importance for governments to balance innovation with protection of personal data. To achieve this, the book reinforces the G20 leaders position,

© Springer Nature Singapore Pte Ltd. 2019

R. Walters et al., *Data Protection Law*,
https://doi.org/10.1007/978-981-13-8110-2_1

and goes a step further, by recommending that an international Model Law be developed, similar to international trade law. Also, consideration to an international treaty or convention will support any model law and go some way to closing the gaps and tensions between country data protection law. Thus, this book calls for greater legal convergence and harmonization in this emerging and complex area of law and policy. The book also identifies that personal data being afforded an intellectual property right. It also highlights the tension between data protection and competition law and cybersecurity/crime law. It will also be argued that data protection can fall within the current transnational international contract legal framework. Adopting data protection within these areas of law, provide valuable tools to strengthen the governance, control and regulation of personal data.

1.1 Problem Definition

1.1.1 Privacy

Privacy and data protection mean different things to industry, governments and the general community. Unlike other areas of law that have well settled legal concepts, norms and principles, it is an area of law that currently is far from settled. More than 40 years ago, Zelman Cowan stated that a man without privacy is a man without dignity.[1] However, privacy and data protection pose a significant challenge for government, society and industry. This challenge is even more pronounced than ever in the new digital economy, with millions of people accessing the Internet daily, and not knowing whether their personal privacy is being infringed.

Privacy has been described in three ways. Firstly, privacy in making certain significant self-defining choices. Secondly, privacy of personal information; and thirdly, privacy as it relates to an individual's personal space and body.[2] Arguably, someone's privacy is compromised when others obtain information about an individual, pay attention to him or her, or, gain physical access. Privacy has therefore protected secrecy, anonymity and solitude.[3] Simon Chesterman suggests that this definition may be too broad, because it would include rights not to be punished. The other element to privacy is through the principal of dignity, which has been expressed as a fundamental human right by the European Union 2000 Charter of Fundamental Rights. However, if privacy is to be solely viewed as a right, the tension with other societal interests such as national security and the economy, arguably dilutes privacy to some degree.[4] That tension becomes even more evident between

[1] Cowan Z *The Private Man* 24 Inst Pub Affairs Rev 26 (1970).

[2] Kang J (1998) *Information Privacy in Cyberspace Transactions*, Stanford Law Review, 1998, pp. 1201–04.

[3] Chesterman S (2012) *After Privacy: The Rise of Facebook, the Fall of WikiLeaks, and Singapore's Personal Data Protection Act 2012*, Singapore Journal of Legal Studies, p 396.

[4] Ibid.

commercial activity and human rights, more generally. This is because personal data defined by the law has become a tradable commodity. Nonetheless, Chesterman believes that the sphere in which privacy relating to personal data can be insulated is the physical confines of one's home, with temporal limits determined by the moment at which one's telecommunication devices are switched off or out of range.[5] However, the ability to insulate one's privacy online is questionable, because most people have very little idea as to the footprint they have personally created when surfing the Internet. Thus, technology focusses on information privacy. It is difficult to define when it relates to personal data, given that data can come in many different forms.

Privacy can also mean different thing to different people. Richard Clarke, [6] believes that privacy can be conceived philosophically,[7] psychologically,[8] sociologically[9] and economically.[10] Clarke goes on to categorize privacy as being interpreted broadly by the individual as a form of personal behavior,[11] personal communication,[12] and personal data, which is often referred to as data privacy and information privacy.

The problem with the contemporary theory of privacy is that it has many deficiencies. Its focus on information means that it excludes many areas widely held to be basic to privacy. These include, but not limited to, the ability to make fundamental decisions about one's personal and family life; insofar as it suggests that personal control is limited to the individual who is the subject of that information.[13] Nevertheless, as a framework through which to view what is loosely termed privacy, the focus on information accurately highlights the overlapping but discrete subject of data protection.[14] Simon Chesterman notes that throughout Asia, in particular, many jurisdictions now embrace data protection laws even in the absence of any formal protection of a more abstract right to privacy. The theory of privacy is also set to evolve and change as technology changes. From its beginning, the scope of

---

[5] Ibid.

[6] Clarke R *What's 'Privacy'?,* Workshop at the Australian Law Reform Commission on 28 July (2006) http://www.cse.unsw.edu.au/~cs4920/resources/Roger-Clarke-Privacy.pdf, accessed 19 April 2018.

[7] Ibid, in Europe, people are regarded as being very important for their own sake. The concepts of 'human dignity' and integrity play a significant role in some countries, as do the notions of individual autonomy and self determination and human rights.

[8] Ibid, people need private space. This applies in public as well as behind closed doors and drawn curtains.

[9] Ibid, people need to be free to behave, and to associate with others, subject to broad social mores, but without the continual threat of being observed.

[10] Ibid, people need to be free to innovate. International competition is fierce, and countries with high labour-costs need to be clever if they want to sustain their standard-of-living.

[11] Ibid, referred to as 'bodily privacy', is concerned with the integrity of the individual's body.

[12] Ibid, including what is sometimes referred to as interception privacy.

[13] Chesterman S (2012) *After Privacy: The Rise of Facebook, the Fall of WikiLeaks, and Singapore's Personal Data Protection Act 2012*, Singapore Journal of Legal Studies, 2012, p. 396.

[14] Ibid.

privacy has evolved to include most elements of modern technology. Thus, privacy can be best described as being culturally sensitive. It can also be described as being culturally biased and is based on Western thought and the Western Liberal Tradition. According to the Western Liberal Legal Tradition, privacy, while considered a principle of data protection, is a fundamental right within itself, and is associated with protecting a person's identity.[15] Privacy has evolved from its traditional notion of the right to be left alone. As a concept it engages the protection of human rights buttressed against the promotion of economic development. In early times, the law gave a remedy only for physical interference with life and property, which has been well understood to be trespass.[16] However, as countries respond to data protection differently the very concept of a privacy right has not been fully accepted, over the Internet. The concept of privacy has meant something different to states outside the Western Liberal Tradition to their counterparts in Western Democratic states.

The challenge is to evaluate how different cultures and legal traditions within the selected nation states influence and regulate these conceptions of privacy. How these concepts are regulated is fundamentally important in the modern world. Indeed, it determines whether a country or region studied in this book is seen as having a competitive disadvantage when compared to other countries or regions that enjoy cultural and legal traditions that are differently attuned to, not only to the development of technology and the Internet, but also in regulating them. The competitive advantage is not only economical, but also operates at a personal level in protecting a person's human rights. One example of the differences in privacy, can in part be summarized when comparing Western thought with other religions in Central Asia, such as Buddhism. Charles Ess states:

> In those countries such as Japan and Thailand where Buddhism plays a central role in shaping cultural values and identity, the Buddhist emphasis on "no-self" (Musi in Japanese) directly undermines Western emphases on the autonomous individual as the most important reality (at least since Descartes), the source of morality (in Kant), the foundation of democratic polity, and in all these ways the anchor of Western emphases on individual privacy. As Buddhism stresses instead the importance of overcoming the ego as the primary illusion at the root of our discontent—it thus provides a philosophical and religious justification for doing away with "privacy" altogether, as in the example of Japanese Pure Land Buddhism (Jodo-shinsyu), which inspires some authors to move towards salvation by voluntarily betraying private, even shameful personal thoughts.[17]

Buddhism is also a major religion, along with Hinduism throughout India. The understanding of privacy in India dates back to 1960s case of *Kharak Singh v. State of UP*[18] where the court noted that privacy was not a fundamental right laid out in the constitution. However, privacy in the law of India is nevertheless a central part of the right to personal liberty, especially as it concerns privacy against arbitrary

[15] Hildebrandt M (2006) *Privacy and Identity*. In: Claes E, Duff A and Gutwirth S (eds) Privacy and the Criminal Law, Oxford, United Kingdom Hart, 2006, pp. 43–60.

[16] Samuel D, Brandeis L (1890) *The Right to Privacy*, Harvard Law Review, Vol. 4, No. 5, pp. 193–220.

[17] Ess C (2005b) *Lost in translation*. Ethics Inf Technol 7, 1 2005b, pp. 1–6.

[18] *Kharak Singh v. State of UP*, (1964) 1 SCR 332, 359 (India).

intrusion. Nonetheless, India's perception of 'privacy' as a concept has been traditionally viewed as subjective in terms of personal space and depends on one's culture, environment and economic condition.[19] It is not about the economic value of that information.[20] However, that view is slowly changing in India. In December 2017, the Indian Government released a White Paper to study various issues relating to data protection. The white paper makes specific suggestions on principles underlying data protection and privacy. India is particularly concerned about the growth in the digital economy, and the need to balance the protection of its citizen's personal data.[21] The committee of experts had not conclusively reviewed community submissions at the time of writing this book. Nonetheless, the development justifies maintaining a watching brief, particularly to see how the committee reacts to those submissions and whether it adopts community concerns selectively.

Viewed broadly, privacy can include protecting all forms of personal communications; the personal body (biometrics and medical)[22]; personal data and personal information (name and address; personal possession such as property).[23] However, the definition of privacy within national and supranational law is rarely defined. It is rather the information that constitutes personal data that is defined. Moreover, the conception of privacy in Islam is worth highlighting. Both Indonesia and Malaysia are predominantly Islamic countries, and privacy has been viewed by many Muslim scholars as a fundamental human right.[24] Privacy stems from the *Maqasid al Shariah,* from which personal rights *(haqq)* are derived. According to the Maqasid, all individual rights are God-given and by their nature not absolute.[25] Even so, there are some exceptions such as witnesses are allowed to give testimony for purposes of law enforcement and the imposition of punishment, even if this means intruding upon another's privacy.[26] In the exercise of such rights, the state is guided by two main functions: *al amr,* or the promotion of certain positive conduct, and *al nahy,* or the prohibition of negative conduct.[27] The establishment of rules and institutions such as the institution of *hisbah* serve as machinery to promote positive conduct. Essential to the prohibition of negative conduct is the creation of a list of offences such as outraging modesty, spying, *'ghibah'* (revealing embarrassing details about

---

[19] Basu S (2010) *Policy-Making, Technology, and Privacy in India*, INDIAN J.L. & TECH. vol 6. p. 66.

[20] Ibid.

[21] White Paper of the Committee of Experts on a Data Protection Framework for India, December 2017.

[22] Westin A (1967) *Privacy and Freedom.* New York: Atheneum, p. 351.

[23] Neethling J, Potgieter M, Visser J (1996) *Neethling's law of personality*, Butterworths. pp. 35–36.

[24] Kamali, H. (2007) *The Right to Life, Security, Privacy and Ownership in Islam* (Cambridge, Islamic Texts Society); Mahmood, T. (ed.) (1993) *Human Rights in Islamic Law* (New Delhi, Institute of Objective Studies).

[25] Madieha Azmi I, *Personal Data Protection Law: The Malaysian Experience*, 16 Info. & Comm. Tech. L. 125 (2007) pp. 130.

[26] Ibid.

[27] Ibid.

others), disclosing matrimonial secrecy, defamation and trespass to property.[28] Therefore, the right of privacy comes in two normative frameworks: the prohibition of intrusion into other's privacy, and instructions and guidance for keeping secrets.[29] Personal privacy (a person is free to conduct their own affairs without interference from outsiders) is guaranteed in the Qur'an in *Surah al Taubah: 105, Surah Fussilat:* 40 and *Surah Saba: 11.* All conduct of a person deserves the highest respect in terms of privacy and secrecy. Any attempt to collect information on their activities would amount to spying ('tajassus'), a conduct forbidden in Islam.[30] Furthermore, Berween argues that privacy is both a very important societal and legal concept to Islam. For Berween, privacy constitutes one of the most precious freedoms, most comprehensive and respected of rights. In Islam, privacy and good manners in public contribute to the highest virtues, and are parts of a Muslim's duty. The right to privacy in Islam includes:

1. the right for every individual to be left alone in their private life;
2. the right to be free from governmental surveillance and intrusion;
3. the right not to have an individual's private affairs made public without their permission;
4. the protection of persons, and places where they live from searches and seizures;
5. the protection of knowledge and thoughts from compulsory self-incrimination; and
6. the right to keep all personal information confidential.[31]

More recently, there has been a hybridization of privacy in which the elements of the West and the East have converged.[32] That convergence has not been by accident. Rather, there has been deliberate attempts at convergence in response to globalization, regionalization, the movement of goods, and services and people in which legal frameworks, concepts, principles and norms require more cohesion. Nonetheless, privacy constitutes a human right, or at least the appropriate use of someone's personal data and information.[33] The level of privacy afforded to an individual is still determined by the national laws of each country or regional entities such as the European Court of Human Rights and the Court of Justice of the European Union. However, the nature if privacy remains divergent across these institutions.[34]

---

[28] Ibid.

[29] Ibid.

[30] Ibid.

[31] Berween M *"The Fundamental Human Rights: An Islamic Perspective"* The International Journal of Human Rights 61 (2002) pp. 70–74.

[32] Ess C (2005b) *Lost in translation*. Ethics Inf Technol 7, 12005b, pp. 1–6.

[33] International Association of Privacy Professionals, https://iapp.org, accessed 20 December 2017. Article 7 and 8 Charter of Fundamental Rights of the European Union C 326/12.

[34] Case –28/08 P *Commission/Bavarian Lager* [2010] ECR I–6055, para. 60.

The ways in which people create, safeguard and enhance their privacy, and the extent to which they exhibit a desire for privacy, vary from culture to culture according to a complex array of factors.[35] In societies which provide little opportunity for physical or spatial solitude, human beings seem to adopt various strategies for cultivating other forms of social distance.[36] Barrington believes that the need for privacy is socially created and is complex, with a strongly felt division between the private realm and public sphere. Privacy, to date, is minimal where technology and social organization is concerned.[37] It has resulted in what can be best described in some regions of the world as a patchwork of law and policy to address privacy issues – on the run. This is certainly true in the contemporary world, where people who operate in the technology sphere are unaware of whether their privacy is being intruded upon. Furthermore, the protection of privacy has a tendency to manifest itself, after the fact, namely, when it is too late to protect privacy that has been violated. It is practically impossible to predict the consequences or the level of harm arising from a violation of a privacy right, notably the misuse of personal data, especially in today's information society.[38] Meg Leta Ambrose and Jef Ausloos argue that privacy is abstract because harms are often concerned with societal and psychological issues. They are distant because many of the consequences will only reveal themselves after a series of reactions to their practical application in specific cases. The impact of privacy breaches is also uncertain because such breaches might never occur, or at least, not occur in a foreseeable way, due to the lack of understanding of the platforms and infrastructure used to capture personal data.[39]

1.1.2 *The Modern History of the Right to Privacy*

The right to privacy itself is not new. In 1890 Samuel Warren and Louis Brandeis wrote an essay titled, "The Right to Privacy," which was published in Harvard Law Review.[40] They proposed recognition of an individual's "right to be let alone" and argued that this right should be protected by existing law, as a matter of human rights. The right has (in Western Liberal Tradition) arisen significantly from the relationship between the individual, society and the nation state. This liberal thought is something that Hobbes and Locke described as protecting rights derived from the

[35] Altman I (1977) *Privacy Regulation: Culturally Universal or Culturally Specific?,* Journal of Social Issues, vol. 33, pp. 66–84.

[36] Barrington M (1987) *Privacy: Studies in Social and Cultural History*, American Journal of Sociology, vol 92, 1987.

[37] Ibid, p. 276.

[38] Ambrose M, Ausloos J *The Right to Be Forgotten Across the Pond,* Journal of Information Policy, Vol. 3 (2013), pp. 1–23.

[39] Ibid.

[40] Samuel D, Brandeis L (1890) *The Right to Privacy*, Harvard Law Review, Vol. 4, No. 5, pp. 193–220.

'state of nature' of mankind, and as forming the basis of the ideals of freedom and liberalization that underpinned the French Revolution.[41] However, scholars have attempted to provide a solid theoretical foundation to the right to privacy. De Boni and Prigmore argue for the protection of a right to privacy from an idealistic, neo-Hegelian philosophy point of view. They see privacy, not as a "human right", but as the logical consequence of the Hegelian idea of free will.[42] This thought is based on traditional Anglo-Saxon philosophy and does not consider the wider world. In particular, it does not consider traditional Central or South East Asian thought.

The right to privacy began to take hold following WWII, when the United Nations had to consider privacy in the context of ideological differences along North–South and East–West ideological lines.[43] Countries in the South-East camp, including mostly developing countries, emphasized the socio-economic benefits of scientific and technological discoveries. In contrast, those in the North-West bloc—mainly industrial nations—argued that priority be given to the negative impact that these technological discoveries may have on human rights, particularly the right to privacy. These divisions could be seen partly as reverberations from a broader debate on the generation of human rights. The resulting effect saw states divided between those that argued for prioritizing civil and political rights, and those that focused on socio-economic and cultural rights. Richard Clayton and Hugh Tomlinson categorize privacy as the:

1. Misuse of personal information. A right to restrict the use of "personal" or "private" information about an individual is central to the right to privacy.
2. Intrusion into the home. The right of the individual to respect for the home is fundamental to any notion of privacy. Unreasonable searches and seizures trigger privacy issues.
3. Photography, surveillance and telephone tapping. The "private sphere" is invaded not only by physical intrusion into the home.
4. Other privacy rights. There is a range of other privacy rights which covers all forms of interference in the "private sphere" including appropriation of a person's image, interference with private sexual behaviour and questions of the sexual identity of transsexuals.[44]

The position taken by Clayton and Tomlinson arguably captures the various cultural, religious and legal thought of privacy in the modern period. Nonetheless, according to Daniel Solove privacy" as a legal concept is challenging at best, and

---

[41] Hobbes T (1981) *Leviathan*, C. B. Macpherson (Editor), Penguin. Locke J (1986) Second Treatise of Government, Prometheus.

[42] De Boni M, Prigmore M (2001) *A Hegelian basis for information privacy as an economic right*, in Roberts M, Moulton M, Hand S, Adams C. (eds) Information systems in the digital world, Proceedings of the 6th UKAIS conference, Manchester, UK, Zeus Pres.

[43] Micheal Yilma K, *The United Nations data privacy system and its limits*, International Review of Law, Computers & Technology (2018).

[44] Clayton R., Tomlinson H *The Law of Human Rights* (2 Ed, Oxford University Press, 2009) [The Law of Human Rights] at 1005.

appears to be about everything, and yet it also appears to be about nothing.[45] On the other side, James Whitman is of the view that our conception of privacy reflects our knowledge of, and commitment to, the basic legal values of our culture and values.[46] Another dichotomy that has arisen in the West, is the view that privacy constitutes an important element of liberty.[47] For instance, the right to freedom from intrusions by the state.[48] Moreover, traditionally, many nation states privacy laws from Western societies have tended to focus on the freedom to control access to one's private life. It is an approach by which one can consent to the loss of privacy. A good example is the entry to one's private property by another, without consent. Importantly, the concept of consent in privacy has been long established, and has also emerged in the modern technological (the Internet) period to also provide a level of control over one's personal data. The resulting effect is that there is a level of privacy protection over the Internet. Yet, the European concept of privacy, by comparison, views it as an aspect of dignity.[49] This subtle difference arguably has little baring on privacy in the context of data protection and the Internet. This has been reinforced by Whitman who argues that privacy today is closely aligned with data protection. However, it is our view that data protection is merely a tool that goes some way to protect privacy over the Internet.

The conceptualization of privacy, along with data protection, is also not subject to a consistent method of expression, or to a particular thought or emotion about the ambit of privacy regulation. This inconsistency arose because the right to privacy could relate to a host of factors, such as signs, paintings, sculpture, music, newspapers,[50] and in the contemporary world, the use of the mobile phone and

---

[45] Solove, D *A Taxonomy Of Privacy,* University of Pennsylvania Law Review, (2006), pp. 477–564.

[46] Whitman, J *The Two Western Cultures of Privacy: Dignity versus Liberty.* Yale Law School (2004) p. 1160.

[47] Ibid. The word 'liberalism' has been used, since the eighteenth century, to describe various distinct clusters of political positions, but with no important similarity of principle among the different clusters called liberal at different times. The roots of liberalism rest in the classical interpretation, that there ought to exist a certain minimum area of personal freedom, which must never be violated. Liberty in this sense is the condition in which an individual has immunity from the arbitrary exercise of authority. It presupposes some frontiers of freedom that nobody should be permitted to cross, and requires the minimum, and demanded a maximum degree of noninterference compatible with the maximum demands of social life.

[48] Ibid.

[49] Ibid. Human dignity plays a role both at the international and state levels. On the international level, the Universal Declaration of Human Rights opens with the statement that "recognition of the inherent dignity and of the equal and inalienable rights of all members of the human family is the foundation of freedom, justice and peace in the world." Other prominent international documents and covenants rely upon human dignity as a leading value. The concept of human dignity also plays a significant role in the debate over the 'universalism' or 'relativism' of human rights. In the contemporary human rights discourse within the international arena, human dignity is highly visible. At the national level, human dignity became a central concept in many modern constitutions. The concept of human dignity now plays a central role in the law of human rights, there is surprisingly little agreement on what the concept actually means.

[50] Ibid.

Internet. Historically, privacy was also used narrowly, such as to identify a person by national identification card, passport, driver's license, birth certificate or bank account. Further accentuating consistent conceptions of privacy is diverse over its ambit in international, European and the laws of nation states. Privacy is also variously conceived as a human right in which the protection of personal data is treated as an important human and economic right.[51]

Privacy, until recently, was largely abstract and not readily applied in relation to technology. However, the scope of a person's privacy is increasingly identified in light of, and in response to, modern day technology. The result is due to an increase in the scope of privacy and the desire to protect a person's human and economic interests from technological invasion. The conception of private in technologically developed countries is further characterised by the exclusion of a person from being subject to undue and invasive publicity.[52] That right to privacy is also closely associated with that person's other rights and freedoms, such as dignity, personal autonomy, and freedom of expression. However, the right to privacy of the individual is not absolute; and governments and organizations can and have the legal power to invade a person's privacy, such as to investigate criminal activity.[53]

This book focusses on privacy as it relates to technology and data (both personal and commercial in nature), rather than with these extended human rights. In other words, data protection law not only protects a person's personal data over the Internet, it also protects a level of the person's privacy over the Internet. It also recognizes that technology, today, more than ever before, allows not only government, but also private organizations to gather private and personal data from its citizens—whether that data is defined by the law or otherwise.

Importantly, while privacy and data protection are distinguished in law, they are commonly used interchangeably to promote comparable legal results. As a result, most scholars conceive of these concepts in light of particular ideological and functional questions, and less according to the distinction between these kinds of rights. These questions include: whether citizens and residents of a nation state receive equal protection of their individual privacy; and how much privacy should be afforded to citizens. They identify the constant challenge that governments face in arriving at viable answers to these questions. This challenge is accentuated by the divergent responses of states to these questions, along both ideological and functional grounds.

This book demonstrates the nature and significance of these divergent responses in which states conceive of the protection of privacy, personal or economic data differently. It demonstrates further, that these differences reflect historical and cultural

[51] Data privacy protection across Asia –A regional perspective, Freshfields Bruckhaus Derringer LLP, October 2008, http://www.freshfields.com/publications/pdfs/2008/oct08/24238.pdf, accessed 2 October 2018.

[52] Neethling J, Potgieter M, Visser J, (1996) *Neethling's law of personality.* Durban: Butterworths. p. 36.

[53] McGarry K (1993) *The Changing Context of Information. An Introductory Analysis.* 2nd ed. London: Library Association Publishing, p. 178.

differences in how states and blocks of states view privacy as a right, and the manner in which they manage data as an economic or social tool.

A perplexing issue for nation states is to redress the tendency to treat privacy as wholly subjective, causing the subject to feel nonessential and politically irrelevant. While the subject remains free to be her or himself, that individual is increasingly unprotected in the public realm.[54] Simon Chesterman stresses that law makers have struggled to ensure that the law remains relevant in response to the changing technological context of privacy and data protection; and that law reform is largely driven by emerging threats, technological breakthroughs, and evolving cultural sensitivities.[55] Chesterman also argues that the pace of change has accelerated, with radical transformations in the way information is produced, stored and shared. This radical transformation serves as a further barrier to a robust theory of privacy. Accentuating that barrier are divergences in how nation states attempt to reconcile differences in the conception of privacy and data protection.[56]

This book demonstrates the divergent conceptions of privacy and data protection adopted in Europe[57] and Asia Pacific. It examines and attempts to harmonize norms and principles at the international level. The book demonstrates that nation states do not treat such conceptions similarly. On the contrary, they often treat data protection and more so privacy law as trade-offs in balancing different societal and economic needs. The book also highlights that the European Union, through the General Regulation on Data Protection in 2018, has set new benchmarks that are likely to significantly influence the privacy and data protection laws of other countries.

1.1.3 Data Protection as a Tool of "Privacy"

Data protection has also been characterized as a tool of 'privacy'. [58] In other words, data protection underpins privacy and constitutes the personal data used to identify a person. Identifying a person by their personal data was historically been achieved through state records, such as, a passport or birth certificate. Data protection as a tool of privacy also helps to facilitate the economic growth in the trade of personal data. This is an important point, because increasingly personal data is being used to develop another area of the economy. Data protection today is increasingly considered as the implementation of appropriate administrative, technical or physical measures that minimize the risk of or harm caused by unauthorized intentional or

[54] Tamás M (2002) *From Subjectivity to Privacy and Back Again,* Social Research, p. 220.

[55] Chesterman S (2012) *After Privacy: The Rise of Facebook, the Fall of WikiLeaks, and Singapore's Personal Data Protection Act 2012*, Singapore Journal of Legal Studies, p. 392.

[56] Ibid.

[57] Ibid.

[58] De Hert p, Gutwirth S, (2006) *Privacy, Data Protection and Law Enforcement. Opacity of the Individual and Transparency of Power*, in Claes E, Duff A, Gutwirth S, *Privacy and the Criminal Law,* Antwerp-Oxford, Intersentia, pp. 61–104.

accidental disclosure.[59] These measures are embodied in the legal and policy frameworks by which nation states protect a person's privacy, including technological systems that collect, store and use data. Regulators also increasingly recognize that the technological use of personal data represents the greatest threat to individuals, accentuating their vulnerability and underscoring the need to protect their rights to privacy.[60]

Paul De Hert and Serg Gutwirth have characterized data protection as a tool of transparency. [61] These tools have been derived from the protection of personal freedoms embodied in the constitutions and laws of democratic societies. De Hert and Gutwirth go on to say that:

> The development of the democratic constitutional state has led to the invention and elaboration of two complementary sorts of legal tools which both aim at the same end, namely the control and limitation of power. We make a distinction between on the one hand tools that tend to guarantee non-interference in individual matters or the opacity of the individual, and on the other, tools that tend to guarantee the transparency/accountability of the powerful.
>
> The tools of opacity are quite different in nature from the tools of transparency. Opacity tools embody normative choices about the limits of power; transparency tools come into play after these normative choices have been made in order still to channel the normatively accepted exercise of power. While the latter are thus directed towards the control and channeling of legitimate uses of power, the former are protecting the citizens against illegitimate and excessive uses of power.[62]

Data protection and privacy have converged primarily as a legal framework to protect people's personal data and is defined by the law and as rights. This convergence also includes finding a balance between economic development and innovation in the digital economy. Even though privacy and data protection is an evolving area of law and economic development, it has not matured as a measure of redressing economic and personal harm comparably to the protection of intellectual property, copyright, criminal procedure and international trade law. What has emerged are fledgling principles directed at regulating privacy and data protection, whether through government regulations, or by courts protecting data on a case by case basis.

In contention for regulators is the very meaning of data protection. Data protection can be used to identify privacy-related laws and regulations. Data protection can also mean the implementation of appropriate administrative, technical or physical means to guard against unauthorized, intentional or accidental disclosure, modification, or destruction of data.[63] Expressed pervasively, data protection consists of

---

[59] International Organisation for Standardisation/IEC 2382-1-1993 and its successors.

[60] Kokott J, Sobotta C, (2013) The distinction between privacy and data protection in the jurisprudence of the CJEU and the ECtHR, *International Data Privacy Law*, Oxford University Press, vol 3, Issue 4, pp. 222–228.

[61] De Hert P, Gutwirth S, (2009) *Privacy, Data Protection and Law Enforcement. Opacity of the Individual and Transparency of Power*. in E. Claes, A. Duff & S. Gutwirth (Eds.), *Privacy and the Criminal Law*, pp. 61–104.

[62] Ibid.

[63] International Organisation for Standardisation/IEC 2382-1-1993 and its successors.

the legal and policy framework established by a nation state or a supranational polity (the EU) to protect someone's privacy. This not only includes the legal framework, but also, the technological systems that collect, store and use data. It also includes data that is used for commercial purposes. However, current data protection laws do not ordinarily regulate the manufacture of these technological systems. That is, a central limitation in many data protection laws. The requirements that personal data must be processed fairly and for a specified purpose is needed to cover many instances in which an interference with privacy occurs. These specific requirements regulating the processing of data can help to focus the debate on data protection in areas that are particularly susceptible to interference with fundamental rights.[64] They can also help to extend data protection and privacy beyond the legal ambit of a single nation state.

1.1.4 Internationalization and Regionalization

Technology knows no national boundaries. This alone, poses one of the greatest challenges to future data protection and privacy law. It is the technology (computers, software systems, servers and other infrastructure) that is used within nation states that can easily transcend national and multiple international borders at any one time. It can also create opportunities to illegally harvest, mine or collect the personal data of individuals, whether accomplished overtly or covertly. Today more than ever, there is the need for global data protection and privacy law. This need is already evidenced in international practice. Data protection and privacy law has largely been directed by concepts and principles established by international, supranational, regional and national institutions, organizations, forums and associations, such as the Organization for Economic Development Cooperation (OECD), the European Union and European Commission, the Association of South East Nations (ASEAN). They also include the laws of nation states. The concepts and principles that have emerged include, but are not limited to, the legal definition of personal data and personal information, as well as consent, accountability, transparency and localization and portability in the use of such data. These concepts have developed a risk management framework to privacy and data protection. Yvonne McDermott identifies this framework with the fundamental right to data protection that includes, not only privacy, but also the principle of autonomy, transparency and nondiscrimination.[65] Scholars also consider consent as one of the most important principle of data protection.[66]

[64] Kokott J., Sobotta C, (2013) The distinction between privacy and data protection in the jurisprudence of the CJEU and the ECtHR, *International Data Privacy Law*, Volume 3, Issue 4, Oxford University Press, pp. 222–228.

[65] McDermott Y (2017) *Conceptualizing the right to data protection in an era of Big Data*, Sage Journals.

[66] Brownsword R, (2009) *Consent in Data Protection Law: Privacy, Fair Processing and*

These principles of consent, accountability, transparency, right to data portability and localization have emerged as core elements of data protection and privacy law. They are all evident in the laws of Australia, the EU, Malaysia, Japan, Singapore, Indonesia and to a lesser extent, the laws of India and Thailand. This book will evaluate these as part of a risk management framework that provides the basis for regulating and protecting personal data, and subsequently privacy over the Internet.

The book emphasizes how EU data protection laws have influenced the development of data protection law around the world. Coupled with the international principles established by the OECD, the EU regulatory framework has also cemented a risk based approach to data protection. Graham Greenleaf reinforces this important point, and argues that approximately 120 countries have adopted, or are considering adopting, some form of data protection or privacy legislation, which are mostly based on European standards.[67] This risk based policy and legislative approach has, not only forced entities to measure the risk from the loss of data or breach of privacy, but also harm to the data subject, and more broadly to society. However, measuring the level of harm arising from violating data protection and privacy laws is dependent on the area of law in issue, such as competition, intellectual property and transnational contract law. Furthermore, the current risk management approach does not go far enough. It focuses on the collector and use of the data, rather than capturing and placing a level of responsibility on the manufacture of the systems and platforms used to collect this data. This book highlights those areas of law that are most in need of reform, including how to measure and redress the risk of harm arising from their violation. Nevertheless, and while it is argued that the EU influence is significant, it hasn't necessarily meant that other nation states have fully embraced the idea of data protection or privacy over the Internet. For instance, Singapore, has not recognized privacy as a right in the same way as its regional or international counter-parts.[68] This is largely based on its historical beginnings. This is no more evident, not only recognizing privacy as a standalone human right, but also the right to be forgotten, which does not exist in every state outside of the EU.

The globalisation of technology has resulted in nation states developing, adopting and applying broad conceptions of data protection and privacy laws to address local economic and social needs. The result has been that data protection and privacy laws have many gaps, variables and tensions across states and regions. The current day laws are adhoc and fragmented. Furthermore, there appears to be little consensus as to what is the most effective model. What consensus there is has only been in relation to the adoption of concepts and principles that have largely come from the OECD. This book argues that a global response is required to redress these tensions, with regional and national institutions and bodies providing the necessary

Confidentiality, In: Gutwirth S., Poullet Y, De Hert P, de Terwangne C, Nouwt S. *Reinventing Data Protection*?. Springer, Dordrecht, pp. 83–88.

[67] Greenleaf G *Global Analysis of Data Privacy Laws and Bills* Privacy Law and Business International Report 145: (2017) pp. 14–24.

[68] Chesterman, S (2018) *Data Protection Law in Singapore, Privacy and Sovereignty in an Interconnected World*, Academic Publishing, p. 4.

resources and information to fill the gaps. This book explores some of the options available to strengthen the regulation of data protection and privacy, globally. It will stress that a greater convergence and harmonisation is required–to address the ongoing economic and social impact arising from the failure adequately to protect personal data. The challenge will be how nation states deal with their sovereign right to regulate in this area of law, based on their local economic and social needs, compared with the wider regional and international needs, and other competing forces.

An even more complex question arises from comparing data protection laws across multiple jurisdictions. What model will serve the international community best in the future? It will be argued throughout this book that by studying these jurisdictions, three models of data protection law now exist. These include Australia, the EU and Singapore. The EU model places privacy as a fundamental right. Secondly, Singapore has created a business friendly model. Thirdly, it is our view that Australia's balanced model sits somewhere between the two (the EU and Singapore) and could also emerge as benchmark. This is becoming even more difficult to determine, if one were to take the position of Douglas Atkin, from AirBnB who believes:

> In the distant future, we'll forget the idea of engaging in technology at all. We'll swallow it, absorb it, and wear it, without us really thinking we're engaging in technology *per se*.[69]

Should this be realised, the narrative, discussion, policy and legal discourse will be very different. Nonetheless, the remaining countries of India, Indonesia, Malaysia, Japan and Thailand while being compared against these countries and their respective models, will also be standalone. They can be best described as either being a combination of the three models, or, they are yet to have specific data protection laws. For instance, both India and Indonesia are in the development stage of their data protection laws. However, it is noted that this book does not consider other models of data protection law such as North America, the Middle East or other ASEAN countries. Nonetheless, it is our view that the models discussed in this book, are likely to be similar to those models in Northern America and Middle East.

1.1.5 Data Protection and Privacy Is Not Limited to One Area of Law

Data protection and privacy law transcends trade and commerce. Data protection and privacy law is emerging in areas of intellectual property, competition and transnational contract law. There is also a greater interest in personal data being considered as part of cybercrime-security and criminology more broadly. What has become evident in recent years, is the increasing ease with which personal data is being

[69] The Wearable Future, Consumer Intelligence, PriceWaterHouseCoopers, http://quantifiedself.com/docs/PWC-CIS-Wearable-future.pdf, accessed 26 October 2018.

illegally obtained to either provide a commercial benefit, or assist in broader criminal activity. Today data protection and privacy law transcends most, if not all, discrete areas of the law. It is unlike, for example, contract law that deals with a specific area of commercial law. This poses another layer of challenges to addressing the many gaps, variables and tensions in comparing data protection and privacy laws of different countries and regions of the world. As already highlighted, this book discusses some of the potential options available to providing a better balance between these tensions and gaps in the law. It proposes a pathway to improved convergence and harmonisation of data protection and privacy law. The book considers how data protection and privacy law has begun to find its way into international (bilateral and multilateral) trade agreements. It evaluates this evolving area of law, recognizing that it is far from being settled and in need of further exploration and development. While out of scope of this book, future data protection law will also need to consider artificial intelligence, quantum and other emerging technologies.

1.2 Structure and Methodology

The methodology adopted in this book to analyze these legal developments extends beyond law *stricto sensu* to include technological developments. Firstly, and due to the varied approach taken by nation states towards data protection and privacy law, this book uses the fundamental principles and concepts of data protection that have evolved from international and regional institutions such as OECD. The OECD has been instrumental in providing the basis for high level concepts and principles such as Collection Limitation, Data Quality, Purpose Specification, Use Limitation, Security Safeguards, Openness, Individual Participation and Accountability. The book highlights the manner in which core differences in culture, history and legal traditions have influenced the data protection laws of states comprising the Asia Pacific Region. It demonstrates that the EUs data protection laws have provided the benchmark for many other countries around the world to develop their own data protection and privacy laws. This is particularly evident when data protection and privacy is conceived as a fundamental right.

Secondly, the book treats the data protection and privacy laws of Singapore, while not being the first to develop such laws in Asia, as a starting point for analyzing the structure of legislation governing data protection in the Region. Thirdly, the book compares the legislative data protection frameworks of states in the Region, primarily through the structure and lens of Australia and Singapore's data protection and privacy laws. Fourthly, the book focuses on the strategic differences across each of the jurisdictions, while refraining from scrutinizing the intricacies of multiple privacy and data protection laws. For instance, it considers the tension between a commercial and rights' focused conception of privacy, the manner in and extent to which states in the Asia Pacific Region have adopted the OECD principles. It also examines the extent to which they have gone further by embodying data protection strategies and laws adopted by the EU.

Even though the book focuses predominantly on personal data protection and privacy laws, it also considers economic and commercial aspects of data protection that have recently surfaced and are not extensively explored in the existing literature. The book also considers intellectual property law, competition law, and transnational contract law. Additionally, the book also examines elements of cybersecurity law.

Due to the breadth, depth and variance of laws across the Asia Pacific Region, this book discusses and analyses core principles related to data protection and privacy law. These include, but are not limited to, the collection of data, the purposes underlying its use, and security, openness and individual participation in its use. The book stresses the importance of consent to the use of personal data, the requirement of accuracy and accountability in that use. The book considers how each country in the Region has approached key principles governing the definition of privacy, data protection or data/personal information, and rules governing the collection, retention, accuracy, and breach of privacy rights. It also stresses differences in their approaches, and in their provision for oversight by a Regulator, Commission or Commissioner.

The book applies the term 'jurisdiction' frequently to identify the Region as a whole, rather than a particular country in it, not unlike the jurisdiction of the EU as a supranational polity and not as an independent nation state. It also considers different international and regional frameworks for the protection of personal data to determine whether they are adequate for the future digital economy, particularly in relation to international trade. The book also recognizes that the framework for data protection and privacy laws of countries like Australia derive from their long standing trade relations with the EU, but also from their geographic proximity to India, Indonesia, Japan, Malaysia, Singapore and Thailand.

It is well understood that Central and South-East Asia are developing at different rates, but are developing at a faster pace than most developed countries. The digital economy will likely enhance geometrically as it has in the last decade. That development will sometimes strengthen regulatory regimes across countries, as they jointly address legal and policy issues relating to data security and protection, as they do today in regulating trade and immigration. The comparative framework adopted by the book is also intended to provide a unique opportunity for the reader to better understand the differences between Australia's approach to data protection and privacy law with the laws in other Asian countries, as well as in the EU.

This book compliments the important work of Professor Graeme Greenleaf, University of New South Wales, and Professor Simon Chesterman, Dean of Singapore Law School, National University Singapore. Both Professor Greenleaf and Professor Chesterman have written extensively and have intimate knowledge of data protection and privacy law.

1.3 Limitation of this Research

There are limitations to this research. Firstly, making accurate comparisons of the degree to which given countries or cultures respect privacy is fraught with difficulty. One of the greatest challenges is the problem of comparing various countries' legal regimes and frameworks for privacy and data protection.[70] Due to the fluid and unsettled nature of the current data protection and privacy laws, some countries have adopted a sectorial approach. Furthermore, other countries have specific privacy and data protection laws that are underpinned by codes of practice, guidelines, standards and procedures. Therefore, the structure of each Chapter will differ, along with breadth of law. In other words, no Chapter will be the same because the concepts and principles adopted by each jurisdiction vary, at least in title headings used. It is beyond the scope of this book to compare the codes of practice, guidelines, standards and procedures that underpin and support these substantive law and statutes. Even though each country specific chapter discusses many of these concepts and principles, the structure of each chapter will vary according to the legislation that each jurisdiction has established.

Another limiting factor is the ability to report on how each country is implementing and enforcing its data protection and privacy laws. Some countries have left these issues entirely to the judiciary, while others have empowered their respective Commission or Commissioner to impose penalties for breaches of the law. Yet, other countries have reported very little, if at all on breaches of their laws, particularly those countries that take a sectorial approach. As the book highlights, there are varying approaches taken by each country in the Asia Pacific Region as well as by the EU in relation to regulating the publication, use and abuse of personal data. The book does not explore the reasons or the extent of transparency in regulation in each nation state or jurisidiction. Rather, it focusses on their regulatory schemes, such as whether they rely on a dedicated agency, Commission, or Commissioner to regulate the use of personal data, and the powers they accord to such agencies or officials. Finally, the book does not purport to explore the implementation and effectiveness of data protection and privacy laws of the jurisdictions analysed in this book.

1.4 Chapters

There are six parts to this book, with seventeen chapters. Part I of this chapter defines the problem. It recognizes that the data protection and privacy laws of each country and region discussed in this book are far from settled. This chapter also outlines the structure and methodology of the book. Part II, Chap. 2 highlights the different approaches that each jurisdiction has adopted to date, in relation to data

[70] Bennett C (2008) *The Privacy Advocates: Resisting the Spread of Surveillance*, M.I.T. Press, Cambridge, Massachusetts, p. 221.

protection and privacy through technology, the law and the digital economy. Part III discusses each jurisdiction's (country) data protection and privacy laws. Part IV compares these laws across each jurisdiction. Part V highlights the commercial issues associated with personal data in relation to intellectual property, competition law and international contract law. It calls for personal data to be afforded an intellectual property right and can fall within the transnational contract law legal framework. This part also identifies how the international framework of contract law can assist in the management of data protection and privacy in the digital economy. Part VI identifies the international and regional institutions and forums that can assist in the management of data and privacy. More importantly, it discusses policy and legal issues relating to data protection and privacy, along with identifying a possible way forward. This part will also conclude the book.

The book includes country specific chapters. The jurisdiction/country chapters commences with the European Union in Chap. 3, because they are the most recent updated data protection laws. The European Union (EU) is considered a leader in many areas of human rights law and, data protection and privacy law is no different. The EU has taken a different approach to many other countries and regions of the world in this important area of law and policy. The EU has effectively forced other countries to redress these issues.

The book then moves to South East Asia and looks at Singapore, in Chap. 4. The book then moves to the Pacific-Oceania. Chapter 5 discusses the data protection laws in Australia. Chapter 6 highlights the approach taken by India. Chapter 7 will discuss Indonesia's sectorial laws. Indonesia, is one of Australia's closest neighbours and a country with one of the largest populations in South East Asia. Malaysia is a commonwealth country, along with Singapore, India and Australia, and Chap. 8 discusses their data protection laws, and their proposed future approach. Thailand is another developing country that makes up the community of ASEAN countries, and Chap. 9 identifies what privacy and data protection laws apply in this country, and provides an overview of the proposed Personal Data Protection laws. Chapter 10 moves to East Asia and looks at Japan's laws. Chapter 11 brings together the jurisdictional chapters by comparing key concepts and principles from the respective laws. However, as will be demonstrated throughout this book, the respective laws vary greatly and therefore, Chap. 11 will only compare key concepts and principles from each Chapter such as consent, definition of personal data, amongst others.

Chapters 12, 13 and 14 digress from the previous chapters and examines the commercial side of personal data, as it is now viewed as a tradable commodity. Chapters 12 and 13 deal with data protection in intellectual and competition law, highlighting some of the key issues related to data protection in these areas of the law. Chapter 14 examines whether personal data can form part of transnational contract law and the conflict of laws involved with online terms and conditions. Furthermore, Chap. 15 discusses briefly the issues arising out of personal data used in criminal activity, highlighting the need for vigilance in the criminology discipline (cyber security).

Chapter 16 highlights the current international and regional institutions, organizations and bodies that assist and influence national data protection and privacy laws. This Chapter also highlights the extensive work that has, and is currently being undertaken by the many international and regional institutions and bodies identified in the book. One of the most important piece (s) of the data protection and privacy framework which has evolved at the international and regional level is the development of the current day legal concepts and principles that can be found, to varying degrees, in EU law and the law of its member states. Some of the concepts and principles include, but are not limited to consent, transparency, accountability, data portability and the right to be forgotten. Responses to these issues have paved the way for greater protection of personal data as is defined in law. However, this approach has created gaps, variables and tensions, not only in data protection and privacy laws, but also in other areas of the law in which data protection and privacy have an impact. These areas overlap with other areas of law, such as in relation to intellectual property rights. The book argues that the variables, gaps and tensions in the law are significant. To overcome these tensions and gaps will be formidable, because the issues no longer only reside in the technology sector, they are political and this was recent highlighted at the 2019 G20 meeting in Japan. The 2019 G20 meeting Declaration (10-12) recognises the urgency and challenges ahead for technology innovation and regulation. The declaration calls for legal frameworks, both national and international, to be respected. (need the following footnote here - G20 Osaka leaders' declaration, Japan Times, 2019, www.japantimes.co.jp That said, this book reinforces what the G20 leaders are seeking to achieve. By establishing an internationally agreed Model Law, and supporting treaty or convention, will go a long way to building consistency and respect of the regulatory framework. Chapter 17 highlights attempts to resolve the gaps, variables and tensions in the law. However, it must be noted that, due to the internationalisation and regionalisation of technology and it interrelationship with data protection and privacy law, there is no single silver bullet available to address all the issues.

1.5 Conclusion

This Chapter has outlined the problem related to privacy and data protection law across the Asia Pacific and the European Union. Privacy has evolved from the long standing notion that people have the right to protect their personal private space. Privacy has been viewed subtly differently by the many different religious and cultural groups from the EU, Asia to the Pacific. In Western society there has been subtle differences between the alignment of privacy, whether privacy should be aligned with dignity or liberty. This makes for a fascinating study, to better understand the differences in data protection and privacy laws, that relate today to the modern economy and changing society.

Nonetheless, no matter how privacy as a right is aligned to these two concept, there is little impact to privacy in the modern world. In other words, privacy today, through and over the Internet can constitute dignity or liberty, or both. What has emerged is that data protection appears to be a tool of privacy. That is, data protection laws have been established to provide individuals with a level of control over their personal data and personal information, and at the same time, the laws provide a level of privacy protection. Thus, privacy over the Internet, is a by-product of data protection. More pervasively, data protection has assisted in establishing a new economic activity - the trade in personal data. How the data protection laws protect privacy differs from country to country, region to region—across the world. This is because they are largely influenced by local national needs, such as, economic, social, cultural and in some cases religious. Yet, it is argued that the concept of privacy even across different cultures, legal systems and religions have converged to mean similar.

One of the major challenges that has arisen from the globalization and internationalization of the Internet and the movement of personal data, is the fact that technology knows no national or supranational (in the case of the European Union) borders. The internationalization of the Internet will continue to challenge nation states in the development of their data protection and privacy laws. However, to some extent the EU are dragging other states into the same regulatory framework as they have established.

The development of data protection and privacy laws is transcending many other areas of the law. The impact to trade and commerce is not fully understood. Even so, this book identifies three areas of commercial law that require a better understanding of how data protection law is shaping and challenging intellectual property, competition and transnational contract law. Moreover, personal data is being used in criminal activity to provide a commercial or financial benefit to individuals and entities. At an international level governments are beginning to consider the inclusion of data protection and privacy into bilateral and multilateral [trade] agreements. Chapter 2 introduces technology and the law. It also highlights the expanding legal and economic significance of personal identity in a constantly evolving digital world.

References

Altman I (1977) *Privacy Regulation: Culturally Universal or Culturally Specific?,* Journal of Social Issues, vol. 33, p. 66–84

Ambrose M, Ausloos J (2013) *The Right to Be Forgotten Across the Pond,* Journal of Information Policy, Vol. 3, pp. 1–23

Basu S (2010) *Policy-Making, Technology, and Privacy in India*, INDIAN J.L. & TECH. vol 6. p. 66

Barrington M (1987) *Privacy: Studies in Social and Cultural History*, American Journal of Sociology, vol 92

Bennett C (2008) *The Privacy Advocates: Resisting the Spread of Surveillance*, M.I.T. Press, Cambridge, Massachusetts, p. 221

Berween M (2002)*"The Fundamental Human Rights: An Islamic Perspective"* The International Journal of Human Rights 61 pp. 70–74

Cowan Z (1970) *The Private Man* 24 Inst Pub Affairs Rev 26

Chesterman S (2012) *After Privacy: The Rise of Facebook, the Fall of WikiLeaks, and Singapore's Personal Data Protection Act 2012*, Singapore Journal of Legal Studies, p 396

Chesterman, S (2018) *Data Protection Law in Singapore, Privacy and Sovereignty in an Interconnected World,* Academic Publishing, p. 4

Clarke R (2006) *What's 'Privacy'?,* Workshop at the Australian Law Reform Commission on 28 July http://www.cse.unsw.edu.au/~cs4920/resources/Roger-Clarke-Privacy.pdf, accessed 19 April 2018.

De Boni M, Prigmore M (2001) *A Hegelian basis for information privacy as an economic right*, in Roberts M, Moulton M, Hand S, Adams C. (eds) Information systems in the digital world, Proceedings of the 6th UKAIS conference, Manchester, UK, Zeus Pres.

De Hert p, Gutwirth S, (2006) *Privacy, Data Protection and Law Enforcement. Opacity of the Individual and Transparency of Power*, in Claes E, Duff A, Gutwirth S, *Privacy and the Criminal Law,* Antwerp-Oxford, Intersentia, pp. 61–104

De Hert P, Gutwirth S, (2009) *Privacy, Data Protection and Law Enforcement. Opacity of the Individual and Transparency of Power.* in E. Claes, A. Duff & S. Gutwirth (Eds.), *Privacy and the Criminal Law,* pp. 61–104

Ess C (2005) *Lost in translation.* Ethics Inf Technol 7, 1 2005b, pp. 1–6

Greenleaf G (2017) *Global Analysis of Data Privacy Laws and Bills* Privacy Law and Business International Report 145, pp. 14–24

Hildebrandt M (2006) *Privacy and Identity.* In: Claes E, Duff A and Gutwirth S (eds) Privacy and the Criminal Law, Oxford, United Kingdom Hart, 2006, pp. 43–60

Hobbes T (1981) *Leviathan*, C. B. Macpherson (Editor), Penguin. Locke J (1986) Second Treatise of Government, Prometheus.

Kamali, H. (2007) *The Right to Life, Security, Privacy and Ownership in Islam* (Cambridge, Islamic Texts Society); Mahmood, T. (ed.) (1993) *Human Rights in Islamic Law* (New Delhi, Institute of Objective Studies).

Kang J (1998) *Information Privacy in Cyberspace Transactions*, Stanford Law Review, 1998, pp. 1201–04

Kokott J, Sobotta C, (2013) The distinction between privacy and data protection in the jurisprudence of the CJEU and the ECtHR, *International Data Privacy Law*, Oxford University Press, vol 3, Issue 4, pp. 222–228

Madieha Azmi I, (2007) *Personal Data Protection Law: The Malaysian Experience*, 16 Info. & Comm. Tech. L. 125 pp. 130

McGarry K (1993) *The Changing Context of Information. An Introductory Analysis.* 2nd ed. London: Library Association Publishing, p. 178

Neethling J, Potgieter M, Visser J (1996) *Neethling's law of personality*, Butterworths. pp. 35–36

Samuel D, Brandeis L (1890) *The Right to Privacy*, Harvard Law Review, Vol. 4, No. 5, pp. 193–220

Tamás M (2002) *From Subjectivity to Privacy and Back Again,* Social Research, p. 220 Neethling J, Potgieter M, Visser J, (1996) *Neethling's law of personality.* Durban: Butterworths. p. 36

Westin A (1967) *Privacy and Freedom.* New York: Atheneum, p. 351

Part II

Chapter 2
Law, Technology and Digital Economy

Abstract This Chapter introduces technology and the law. This Chapter highlights the importance of the digital economy to Asia Pacific and Europe. The world is living in a period of rapid technological change, which is creating complex policy, regulatory and legal issues for governments and the broader community. The explosion in connectivity, data volumes, digital communication, e-commerce and use of the Internet is challenging government and the business community. The digital economy is expected to be a key economic driver over the next decade.

The Internet has created pressures in the law, and today more than ever the law of the Internet in new technologies has become institutionalized. However, there is a lack of regulation of the manufacture of Internet systems and infrastructure. This is a major concern. It is like no other industry, which is regulated by government through a minimum set of standards. At issue is that society generally knows next to nothing about the extent of injuries on and from the Internet. Doing business over the Internet raises a host of challenging issues, such as how to deal with electronic cash, online banking, commercial transactions in digital information and digital signatures. In addition, the rise of intellectual property, data protection (commercial and personal) and privacy on and over the Internet poses new legal dilemmas for preserving an asset which is challenged by the digital capture of data.

The commercial use of personal and business data will form a key component of the digital economy. Identity in the digital world will also expand. Given this expected growth in the ability to capture personal data, personal information and privacy is likely to be increasingly compromised. This Chapter briefly highlights the origins of and present day constituent elements of personal identity.

2.1 Introduction

In the contemporary world, we are living in a period of rapid technological change, which is creating complex regulatory and legal issues for governments and the broader community. The explosion in connectivity, data volumes, digital communication, e-commerce and use of the Internet is challenging government and the business community.

© Springer Nature Singapore Pte Ltd. 2019

R. Walters et al., *Data Protection Law*,
https://doi.org/10.1007/978-981-13-8110-2_2

The Internet has created pressures in the law. Michael Rustad believes that cyberspace law in emerging technologies has become institutionalized.[1] As an example, Rustad beleives that we know next to nothing about the extent of injuries on and from the Internet. Doing business over the Internet raises a host of challenges, such as how to deal with electronic cash, online banking, commercial transactions in digital information and digital signatures. In addition, the rise of intellectual property, data protection (commercial and personal) and privacy on the Internet poses new legal dilemmas for preserving an asset which is challenged by the digital capture of data.

The development of a global network of interconnected computers has simplified access, storage, process, and transmission of vast amounts of information and data in digital form. In the decades to come, this is likely to transform many of our assumptions about communication, knowledge, invention, information, sovereignty, identity and community.

In 1999, Lawrence Lessig released *Code and Other Laws of Cyberspace*.[2] For Lessig, Internet law is about the intersection of laws, norms, architecture and market forces. The complexity surrounding Internet law is not simply knowing the law, but also understanding the architecture and systems sitting behind the technology. Understanding the interrelationship between the law and technology will increasingly become a challenge. A report released in 2017 estimates that digital activity will grow nearly 20 times between 2015 and 2025 to 180 zettabytes (or 180 × 10 bytes – 180 trillion gigabytes).[3] As digital activity takes hold in countries such as China, India and Brazil, it is estimated that these countries alone will contribute up to two thirds of total digital output.[4]

Thus, four constraints exist in relation to cybersecurity and the world of technology: (1) laws or legal sanctions, (2) social norms, (3) the market, and (4) code.[5] However, and while they might be considered constraints, they are essential to the overall framework of managing the current and future development and use of technology. They contribute in some way to the management and protection of data and privacy. No different to any other industry or profession, the law lags behind today's technology, and over the past two decades' governments and industry have been scrambling to regulate technology. Governments have not fully understood data protection and privacy issues associated with new technology, which has also seen them establishing laws in this area.

---

[1] Rustad M (1996) *Legal Resources for Lawyers Lost in Cyberspace*, Suffolf, U. L. Review, pp. 317–18.

[2] Lessig L (1999) *Code and Other Laws of Cyberspace*, http://codev2.cc/download+remix/Lessig-Codev2.pdf, accessed 20 July 2017.

[3] Smart C (2017) Regulating the Data that Drive twenty-first- Century Economic Growth The Looming Transatlantic Battle, US and the Americas Programme, 2017, https://www.chathamhouse.org/sites/files/chathamhouse/publications/research/2017-06-28-regulating-data-economic-growth.pdf accessed 5 December 2017, accessed 10 November 2018.

[4] Ibid.

[5] Ibid.

Lawrence Lessig points out that historically it has been very easy for governments to regulate the activity of people and industry.[6] His concerns in relation to code are that it:

> will present the greatest threat to both liberal and libertarian ideals, as well as their greatest promise. We can build, or architect, or code cyberspace to protect values that we believe are fundamental. Or we can build, or architect, or code cyberspace to allow those values to disappear. There is no middle ground. There is no choice that does not include some kind of building. Code is never found; it is only ever made, and only ever made by us.[7]

Lessig is advocating that it is imperative that a clear set values are established that each country can adopt. There are concerns with the current approach as arguably it is failing and will continue to fail well entrenched legal principles that span both the commercial and private sphere (privacy, data protection, intellectual property, transnational contracts and competition).

Regulation and the law of the Internet is likely to become even more complex. The code itself may become the regulator by default, and be used to lock individuals and entities out of the Internet all together.[8] Moreover, when the courts have to decide on areas of law such as contract, property rights, copy right and competition law that involve developing and future technology such as code. Nevertheless, the overall values and direction are being set by nation states, supranational polities and the international community. Although there are varying approaches, there will need to be continued collaboration and agreed standards set by all. This can be achieved through government, industry and community regulation.

More than at any other time in history, there are many more devices and applications available to identify a person. Apart from the traditional modes mentioned, which are now mostly electronic, there is CCTV, computer usage email and Internet, and the mobile phone. The advances have recently seen the introduction of facial recognition, which can been found at most airports, and starting to develop within organizations for employees to enter the premises, and places where people would not even expect. This includes, but not limited to public places, such as courts. The most recent concern for privacy advocates has been the way that technology can not only trace an individual's consumer and computer use, but also a person's emotions. Some, technology today is so advanced that it can scan up to 30,000 points of the face. That being the case, it will only become even more advanced and precise into the future. This information can be used to detect if you are depressed, a person's age, sexuality and race.[9] Some of this information constitutes personal data, as will be discussed in each of the jurisdictional chapters throughout this book.

Importantly, data collection (personal or commercial), storage and its use serves both the private and commercial. The private is the data that identifies a person. The commercial is the personal data of an individual that is used for commercial purposes

---

[6] Lessig L (2006), *"CODE" version 2.0*, A Member of the Perseus Books Group New York, p. 23.

[7] Ibid, 6.

[8] Ibid, 82.

[9] Fowler G (2017), *New privacy worry: Apple sharing your face with apps*, The Strait Times, p. 10.

or data that is used for the development of products. That is, data in relation to individuals can assist to improve its services to customers (competition) and used for internal marketing purposes, including licensing of third parties (intellectual property).[10] Furthermore, the trade in personal data in the new digital economy constitutes the trade in goods and services along with data used as part of international investments. It is forming an important element of economic activity, which is expected to continue to grow. Protection of commercial and personal data is paramount to ensure certainty, trust and confidence, by consumers and business in the market. Therefore, nation states have, and will need to continue to consider data protection along with privacy across economic and social boundaries. This balance may in fact become increasingly more difficult to control. In addition, as the digital economy continues to expand, data protection will need to be considered as part of regional and international laws. There is a need to foster greater collaboration and capability which is important for global trade and investment flows that will increasingly be undertaken online, and will include personal data.

In 2016, 6.4 billion devices were connected worldwide, and by 2020 it is predicted this will extend to more than 20 billion intelligent connections. [11] These devices are generating more and more personal and commercial data, with software being embedded in more and more products.[12] There is a greater potential for that data to be transferred and used in a way that could compromise a person's privacy, business secrets and creating opportunities for anti-competitive behavior. It was estimated that the cloud computing industry alone is estimated to be worth between $107 and $127 billion in 2017.[13] One report estimates that value-added services related to the 'Internet of Things' will grow from about $50 billion to approximately $120 billion by the end of 2019.[14] Another report forecasts a potential economic impact of between $3.9 and $11.1 trillion per year by 2025.[15] Therefore, data protection regulation must carefully correspond to the evolving needs and possibilities associated with these changes in order to facilitate potential benefits to local and international economy, business community and citizens. In 2014, approximately $30 trillion worth of goods, services and finance was transferred across borders. Around 12% of international trade in goods has been

[10] Smarajiva R (1995), *Interactively As Thought Privacy Mattered, Technology and Privacy*, supra, note 4, pp. 277–279. United States Department of commerce, Privacy and the Nil, Safeguarding Telecommunications-Related Information, p. 15.

[11] Ibid.

[12] Ibid.

[13] Top Markets Report: Cloud Computing, U.S. Department of Commerce. International Trade Administration, July 2015. http://trade.gov/topmarkets/pdf/Cloud_Computing_Top_Markets_Report.pdf. See also http://openviewpartners.com/news/global-cloud-computing-services-market-to-reach-us127-billion-by-2017- according-to-new-report-by-global-industry-analysts-inc, accessed 2 December 2017.

[14] The Internet of Things *"Smart" Products Demand a Smart Strategy Using M&A for a Competitive Edge."* Woodside Capital Partners, March 2015. http://www.woodsidecap.com/wp-content/uploads/2015/03/WCP- IOT-M_and_A-REPORT-2015-3.pdf, accessed 2 December 2017.

[15] Ibid.

estimated to occur through global e-commerce platforms such as Alibaba and Amazon. The international dimension of flows has increased global GDP by approximately 10%, equivalent to a value of $7.8 trillion in 2014. Data flows represent an estimated $2.8 trillion of this added value.[16]

The contribution of digital technologies to the Australian economy is forecast to be $139 billion by 2020.[17] The ICT workforce is expected to increase to 722,000 workers by 2022. Australia's IT services exports were estimated at $2.8 billion in 2015–16.[18] This increasing digital activity also brings risks, such as the cyber security risks from the digitization of more consumer and business transactions. However, the cost and impact from cyber-attacks to an Australian business is around $419,000.[19] Economic modelling suggests that a greater focus on cyber security by Australian businesses could increase business investment by 5.5% and wages by 2.0%. It could also employ an additional 60,000 people by 2030.[20]

India like other countries are on the path to transform their economy by embracing the digital world. There is currently a push by the Indian government to grow the digital economy to $1 trillion over the next 7 years.[21] India is still considered a developing country, and the digital economy could be a way to fast track the country's progress to the first world over the next decade. Even so, India has the largest outsourcing economy in the world, which deals with extensive amounts of data over the Internet. Indonesia is also moving towards a digital economy, similar to its neighbours such as Australia and Singapore. Should Indonesia fully embrace digitization, it can realize an estimated USD $150 billion ingrowth 10% of GDP by 2025.[22] The Indonesian Government is focused on creating a conducive business environment and in 2017 released the 'Economic Policy Package', which is aiming to attract large scale foreign investment into the marketplace.[23]

---

[16] Digital Globalization: The new era of global flows, McKinsey Global Institute https://www.mckinsey.com/business-functions/digital-mckinsey/our-insights/digital-globalization-the-new-era-of-global-flows, accessed 2 December 2017.

[17] Australian Bureau of Statistics (ABS) occupation and industry classifications, based on the methodology used in previous editions of Australia's Digital Pulse. This methodology draws on definitions and nomenclature developed by Ian Dennis FACS, lead researcher from the Centre for Innovative Industries Economic Research (CIIER), and used in the Australian Computer Society's 2008–13 statistical compendiums and other CIIER analysis.

[18] Australia's Digital Pulse Policy priorities to fuel Australia's digital workforce boom Australian Computer Society, 2017, Delloitte Access Economics, https://www.acs.org.au/content/dam/acs/acs publications/Australia's%20Digital%20Pulse%202017.pdf, accessed 24 November 2017.

[19] Ponemon Institute (2015), 2015 Cost of Cyber Crime Study: Australia, https://ssl.www8.hp.com/ww/en/secure/pdf/4aa5-5210enw.pdf.

[20] Australia's Digital Pulse Policy priorities to fuel Australia's digital workforce boom Australian Computer Society, 2017, Delloitte Access Economics, https://www.acs.org.au/content/dam/acs/acs publications/Australia's%20Digital%20Pulse%202017.pdf, accessed 24 November 2017.

[21] Shankar Prasad R (2017), *Union Minister for IT, Electronics, Law and Justice*, The Times of India Business.

[22] Unlocking Indonesia's digital opportunity, McKinsey Indonesia Office, October 2016.

[23] Digital Economy of Indonesia, http://www.cicc.or.jp/japanese/kouenkai/pdf_ppt/pastfile/h28/161026-03id.pdf, accessed 25 November 2017.

Japan, on the other hand, is considered to have the world's third largest economy. In 2013, Internet usage across Japan was at saturation point, being an estimated 99% for the business sector, and more than 70% of the population using the Internet.[24] As e-commerce has steadily increased in 2015 it was projected that the market size was worth USD $134 billion and growth of 7.1%. Purchasing or trading goods and services was the second most common purpose of Internet use at home with 57.2%, which is anticipated to steadily increase over the next decade.[25] Malaysia's digital economy in 2014 contributed 17% to national GDP and is expected to grow at 9.5% through to 2020.[26] Interestingly, Malaysia when compared to its neighbors, particularly Singapore, has not fully deregulated the technology (communications) sector. On the other side, Singapore has managed to get the jump on its neighbor and other countries in the region to strengthen its "hub" status in the technology sector.

Since the independence of Singapore more than 50 years ago, the island state and its people have transformed themselves from the third world to the first. Singapore prides itself on being a central business hub that is based on promoting the rule of law. Its recent implementation of the 'future economy'[27] and economic strategy will see Singapore deepen its diversification and international connections. Singapore is rapidly pushing to develop strong digital capabilities and support the pervasive adoption of digital technologies across all sectors of the economy.[28] This push will pose many challenges in the management of data and protecting individual's privacy. In 2014, approximately 35% of Thailand's population had access to some form of Internet.[29] The digital economy in Thailand today is estimated to be worth TBT $11.5 trillion, equivalent to 15.5% of global GDP.[30] Thailand have a Digital Plan, which aims to assist business to adopt and embrace digital technology.[31] This is no different to other nation states, except they are, to date, focused on the Thai economy. Thailand's economy is growing and the country continues to develop, however, they are not as advanced as other Asian countries in the area of data protection and privacy.

---

[24] Digital Economy in japan and the European Union, An Assessment of the Common Challenges and the Collaboration Potential, 2015, https://www.eu-japan.eu/sites/default/files/publications/docs/digitaleconomy_final.pdf, accessed 5 December 2017.

[25] Ibid.

[26] Malaysia Digital Economy, https://www.mdec.my/assets/migrated/pdf/2015-MSC-Malaysia-Annual-Industry-Report-final.pdf, accessed 25 November 2017.

[27] Report of the Committee on the Future Economy Pioneers of the next generation, https://www.gov.sg/~/media/cfe/downloads/cfe%20report.pdf?la=en, accessed 20 October 2017.

[28] Ibid.

[29] Thailand's Digital Economy, http://www.asean-sme-academy.org/wp-content/uploads/Presentation_20151012_DE_Introduction_Kasisitorn.pdf accessed 25 November 2017.

[30] Thailand Economic Monitor– August 2017: Digital Transformation, http://www.worldbank.org/en/country/thailand/publication/thailand-economic-monitor-august-2017-digital-transformation, accessed 25 November 2017.

[31] Thailand's 3 Year Digital Government Master Plan, https://www.ega.or.th/upload/download/file_49732b080dd2dc0a2125b5288c63c2c5.pdf, accessed 2017.

Even with a greater focus on protecting personal data and information as a 'fundamental right',[32] which is front and center of EU policy making, the EU continues expanding the 'single market' concept. The EU plans to mobilize €50 billion of public and private investments in support of the digitization of industry, to boost digital innovation by establishing digital innovation hubs.[33] The digital economy is growing at seven times the rate of the rest of the economy, but this potential is currently held back by a patchy pan-European policy framework.[34] Europe is lagging behind many other regions when it comes to the fast, reliable and connected digital networks which underpin economies and every part of our business and private lives. There is approximately 250 million Europeans use the Internet daily and there are still millions of people that have never used the Internet at all.[35] The number of jobs that require information and communications technology skills is expected to rise by 16 million by 2020. 90% of jobs will require basic information technology skills by 2015.[36] The increase in economic activity related to data is likely to result in more breaches and unauthorized use, rather than less. Therefore, it is paramount that governments, industry, the legal profession work together to overcome the challenges that lie ahead. Moreover, identity in the modern world has changed from its traditional past. Identity in the modern technological Internet has changed the way people view their own personal identity. The next section highlights what identity means in the modern Internet world.

2.1.1 *Identity in the New World*

Identity – what is it? There are various aspects associated with identity. The identity of a person, group of people, a community, an organization, political party, business, nation state, corporation, sports organization and non-government organization. In the 1600s Leibniz defined identity in terms of whether one thing can be distinguished from another.[37] If object A shares absolutely every characteristic of object B, including its shape, extent, position in time and space, then A and B are identical: they have the relationship of identity.[38] The most common identity has

[32] Article 1. 2. "Regulation (EU) 2016/679 of the European Parliament and of the Council," April 27, 2016.

[33] Digitizing European Industry, https://ec.europa.eu/digital-single-market/en/policies/digitising-european-industry, accessed 25 November 2017.

[34] European Commission, The European Union Explained, *Digital Agenda for Europe, Rebooting Europe's economy*, The digital agenda for Europe will help Europe's citizens and businesses to get the most out of digital technologies.

[35] Ibid.

[36] Ibid.

[37] Wilton R (2008), *Identity and Privacy in the Digital Age*, Int. J. Intellectual Property Management, Vol. X, No. Federated Identity Chief Technology Officer c/o Sun Microsystems Guillemont Park, Camberley, Surrey GU17 9QG, United Kingdom, pp. 1–15.

[38] Ibid.

been a person's date of birth, residential address, picture, gender and birth place in the case of a passport or identity card. Moving forward to today, and the future digital economy, what and how will identity at the personal level be defined and viewed. Wilton highlights, 'identity' is also commonly used in the sense of 'digital footprint'. In the context of this book 'identity' constitutes the personal data and information that forms part of the current economy and the future digital economy. Identity also, as highlighted throughout the book, constitutes the personal data and information that has been defined by the law.

The rise of the Internet and access by people has risen at such a rate that today individuals have a digital identity. The graph below highlights the use of the Internet by regions across the world. In the context of this book, and the countries, jurisdictions discussed Internet use accounts for about 40% of total Internet use.[39]

Internet Users in the World by Regions - June 30,2017

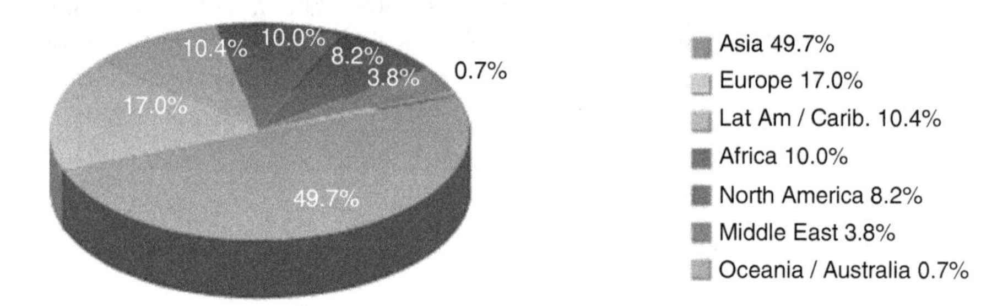

Source: Internet World stats - www.internetworldstats.om/stats.htm
Basis: 3,885,567,619 Internet users in June 30, 2017
Copyright © 2017, Miniwatts Marketing Group

This is an interesting point because a person's identity will at some stage become available through the Internet. For instance, health records, banking details, driving and other licensing details, social media and email. Even though these may be protected, they create a digital identity. It is that digital identity that can be traced to an individual person. Another example is the lawyer or legal academic who has his or her profile on the law firm webpage or the university institutions website. That in itself creates a digital identity of the lawyer or academic. This type of identity is a recent phenomenon and can only be found with the introduction of the Internet. It provides an outline of the person's, profile, eye colour, facial contours, hair colour, skin colour, race and sex. This data and information can be seen in the current data protection laws.

[39] Internet Users in the World by Regions, (2017) http://www.internetworldstats.com/stats.htm, accessed 5 December, 2017.

The online identity is analogous to personal identity as a social construct.[40] This is because, our notion of identity has been traditionally focused (except for media communication) on the personal human interaction. However, people's social existence is becoming increasingly online through email and other social networks. People today can quickly have access to the shared knowledge of a person and their identity, through the online world. People can, if they choose to do so be public. Even, if people want to hide who they are, they can be identified simply by their Internet use. For instance, a person who is undertaking higher education and completes doctorate level studies. It is common place for the university to place that research online so as anyone can access the information. Furthermore, the individual through their research accesses of various websites for information, in the same way as people do so for business, are creating an online identity. As such, they create a digital footprint, and in the case of social media like Linkedin they create a network identity.[41] That identity can also be, as Wilton argues, through web browsing patterns. Another example is how an online retailer collects your information from their website. The data held about you by credit reference agencies and disclosed to third parties, from the use of a credit or debit card, leaves an audit trail of expenditure, which is another form of a digital identity. Therefore, the online identity is growing, and is likely to continue to expand, as technology develops.

2.1.1.1 Personal Identity

Personal data and personal information that identifies a person can come in various forms, within and over the Internet. Personal information and personal data has been defined by law, by various jurisdictions, including those discussed in this book to varying degrees, and is also used to identify an individual. However, the personal data and personal information not defined by the law, which can also be used to identify a person over the Internet, also creates a personal identity. In other words, the ability for tracing a user's identity can be easily undertaken through tags or digital water marks. A similar process can be undertaken to track the misuse of protected information, which is used to process personal data. Furthermore, IP addresses can be personal data. In 2016, Advocate General of the Court of Justice of the European Union delivered an opinion that businesses will need to ensure the collection and processing of "dynamic IP addresses" complies with the EU data protection law.[42] A dynamic IP address is an address that is provided to a user for a period that allows communication between its device and the server provider.[43] However,

---

[40] Ibid.

[41] Ibid.

[42] Case 582/14 – *Patrick Breyer v Bundesrepublik Deutschland*, The Court of Justice of the European Union, 2016, 30–34.

[43] Google Fibre, https://support.google.com/fiber/answer/3547208?hl=en accessed 16 December 2017.

the 'static IP address' does not change.[44] In *Scarlet Extended SA v Société belge des auteurs, compositeurs et éditeurs SCRL*, the court was required to balance the right to property against other fundamental rights and what constitutes personal data. In 2004, SABAM requested that Scarlet, a Belgian Internet service provider, filter the online activities of its users to avoid illegal downloads, that were affecting copyright holders. SABAM was concerned with copyright infringement, arising from file sharing and peer-to-peer networks, whilst Scarlet was concerned about data protection. IP addresses constitute personal data, Scarlet argued that the measures required by the court on SABAM's application constituted a general obligation to monitor communications on its network, inasmuch as any system for blocking or filtering peer-to-peer traffic would necessarily require general surveillance of all the communications passing through its network.[45] The court ruled:

> for an IP address is to be considered personal data. Under the data protection laws of the EU, an identifiable person is one who can be identified, directly or indirectly. The rights to run a business as significant (Article 16) before turning to those insignificant civil and political rights comprising right to protection of their personal data and their freedom to receive or impart information and found in Articles 8 and 11 of the Charter respectively.[46]

The implementation of the GDPR does not provide any guidance on whether an IP address constitutes personal data. The use by the EU legislature of the word "indirectly" suggests that, in order to treat information as personal data, it is not necessary that the information alone allows the data subject to be identified.[47] The court also referred to recital 26 of the Directive, which reads as follows:

> to determine whether a person is identifiable, an account should be taken of all the means likely to be used either by the controller *or by any other person to identify the said person.*[48]

In conclusion, the court ruled that 'for information to be treated as "personal data" it is not required that all the information enabling the identification of the data subject must be in the hands of one person.'[49]

The issue arguably centers around the definition of personal data and information, and secondly the application of the law. For instance, the United Kingdom, while still a member of the EU, implemented the data protection laws by limiting the definition of personal data to information that the controller holds or is likely to come into its possession.[50] Most other EU member states have not restricted the definition. This provides a clear example where harmonization of laws and rules are

[44] Case C-70/10 *Scarlet Extended SA v Société belge des auteurs, compositeurs et éditeurs SCRL*, The court of justice of the European Union, 2011.

[45] Ibid.

[46] Ibid, para 40–41.

[47] Ibid, 41.

[48] Ibid, 42.

[49] Ibid, 43.

[50] Library of Congress, Online Privacy Law: United Kingdom, https://www.loc.gov/law/help/online-privacy-law/uk.php, accessed 16 December 2017.

not consistent. Most noteworthy is the use of the term personal data, whereas, in other countries use the term personal information. Nonetheless, they constitute the identity of the person. The Australian example is very different to the EU. The Federal Court of Australia had to decide on what constituted personal information. In *Ben Grubb and Telstra Corporation Limited,*[51] the court was required to determine whether personal information defined under the Privacy Act 1988 included 'all the metadata information Telstra has stored' about him in relation to his mobile phone service, including (but not limited to) cell tower logs, inbound call and text details, duration of data sessions and telephone calls and the URLs of websites visited. On 8 August 2013, the complainant lodged a complaint with the Office of the Australian Information Commissioner (OAIC) against Telstra under s 36 of the Privacy Act. The complainant claimed that Telstra had breached his privacy by refusing him access to the personal information it holds about him. The complainant seeks a declaration by me that Telstra meet its access obligation under the Privacy Act and provide the complainant with access to all the information he has requested. The complainant has not sought an apology or compensation. The court stated that:

> meta data under Australia's privacy laws only protects personal information when the person is identified from that data. Personal information is defined as information or an opinion (including information or an opinion forming part of a database), whether true or not, that is recorded in a material form or not, about an individual whose identity is apparent, or can be reasonably ascertained, from the information or opinion. [52]

The court went onto say that data is only personal information if a person is the subject matter of that information. Thus, data held by Telstra and other organizations including but not limited to URLs and IP addresses are not protected under Australian law.

Furthermore, in *Privacy Commissioner v Telstra Corporation* Limited,[53] Telstra, Australia's leading telecommunications organization believed that an individual's meta data does not constitute personal information because meta data is not information about a particular person, which can reveal a persons' identity. Telstra contended that it is not in breach of its access obligation under NPP 6.1 because: the metadata in dispute, which sits on its network management systems, is not personal information as defined under the Privacy Act 1988 as the complainant's identity is not apparent nor can it reasonably be ascertained from that data incoming call records are not the personal information of the complainant. Rather they were of the view that personal information of third parties if disclosed would have an unreasonable impact on the privacy of individuals (those incoming callers, and/or potentially place Telstra in breach of the provisions of the *Telecommunications Act 1997 (Cth),* which regulates the use and disclosure of telecommunications data). The complainant's position is that the metadata stored about him is his personal information in relation to inbound call numbers, would

[51] [2015] AICmr 35.

[52] *Ben Grubb and Telstra Corporation Limited* [2015] AICmr 35.

[53] [2017] FCAFC 4.

not have an unreasonable impact on the privacy of other individuals in cases where the calling number display has not been blocked or the option of a silent line not taken.[54] The Privacy Commissioner[55]confirmed the earlier position of the Federal court in Grubb, and provided little guidance on whether the earlier decision was correct in evaluating what constituted personal data.[56] Australia, appears to have not considered other jurisdictions when deciding on this issue.

A personal identity over the Internet, is being created by all users of the Internet, whether the individuals realizes it or not. Arguably, most people will not care or even know that this personal identity has been created or even exists. On the other hand, people will have concerns about their privacy as a result of Internet systems and platforms creating personal identifiers which include that personal data and information, which is defined by law. It is this personal data and personal information, which is protected. That protection is afforded so as a persons' privacy is protected, even though the level of privacy protection will vary from jurisdiction to jurisdiction. Privacy and data protection has not had a lot of consideration by the courts. The future concern is that different jurisdictions adopt very different laws and courts apply standard principles differently, and therefore, the impact on organizations operating in multiple jurisdictions will be enormous. One way to minimize this type of impact is for the international community to develop legal and regulation mechanisms and frameworks that converge and harmonize. The convergence and harmonization is not just limited to government regulation. Regulation can come in different forms. For example, there is regulation imposed by government, or, industry self-regulation or co-regulation (government and industry).

2.1.2 Co-regulation [Government and Industry]

The idea of government regulation, self or co-regulation is not new. Governments and industry sectors have been doing this for decades. An excellent example is the airline industry. Even though law is a necessary part of society, it has to be underpinned and supported with industry sector policy and procedural instruments. Government alone cannot regulate every part of the privacy and data protection chain. In the same way as the airline industry is global, Internet technology is global and knows no national borders. The development of data protection and privacy as a national and international issue has seen the rise in co-regulation. Industry sectors have established a regulatory tool box, which has been widely accepted and encouraged. However, some jurisdictions have imposed obligations on industry to adopt self-regulation through the establishment of Codes of Practice. The EU, even prior

[54] Ibid.

[55] *Privacy Commissioner v Telstra Corporation* Limited [2017] FCAFC 4.

[56] Ibid, at 61 to 64.

to the GDPR adopted a co-regulatory approach to data protection and privacy.[57] Australia, Japan, Malaysia and Singapore along with the EU has also adopted a comprehensive legal framework into a co-regulatory model.[58] India, Indonesia and Thailand are slowly moving towards a similar model. There are well established international institutions that are well equipped to assist organizations and industry to develop, implement and oversee their co-regulatory obligations.

2.1.2.1 ISO – IEC – Cobit

To assist entities in developing self and co regulation data protection frameworks, the International Organization for Standardization (ISO) and International Electrotechnical Commission (IEC) have formed the specialized system for worldwide standardization. National bodies that are members of ISO or IEC participate in the development of International Standards through technical committees established by the respective organization to deal with particular fields of technical activity. Even though the ISO Codes do not specifically[59] relate to data protection, they provide the basis for an organization to manage their actual and potential IT systems risks. The range of ISO standards relating to information security includes ISO 27001, ISO 27002 and ISO 27005.[60] These are designed for voluntary use, and in some cases are imposed by governments, as stated later in Chap. 1. ISO 27001 and ISO 27002, are referred to as the closest thing to a universal information security standard. These standards were referred to as the benchmark for reasonable security in those Australian texts that have considered the issue.

Moreover, ISO 27001 provides the specification for an information security management system (ISMS) against which certification by an ISO Certification body, based on the recommendation of an authorized third party auditor, can be granted. The standard provides a model for establishing, implementing, operating, monitoring, reviewing, maintaining, and improving an Information Security Management System.[61] On the other side, ISO 27005 defines a security risk as the as the potential that a given threat will exploit vulnerabilities of an asset or group of assets and thereby cause harm. Identifying the risk and prioritizing the action allows for controls to be established so as to manage the risk. In the context of business risk (s), they can be selected from the list in ISO 27001 Annex A and ISO 27002 for the

[57] Banisar D, Davies S (1999), *Global Trends in Privacy Protection: An International Survey of Privacy, Data Protection, and Surveillance Laws and Developments*, 18 J. Marshall Computer & Info L. 1, (1999) pp. 108–09.

[58] Ibid, 13.

[59] ISO 27001 and 27002.

[60] International Standards Organization, *ISO/IEC 27005: 2008 Information Technology – Security techniques – Information Security Risk Management* (2005) ('*ISO 27005*'), http://www.27000.org/iso-27005.htm.

[61] ISO 27001.

sake of completeness. ISO 27001 Annex A and ISO 27002 contains 14 control areas and 114 separate controls.[62]

The most commonly used of the ISO systems is 27,001. Cath Everett believes that ISO 27001 is outdated because it was developed more than 20 years ago. It focuses on outputs rather than inputs and does not take into consideration organizational or societal factors.[63] However, this form or regulatory standard has been widely accepted by the OECD and formed part of the 2002 Security guidelines, to enhance the risk-based regulatory approach.

In addition to ISO, the Control Objectives for Information Technology (Cobit). Cobit is a security framework that is used and developed by the Information Systems Audit and Control Association (ISACA). It provides controls for domains that include Planning and Organization, Acquisition and Implementation, Development and Support and Monitoring and Evaluation. Cobit compliments ISO 27001, and is broader in scope focusing on the governance of information technology. This is because, ISO 27001 is restricted and only focuses the governance rather than procedures.[64] Cobit ensures information security is managed at the highest level of the organization and in line with business requirements. It also identifies the necessary requirements based on a risk analysis and compliance requirement. Furthermore, the framework ensures it is effectively implemented and communicated to all users, including shareholders of an organization. This approach has the ability to strengthen an organizations risk-based approach to data protection, and serves to demonstrate to clients, customers, investors and shareholders that the organization is able to establish best practice management systems to manage the security of data, and privacy. Nevertheless, analyzing the extent to which this form of regulation has been adopted and implemented by each jurisdictions is outside the scope of this book.

2.2 Conclusion

This Chapter has discussed the economic importance the digital economy is likely to have to Australia, India, Indonesia, Japan, Malaysia, Singapore, Thailand and the European Union. The rise of the Internet has created pressures and tensions in the law. One of the challenges facing government, business community and broader

---

[62] There are 14 risks that include: Security policy; Organization of information security; Asset management (which includes classification); Human resources security; Physical and environmental security; Operations management; Access controls; Cryptographic controls; Information systems acquisition, development and maintenance; Communications security; Supplier management; Information security incident management; Business continuity management; and Compliance.

[63] Everett C (2011), 'Is ISO 27001 *worth it?' Computer Fraud & Security* 5; and Alan Gillies, 'Improving the quality of information security management systems with ISO 27000' 23(4) The TQM Journal, pp. 364–368.

[64] Gifford N (2009), *Information security: managing the legal risks,* CCH Australia Limited, pp. 196–198.

society is that we currently know very little about the extent of injuries or harm on and from the Internet. Doing business over the Internet raises a host of challenging issues, such as how to deal with electronic cash, online banking, commercial transactions in digital information and digital signatures. In addition, the rise of intellectual property, data protection (commercial and personal) and privacy in the Internet poses new legal dilemmas for preserving an asset which is challenged by the digital capture of data. Some of these challenges will be explored later in the book.

This Chapter has highlighted how a person's identity has been transformed as a result of the Internet. Prior to the Internet a person's identity was largely constrained to government records and long standing personal information such as ones' name, date of birth and place of residence. The Internet has introduced an online personal identity, which many people would be unaware exists. This new identity can be obtained from an individual's use of the internet, to more sophisticated systems that monitor people's Internet behavior. This new personal identity has begun to be defined by the law. However, as this book will highlight the definition of personal information or otherwise referred to as personal data is not always clear, and the courts are far from settled on what personal identity in and on the Internet finally constitutes.

This Chapter briefly introduced the co-regulatory approach that is currently applied, and needed in data protection and privacy, because government is not equipped to solely regulate every aspect of data protection and privacy. It is argued that this is a complex area, and Government alone cannot regulate every part of the privacy and data protection supply chain, or, life-cycle. The rise of data protection and privacy as an issue both nationally and internationally has resulted in the development of a co-regulatory framework. Industry sectors have established a regulatory tool box, which has been widely accepted and encouraged. However, along with most if of all other areas within data protection and privacy law, this co-regulatory model is also far from being settled. It is unlikely to be settled for some time because of the sheer growth and transition in the digital economy and Internet technology.

Finally, it is our view that, the co-regulatory approach is the way forward in data protection and privacy, no different to many other industries. Combine the co-regulatory model with legal convergence and harmonisation, and some of the gaps in data protection law may be addressed. Furthermore, combine this with an international Model Law will strengthen the governance of personal data. The next Chapter begins looks at the current day privacy laws of Australia.

References

Banisar D, Davies S (1999), *Global Trends in Privacy Protection: An International Survey of Privacy, Data Protection, and Surveillance Laws and Developments*, 18 J. Marshall Computer & Info L. 1, (1999) pp. 108–09

Everett C (2011), 'Is ISO 27001 *worth it?' Computer Fraud & Security* 5; and Alan Gillies, 'Improving the quality of information security management systems with ISO 27000′ 23(4) The TQM Journal, pp. 364–368

Fowler G (2017), *New privacy worry: Apple sharing your face with apps*, The Strait Times, p. 10

Gifford N (2009), Information security: managing the legal risks, CCH Australia Limited, pp. 196–198

Lessig L (1999) *Code and Other Laws of Cyberspace*, http://codev2.cc/download+remix/Lessig-Codev2.pdf, accessed 20 July 2017

Rustad M (1996) *Legal Resources for Lawyers Lost in Cyberspace*, Suffolf, U. L. Review, pp. 317–18

Smarajiva R (1995), *Interactively As Thought Privacy Mattered, Technology and Privacy*, supra, note 4, pp. 277–279

Wilton R (2008), *Identity and Privacy in the Digital Age*, Int. J. Intellectual Property Management, Vol. X, No. Federated Identity Chief Technology Officer c/o Sun Microsystems Guillemont Park, Camberley, Surrey GU17 9QG, United Kingdom, pp. 1–15

Part III

Chapter 3
European Law

Abstract This Chapter highlights the importance of data protection law to the European Union (EU) and its member states. Chapter 3 will discuss the General Data Protection Regulation (GDPR) (Regulation 2016/679 Of the European Parliament and the European Council, on the protection of natural persons with regard to the processing of personal data and on the free movement of such data and repealing Directive 95/46/EC (General Data Protection Regulation), Official Journal of the European Union L 119/1), which has strengthened data protection law across member states of the EU. This Chapter does not discuss the GDPR's predecessor, the 1995 Directive on data protection. The EU has placed the concept of privacy as a high priority, no matter where and how the concept is applied, and this includes the protection of personal data. The European Charter of Fundamental Rights 2000 (Articles 7 & 8, Charter of Fundamental Rights of the European Union, Official Journal of the European Union, 2000/C 364/01) protects European citizen's private life and personal data. Arguably, the EU have led the way in the area of data protection and privacy. The European Convention of Human Rights (ECHR), ratified in 1953, has for more than 60 years provided the basis for human rights across Europe. The European Charter of Fundamental Rights, promulgated in 2000, enhanced citizen's right to privacy and personal data protection. The European Court of Justice has discussed the right to be forgotten and this Chapter will briefly explore this concept (The Court of Justice of the EU and the "Right to be Forgotten", http://ec.europa.eu/justice/data-protection/files/factsheets/factsheet_rtbf_mythbusting_en.pdf, accessed 7 June 2018).

However, there is a fine line between these rights and having to balance those in conjunction with the single market concept. This Chapter also highlights how the EU and GDPR, are making a significant impact beyond the borders of the EU. In other words, it is argued that the GDPR while considering the single market has placed the right to privacy over the Internet ahead of most other nation states data protection laws. Due to the depth of EU data protection law, Chapter 3 begins with an overview of the GDPR, the Definition of Personal Data, and then highlights the key concepts embodied within the law. This includes, Controller, Processor and Officer, the Right to be Forgotten, the Agency [Regulator] – Authority, whether the laws apply to both the Public and Private sectors, Consent, the Extra-Territorial Reach of the law, Retention Principles and Codes of Practice, the Cross Border Transfer of personal data and Breaching the laws. This Chapter concludes by argu-

© Springer Nature Singapore Pte Ltd. 2019

R. Walters et al., *Data Protection Law*,
https://doi.org/10.1007/978-981-13-8110-2_3

ing how the EU has to a large degree been able to drag other nation states into the same circle – in order to regulate and protect personal data over the Internet more pervasively. It is our view that the GDPR now sets the overall benchmark for data protection law in the world.

3.1 Introduction

The EU has a long history of dealing with data protection and privacy. Arguably, the EU and its member states have led the way, not only in the area of data protection and privacy, but in respect of human rights more broadly. The European Convention of Human Rights (ECHR) 1950 came into force in 1953, established the beginnings of a more comprehensive legal framework to protect individuals human rights.[1] Article 8 provides the right to a private life, but there is no specific mention of data protection.[2] Since this period, the Council of Europe has arguably been at the forefront of human rights protection in Europe.[3] Simon Bronnitt argues the ECHR adopted a court centered model. That is, the ECHR establishes the European Court of Human Rights, which are part of the Council of Europe.

In 2000 the European Charter of Fundamental Rights[4] was introduced and strengthened many areas of human rights for European citizen's. The Charter

[1]The European Convention of Human Rights (ECHR) 1950. It was the first instrument to give effect and binding force to certain of the rights stated in the Universal Declaration of Human Rights. It established the European court of human Rights. More importantly the ECHR has been used to successfully to strengthen human rights. That is, in order to join the Council of Europe, a State must first sign and ratify the European Convention on Human Rights, thus confirming its commitments to the aims of the Organization, namely the achievement of greater unity between its members based on human rights and fundamental freedoms, peace and respect for democracy and the Rule of Law.

[2]Article 8, Convention for the Protection of Human Rights and Fundamental Freedoms, 1950. Rome, 4.XI.1950, European Treaty Series-No. 5. Everyone has the right to respect for his private and family life, his home and his correspondence. There shall be no interference by a public authority with the exercise of this right except such as is in accordance with the law and is necessary in a democratic society in the interests of national security, public safety or the economic well-being of the country, for the prevention of disorder or crime, for the protection of health or morals, or for the protection of the rights and freedoms of others.

[3]Bronitt, S *A Tale of Two European Charters of Rights: Comparing the European Convention on Human Rights and the EU Charter of Fundamental Rights*, Professor, Director, National Europe Centre and Australia National University College of Law, http://www.academia.edu/6410839/A_Tale_of_Two_European_Charters_of_Rights_Comparing_the_European_Convention_on_Human_Rights_and_the_EU_Charter_of_Fundamental_Rights, accessed 13 December 2018.

[4]Articles 7 & 8, Charter of Fundamental Rights of the European Union, Official Journal of the European Union, 2000/C 364/01. Article 7 states that everyone has the right to respect for his or her private and family life, home and communications. Article 8 provides that everyone has the right to the protection of personal data concerning him or her. Such data must be processed fairly for specified purposes and on the basis of the consent of the person concerned or some other legitimate basis laid down by law. Everyone has the right of access to data which has been collected concerning him or her, and the right to have it rectified. Compliance with these rules shall be subject to control by an independent authority.

reinforces the need to protect an individual's private life, but more importantly, protect people's personal data. Bronnitt highlights that the Charter adopts a legislative centered model.[5] The Charter was the first legal document to include data protection as a fundamental human right.[6] The Charter is binding primarily on the institutions of the European Union themselves and all member states. The binding effect of the 2000 Chapter has paved the way for personal data and privacy over the Internet, to be protected through the European courts.[7]

It is well understood that the EU have human rights front and center of their policy and legal considerations, across all areas of society. In other words, the respect of human rights together with the principles of freedom, democracy, equality and the rule of law, are values common to all European Union (EU) member states.[8] In the case of *Nold Kohlen- und Baustoffgroßhandlung v Commission of the European Communities*[9] the CJEU asserted its commitment to human rights in the strongest of terms. The CJEU ruled that:

> in safeguarding human rights, the court is bound to draw inspiration from constitutional tradition common to member states, and it cannot therefore uphold measures which are incompatible with fundamental rights recognized by the constitutions of those states. Similarly, international treaties and conventions for the protection of human rights on which the member state have collaborated or of which they are signatories', can supply guidelines which should be followed with the framework of Community law.[10]

Furthermore, another international treaty that relates to human rights which has been referred to by the CJEU in *Gabrielle Defrenne v Société anonyme belge de navigation aérienne Sabena*, as constituting a source of general principles is the European Social Charter 1971.[11] The European Social Charter is a Council of

[5] Bronitt, S *A Tale of Two European Charters of Rights: Comparing the European Convention on Human Rights and the EU Charter of Fundamental Rights*, Professor, Director, National Europe Centre and Australia National University College of Law, http://www.academia.edu/6410839/A_Tale_of_Two_European_Charters_of_Rights_Comparing_the_European_Convention_on_Human_Rights_and_the_EU_Charter_of_Fundamental_Rights, accessed 13 December 2018.

[6] Ibid, Article 8.

[7] Emmert, F, Carney, CP *The European Union Charter of Fundamental Rights vs. The Council of Europe Convention On Human Rights And Fundamental Freedoms – A Comparison*, Fordham International Law Journal *Volume* 40, *Issue* 4 (2017). Within their own sphere of authority, the Member States remain accountable under their own constitutional provisions, as well as the European Convention. However, to the extent the Charter might provide substantially better protection, the Member States are not bound by it when adopting or implementing their own law outside of the sphere of application of EU law.

[8] EUR-Lex Access to European Union Law, Human Rights, Respect for human rights and dignity, together with the principles of freedom, democracy, equality and the rule of law, are values common to all European Union (EU) countries, https://eur-lex.europa.eu/summary/chapter/human_rights.html?root_default=SUM_1_CODED%3D13, accessed 13 December 2018.

[9] Case 4/73 J *Nold Kohlen- und Baustoffgroßhandlung v Commission of the European Communities [1974]*, Court of Justice of the European Union.

[10] Ibid.

[11] Case 149/77 *Gabrielle Defrenne v Société anonyme belge de navigation aérienne Sabena [1977]*.

Europe treaty that guarantees fundamental social and economic rights as a counterpart to the European Convention on Human Rights, which refers to civil and political rights. It guarantees a broad range of everyday human rights related to employment, housing, health, education, social protection and welfare.[12]

Notwithstanding the above, the CJEU continues to reinforce its commitment to privacy in data protection, whereby it expressed the importance of the concept of the right to be forgotten.[13] This is what sets the EU apart from any other region and country in the world. Human rights have emerged following WWII as a key policy pillar within the EU legal framework. The EU has had decades of experience in balancing human rights with economic and other social policy issues. It is argued that the EU data protection and privacy law, including the right to be forgotten, is having a significant influence on the same laws of other jurisdictions discussed throughout this book. The right to be forgotten alone has today set a benchmark, and to varying degrees is forcing other countries to consider the concept.

In addition to the above, the Council of Europe, has been, and remains responsible for the Convention for the Protection of Individuals with regard to Automatic Processing of Personal Data which was introduced in 1981 (Convention 108).[14] Convention 108 provides for the setting up of national supervisory authorities who are responsible for ensuring compliance with laws or regulations adopted in pursuance of the convention, concerning personal data protection and transborder data flows,[15] to third countries. Data may only be transferred if the recipient state or international organization is able to afford an adequate level of protection. Furthermore, Convention 108 is a binding multilateral instrument that has been an international benchmark[16] for data protection. Convention 108 is important for countries across Asia and allows countries outside the EU to also adopt and ratify the Convention.[17] Bu-Pasha is of the view that the Convention is regarded as the first and only legally binding international instrument that is not only accessible within the EU, but also outside the EU, and third countries.[18] Firstly, Convention 108,

[12] The European Social Charter, https://www.coe.int/en/web/turin-european-social-charter, accessed 13 November 2018.

[13] The Court of Justice of the EU and the "Right to be Forgotten", http://ec.europa.eu/justice/data-protection/files/factsheets/factsheet_rtbf_mythbusting_en.pdf, accessed 7 June 2018.

[14] Council of Europe, Data Protection Commissioner, is in charge of overseeing that data protection rules are respected for all personal data collected and processed by the council of Europe, www.coe.int, accessed 12 December 2018. The Committee of Ministers has responsibility for the adoption, review and change of the Convention and its Protocols.

[15] European Council, Details of Treaty No.181 *Additional Protocol to the Convention for the Protection of Individuals with regard to Automatic Processing of Personal Data, regarding supervisory authorities and transborder data flows.*

[16] Greenleaf, G *Asian Data Privacy Laws: Trade and Human Rights Perspective*, Oxford University Press, (2014), pp. 35–42.

[17] Ibid.

[18] Bu-Pasha S, *Cross-border issues under the EU data protection law with regards to personal data protection*, Taylor & Francis, (2017) pp. 213–228.in p.

along with the OECD principles of 1981[19] provide a minimum set of privacy principles that serve as an international standard. Secondly, Convention 108, Protocol 2001, along with the EU Directive 95/46/EC [20] on the processing of personal data created standards that have not only been influential across Europe, but also across the world. To date, no country in the Asia Pacific Region, or country discussed in this book, has signed or ratified Convention 108. The EU is now more strongly supporting Convention 108 as a global privacy treaty.[21] In other words, the globalization of Convention 108 is accelerating, with Mauritius, Senegal, and Tunisia invited to accede in 2015. Moreover, in 2018 the modernization of Convention 108 pursued two main objectives. The first is to deal with challenges resulting from the use of new information and communication technologies and the second is to strengthen the Convention's effective implementation.[22] However, Convention 108 only requires a Party to 'take the necessary measures in its domestic law to give effect' to the principles in the Convention.[23]

The European Union Data Protection Directive[24] in 1995, was a major step towards harmonizing member states law in regard to data protection and privacy law. The Directive required member states to protect the privacy of personal information. The Directive has now been replaced by the 2018 GDPR. This is an important point because the GDPR is an EU Regulation, which under the supranational legal framework has a higher status than the former Directive.[25] That is, member states are obliged to fully implement and fully transpose EU Regulation into national law. Arguably, the EU has strengthened the uniform standards that are directly related to EU member states, who are responsible for protecting an individual's personal data over the Internet.

---

[19] OECD *Guidelines on the Protection of Privacy and Transborder Flows of Personal Data, https://www.oecd.org/sti/ieconomy/2013-oecd-privacy-guidelines.pdf*, accessed 5 December 2018.

[20] Directive 95/46/EC of the European parliament and European Council on the protection of individuals with regard to the processing of personal data and on the free movement of such data, Official Journal of the European communities, L281/31.

[21] Greenleaf, G *Asia Data Privacy Laws – Trade and Human Rights Perspectives*, University New South Wales, (2017).

[22] Council of Europe, Modernisation of Convention 108, https://www.coe.int/en/web/data-protection/convention108/modernised, accessed 13 November 2018.

[23] Chart of signatures and ratifications of Treaty 108, *Convention for the Protection of Individuals with regard to Automatic Processing of Personal Data,* Status as of 24/04/2018 https://www.coe.int/en/web/conventions/full-list/-/conventions/treaty/108/signatures?p_auth=VbbXiWQw, accessed 24 April 2018.

[24] Directive 95/46/EC of the European parliament and European Council on the protection of individuals with regard to the processing of personal data and on the free movement of such data, Official Journal of the European communities, L281/31.

[25] Note, Regulations are legal acts that apply automatically and uniformly to all EU countries as soon as they enter into force, without needing to be transposed into national law. They are binding in their entirety on all EU countries. Directives require EU countries to achieve a certain result, but leave them free to choose how to do so. EU countries must adopt measures to incorporate them into national law (transpose) in order to achieve the objectives set by the directive.

The GDPR has provided wider power and therefore a broader reach. *van der Sloot* highlights that the GDPR has direct effect and need not be implemented in the national legal frameworks of the different member countries.[26] Besides extended rules for cross-border data processing,[27] the GDPR introduces the possibility of a leading supervisory authority investigating an EU-wide data processing activity.[28] The former Working Party[29] has been replaced by a European Data Protection Board (EDPB), and is granted wider powers.[30] However, it is out of scope to fully explore the wider powers afforded to the EDPB. The GDPR has also introduced fines and sanctions of up to €20,000,000 or, in the case of a commercial entity, up to 4% of its annual worldwide turnover, which will also be discussed later in this Chapter. This new addition significantly raises the stakes for entities breaching the law. It must be noted that many entities operating in this area of the economy are attracting excessive profits, that far outstrip the level of penalty that could, and can be imposed by the regulator.

The CJEU has also reaffirmed[31] the notion that the fundamental rights and freedoms across the EU are weighted highly, which arguably has been used to advance the right to privacy over the Internet, and more broadly citizens' rights in data protection. The court has also had to strike a balance between the right of data protection and privacy with that of the right to property to be equally protected.[32] For instance, the court had to interpret EU law protecting the intellectual property and copyright ("sketching the outline of a ghost: the fair balance between copyright and fundamental rights in intermediary third party liability"), to that of the individual's right to privacy and data protection.[33] However, it must be noted that the early developments in jurisprudence by the courts were at a time when the previous Directive 95/46/EC on data protection was in force.

---

[26] Bart van der Sloot & Frederik Zuiderveen Borgesius The EU General Data Protection Regulation: A New Standard for Information Privacy, https://bartvandersloot.com/onewebmedia/SSRN-id3162987.pdf, accessed 13 December 2018.

[27] Ibid.

[28] Ibid.

[29] Article 29 Working Party", which is the short name of the Data Protection Working Party established by Article 29 of Directive 95/46/EC.

[30] Ibid. Bart van der Sloot further states the legal basis of the Data Protection Directive is the regulation of the internal market.

[31] C-29/69; *Stauder v City of Ulm*, judgment in C-11/70. It was highlighted that the CJEU also contributed through its case-law to the development of and respect for fundamental rights, by affirming that the Treaties also protected those fundamental rights which result from the constitutional traditions common to the Member States as general principles of Community law.

[32] Ibid.

[33] Case C-70/10 (*Scarlet Extended SA v Société belge des auteurs, composi- teurs et éditeurs SCRL (SABAM)*), judgment of 24 November 2011, [2011] E.C.R. I-11959; Cf. Info 17(6), p. 72–74. Case C-275/06 (*Productores de Música de España (Promusicae) v Te- lefónica de España SAU*), judgment of 29 January 2008, [2008] E.C.R. I-00271, para. 51.

Moreover, the CJEU has strengthened the protection of data and privacy as a fundamental human right. In *Productores de Música de España (Promusicae) v Telefónica de España SAU*[34] the CJEU had to ruled that:

> data protection is based on the fundamental right to private life, as it results in particular from Article 8 of the [ECHR]. The Charter of Fundamental Rights confirmed that fundamental right in Article 7, and in Article 8 specifically emphasized the fundamental right to the protection of personal data, including important fundamental principles of data protection.[35]

However, 2 years later the CJEU ruled in *Volker und Markus Schecke GbR, Hartmut Eifert v Land Hessen*[36] that the right to the protection of personal data is not an absolute right. It must be considered in relation to its function in society.[37] The CJEU interpreted the EU Data Protection Directive 95/46/EC as ensuring the effective and complete protection of the fundamental rights and freedoms of the natural person in accordance with Articles 7 and 8 of the Charter.[38] Arguably, the EU reinforced the principle of effective and complete protection whereby the court confirmed the importance of protecting citizens fundamental rights and freedoms to the highest possible level.[39]

The issue of having to balance competing and sometimes conflicting legal principles was also reinforced in *Coty Germany GmbH v. Stadtsparkasse Magdeburg*.[40] In this case the court had to balance the right to information with the right to intellectual property and banking secrecy. The court stated:

> that the communication of the name and address of a banking institution's costumer constitutes the processing of personal data.[41]

The court went on to say that this case highlights the intention to implement the fundamental right to an effective remedy concerning the infringement of the right to property and, on the other hand, the right to protection of personal data.[42] The Court argued that:

---

[34] Case C-275/06 *Productores de Música de España (Promusicae) v Te- lefónica de España SAU*, judgment of 29 January 2008, [2008] E.C.R. I-00271, para. 51.

[35] Ibid.

[36] Cases C-92/09 and C-93/09 *Volker und Markus Schecke GbR, Hart- mut Eifert v Land Hessen*, judgment of 9 November 2010, [2010] E.C.R. I-11063.

[37] Ibid.

[38] Case C-131/12 *Google Spain SL and Google Inc. v Agencia Española de Protección de Datos (AEPD) and Mario Costeja González*, judgment of 13 May 2014, ECLI: EU:C:2014:317, 53.

[39] Case C-362/14 *Maximillian Schrems v Data Protection Commissioner*, judgment of 6 October 2015, ECLI:EU:C:2015:650, 39.

[40] Case C-580/13, *Coty Germany GmbH v. Stadtsparkasse Magdeburg*, 16 July 2015, para.26–34.

[41] Ibid.

[42] Ibid.

> EU law requires that, when transposing Directives, the Member States must take care to rely on an interpretation of them which allows a fair balance to be struck between the various fundamental rights protected by the EU legal order.[43]

The court was saying that it is a constant balancing exercise between protecting and advancing social (rights) and economic policies. The competing and conflicting policy areas, have arguably created tension within the law, because the EU and its citizens, unlike other nation states, demand that their personal data and privacy over the Internet be protected. Chapters 13, 14 and 15 look at the developing law between intellectual property, competition and transnational contracts, with data protection law. These Chapters will highlight the tension and conflicts in the law.

The concerns raised by the EU in relation to privacy over the Internet appear to be valid. That is, the extent to which personal data that can identify the data subject is being captured, stored and used was summarized by the Court of Justice of the European Union. In *Case -698/16, Joined Cases, Court of Justice of the European Union, Grand Chamber*[44] the court was concerned with the way metadata was being held and stored. The court held that meta data:

> "allows very precise conclusions to be drawn concerning the private lives of the persons whose data has been retained, such as everyday habits, permanent or temporary places of residence, daily or other movements, the activities carried out, the social relationships of those persons and the social environments frequented by them [...]. In particular, that data provides the means [...] of establishing a profile of the individuals concerned, information that is no less sensitive, having regard to the right to privacy, than the actual content of communications.[45]

Arguably, and unless one is working within the Internet profession and fully understands the infrastructure used to collect and store data, the use of meta data, which constitutes a level of personal data defined by the law, would not know their privacy is being infringed. Therefore, one argument has to be, the need for greater transparency. Transparency is a fundamental principle of the OECD and enhances the other principle of 'trust' over the Internet. The resulting effect would allow the general public to better understand what they are dealing with, when using the Internet.

However, with the implementation of the GDPR, even the member states of the EU are yet to fully united behind the laws. At implementation, only Germany and Austria had adopted new data protection laws that are in line with the EU GDPR.[46] Graham Greenleaf rightly raises the concern that harmonization amongst EU member states may not have been achieved following the implementation of the GDPR, and this poses complications for third countries who are dealing with organizations

[43] Ibid.

[44] (C-698/16), *Joined Cases, Court of Justice of the European Union, Grand Chamber*, Judgment (21 December 2016).

[45] C-698/16, *Joined Cases, Court of Justice of the European Union, Grand Chamber*, Judgment (21 December 2016).

[46] Greenleaf G, *The Legal and Business Risks of Inconsistencies and Gaps in Coverage in Asian Data Protection Laws* Session II Materials, *Asian Business Law Institute (ABLI) Data Privacy Forum*, Singapore, (2018).

in member states or multiple member states. At the time of writing this book, full and complete harmonization had not been achieved. As highlighted in Chap. 1 of the key questions likely to emerge in the future, which has begun to surface, is what model will serve the international community best in the future?

It is our view that governments, regulators, experts, the general population and nation states have largely, to date, looked to the EU to develop their respective data protection laws. Doing so, has seen the rise in the need to protect a data subject's privacy over the Internet. Today, the EU model can be best described as not only balancing the single market, but also ensuring that privacy over the Internet remains a fundamental right. This rights based approach is setting the direction for data protection law across the world. However, it is also our view that Singapore has created a business friendly model, and could be seen as a benchmark into the future. It is also our view that Australia's balanced model sits somewhere between the two (the EU and Singapore) and could also emerge as a benchmark in the future. No doubt the EU, is likely to argue to the international community that their model must be assured – for the protection of their citizens. In part, this is being achieved through the GDPR's extra-territorial reach. Ultimately, it will come down to whether nation states see that privacy over the Internet is such a public policy issue that their data protection laws will follow the EU. In other words – do the citizens see that their personal data needs to be protected for privacy reasons? However, if people do not care about their personal data and privacy over the Internet, Singapore's model, would be considered as being a valid international option (model).

3.2 General Data Protection Regulation

The GDPR respects all fundamental rights, particularly the right to private and family life, home and communications, the protection of personal data, freedom of thought, conscience and religion, freedom of expression and information.[47] Additionally, the GDPR respects the freedom to conduct a business, the right to an effective remedy and to a fair trial, and cultural, religious and linguistic diversity. The EU has retained the balance of these rights, with the overall object of the Union to retain and strengthen the common economic market. The GDPR reinforces this point within the preamble whereby at (6) it states that 'technology allows both private companies and public authorities to make use of personal data on an unprecedented scale in order to pursue their activities'.[48] The GDPR also provides greater legal certainty, and arguably, an even level playing field for all companies involved with and located in the EU market. Its replacement of the earlier Directive 95/46/

[47] Regulation 2016/679 Of the European Parliament and the European Council, on the protection of natural persons with regard to the processing of personal data and on the free movement of such data and repealing Directive 95/46/EC (General Data Protection Regulation), Official Journal of the European Union L 119/1.

[48] Ibid.

EC, introduced the concepts of data portability, standardized privacy icons along with data protection by design and default.[49] Nonetheless, the notion of privacy by design or privacy by default have become important concepts.[50]

The protection of the rights and freedoms of natural persons with regard to the processing of personal data require that appropriate technical and organizational measures be taken to ensure that the requirements of the Regulation are met.[51] When developing, designing, selecting and using applications, services and products that are used for processing of personal data, should also take into consideration the design aspect – to ensure privacy is protected. In other words, producers of the products, services and applications should be encouraged to take into account the right to data protection, and build products with data protection in mind. Furthermore, and due to the many public tenders provided by the EU, the principles of data protection by design and by default should also be taken into consideration.[52]

Mike Hintze and Gary LaFever argue that the GDPR specifically developed privacy by design,[53] by regulating personal data and information. The resulting effect has seen a level of privacy protection over the Internet, but, that level is not fully understood. It is such a nebulous concept to understand and measure because technology continues to evolve and change. Hintze and LaFever point out that data protection by default supports data protection over the full lifecycle of data by leveraging technical and organizational measures, including pseudonymization, to ensure that, by default, personal data is not made accessible without the individual's intervention to an indefinite number of natural persons.[54] It has enhanced the risk-based regulatory model to strengthen the governance of personal data. However, and

[49] Albrecht J, *How the GDPR Will Change the World, European Data Protection Law Review*, Volume 2, Issue 3, (2016), pp. 287–289.

[50] Regulation 2016/679, Article 25 states that – Taking into account the state of the art, the cost of implementation and the nature, scope, context and purposes of processing as well as the risks of varying likelihood and severity for rights and freedoms of natural persons posed by the processing, the controller shall, both at the time of the determination of the means for processing and at the time of the processing itself, implement appropriate technical and organizational measures, such as pseudonymisation, which are designed to implement data-protection principles, such as data minimization, in an effective manner and to integrate the necessary safeguards into the processing in order to meet the requirements of this Regulation and protect the rights of data subjects. The controller shall implement appropriate technical and organizational measures for ensuring that, by default, only personal data which are necessary for each specific purpose of the processing are processed. That obligation applies to the amount of personal data collected, the extent of their processing, the period of their storage and their accessibility.[3] In particular, such measures shall ensure that by default personal data are not made accessible without the individual's intervention to an indefinite number of natural persons.

[51] Recital 78.

[52] Ibid.

[53] Hintze M, LaFever, G (2018) *Meeting Upcoming GDPR Requirements While Maximizing the Full Value of Data Analytics, Balancing the Interests of Regulators, Data Controllers and Data Subjects, Unlock Big Data Value while Complying with the GDPR,* http://files8.design-editor.com/93/9339158/UploadedFiles/1B4F2EF8-BC8D-A12D-C9B1-7DF644A29C1F.pdf, accessed 24 April 2018.

[54] Ibid.

similar to data protection by design, data protection by default remains fluid. Yet, the principles that allow for a level of intervention, place data protection by default as providing a higher level of certainty – in protecting data subject's personal data.

Notwithstanding the above, data impact and privacy assessments (DPIA) have also been introduced as an important step in regulating the risk associated with personal data on the Internet.[55] The impact assessment is a risk management tool that can be used by an entity to assess the risk of a data breach within a single or multiple projects. Moreover, the impact assessments are one way an entity can assess the potential level of privacy harm to data subjects. The assessment should comprise of a systematic description of the envisaged processing operations and the purposes of the processing, including, where applicable, the legitimate interest pursued by the controller. In addition, there is to be an assessment of the necessity and proportionality of the processing operations in relation to the purposes. Failing to carry out a DPIA when the processing is subject to a DPIA[56], or carrying out a DPIA in an incorrect way,[57] or failing to consult the competent supervisory authority where required,[58] can result in a fine of up to €10 M. The risk management imposed by the GDPR goes some way to better understanding what level of risk might be come from a project or activity undertaken by the entity.

3.3 Definition of Personal Data

The GDPR defines personal data to mean any information relating to an identified or identifiable natural person ('data subject') an identifiable natural person is one who can be identified, directly or indirectly.[59] In particular by reference to an identifier such as a name, an identification number, location data, an online identifier or to one or more factors specific to the physical, physiological, genetic, mental, economic, cultural or social identity of that natural person.[60] Arguably, this is a broad definition that encompasses many different issues and scenarios where a person might be identified by or over the Internet and computer through its search engines, websites, systems and platforms.

---

[55] Regulation 2016/679 Of the European Parliament and the European Council, on the protection of natural persons with regard to the processing of personal data and on the free movement of such data and repealing Directive 95/46/EC (General Data Protection Regulation), Official Journal of the European Union L 119/1, Article 35, Recitals 75, 89, 91 and 93. Guidelines on Data Protection Impact Assessment (DPIA) and determining whether processing is "likely to result in a high risk" for the purposes of Regulation 2016/679.

[56] Ibid, Article 35(1) and (3)–(4).

[57] Ibid, Article 35(2) and (7)–(9).

[58] Ibid, Article 36(3)(e).

[59] Ibid, Article 4.

[60] Ibid.

The definition and characterization of personal data has been tested by the courts. In *ClientEarth and Pesticide Action Network Europe (PAN Europe) v European Food Safety Authority*[61] the CJEU ruled that the:

> 'characterization as personal data cannot be excluded: a) by the fact that the information is provided as a part of the professional activity and b) by the circumstance that the identity of the experts and the comments were previously made public on the EFSA website and c) by the circumstance that the persons concerned do or do not object'.[62]

This case offered an opportunity of ruling on a question whether the interaction between the general or ordinary regulatory scheme relating to access to the documents of the institutions laid down in Regulation 1049/2001[63] and the specific or special schemes laid down in other EU legislative provisions. It was also seen as an opportunity to reconcile the regulatory scheme relating to access under that regulation with the rules on the processing of personal data laid down in Regulation 45/2001.[64] In other words, the case involved the disclosure of personal data of scientific experts commenting on the placing of plant protection products onto the market. At issue was also whether the transfer of the personal data was necessary and determine whether that transfer was prejudicial to the legitimate interest of the data subject. The court held that:

> the fact that information is provided as part of a professional activity does not mean that it cannot be characterized as a set of personal data. Personal data within the meaning of Article 2(a) of Regulation No 45/2001, and of 'data relating to private life' are not to be confused. Personal data may, as a general rule, be transferred only if the recipient establishes the necessity of having the data transferred and if there is no reason to assume that that transfer might prejudice the legitimate interests of the data subject.[65]

Therefore, what can be seen from this case was that in general, there is no automatic priority to be conferred when having to balance other norms and principles with the objective of transparency over the right to the protection of personal data. The case is an example of where the court had to balance four elements (1) privacy (2) the market, (3) the concept of personal data and (4) the consideration of the definition of personal data.

Nevertheless, the GDPR does not specifically define sensitive data. Article 9 deals with processing special categories of personal data.[66] The processing of personal data revealing racial or ethnic origin, political opinions, religious or

---

[61] Case C-615/13, *ClientEarth and Pesticide Action Network Europe (PAN Europe) v European Food Safety Authority*, 16 July 2015, para 29–30.

[62] Ibid.

[63] Ibid, when referring to Regulation 1049/2001, regarding public access to European Parliament, Council and Commission documents, Official Journal of the European Union, L 145.

[64] Ibid, when referring to Regulation 45/2001, on the protection of individuals with regard to the processing of personal data by the Community institutions and bodies and on the free movement of such data Official Journal of the European Union, L 8.

[65] Ibid, 38–46.

[66] Regulation 2016/679, Article 9.

philosophical beliefs, or trade union membership is illegal.[67] Furthermore, prohibitions also exist in relation to processing genetic data, biometric data for the purpose of uniquely identifying a natural person, data concerning health or data concerning a natural person's sex life or sexual orientation – is prohibited.[68] However, there are exemptions to this and Article 9 (2) states that sensitive data can be processed where consent has been obtained. Furthermore, sensitive data can be processed for employment and social security and social protection law.[69] This also extends to areas that are in the public interest such as health and national security. Capacity has been provided to member states that allows them to introduce additional conditions regarding the processing of genetic data, biometric data or data concerning health.

The exemptions introduced by Article 9(2) are considered far reaching and extend to genetic data used in research. Thus, for Kart Pormeister, the question has been whether specific or broad consent is required under EU or national laws?[70] Pormeister argues that the research exemption creates a situation where once genetic data has been obtained from the data subject, it can be further processed for any research purposes, and stored for an unspecified time to enable such processing.[71] The obligation to inform the data subject is likely to not apply in cases of processing genetic data for research purposes, the data subject will not be aware of the processing. Therefore, the right to object to the processing of sensitive data for research purposes could be excluded by a member state under Article 89(2) or Article 21(6). The broad approach that can be applied under Article 9 provides that consent is largely left to other EU and national laws. It is outside the scope of this Chapter to examine the other EU laws and national laws.

This current definition also includes IP addresses. Arguably, it is the most advanced definition of personal data and information in the world. As will be shown in this Chapter, this definition coupled with the other key concepts and principles such as consent and data portability, provide data subjects with greater control and ownership over their personal data. Not having a clear and broad definition will significantly restrict a data subject's ability to identify what personal data is protected over the Internet. The challenge will be whether the current definition adequately provides for and allows the courts to deal with all areas of the law that data protection law transcends. Another challenge is how this definition has or will, if at all, influence the laws of other countries to expand their definition of personal data and personal information.

---

[67] Ibid.

[68] Ibid.

[69] Regulation 2016/679, Recital 10, 34, 35, 51.

[70] Pormeister K, *Genetic data and the research exemption: is the GDPR going too far?,* International Data Privacy Law, Volume 7, Issue 2, (2017) pp. 137–146.

[71] Ibid, for example, the exemption is not subject to purpose or storage limitation.

3.4 Controller, Processor and Officer

The EU have adopted a multi-layered approach to ensure organizations are responsible for the management or personal data. The legal basis that requires the appointment of a Data Controller has become an important element of the EU data protection framework. Unlike other countries, who do not have this legal requirement, it places a level of responsibility on an organization that is handling personal data to appoint such a position. The data Controller is responsible for implementing an organizations technical, systems and processors that will collect, store and use personal data. They are required to protect the organization from the illegal collection, use and processing of data.[72] This has been reinforced by Article 25, which requires that the controller to establish measures to ensure personal data is limited to the required recipient.[73] The Controller has an added layer of responsibility as they are also required to comply with the principle set out in Article 5. For instance, the Controller is accountable for an organizations data protection policies and procedures. The principle of accountability requires a data protection impact assessment to be undertaken to determine the level of risk to the rights and freedoms of a person, to whom the data applies.

For larger organizations, where appointing a single Controller would not be adequate, there is the ability under the law to appoint joint controllers. That is, Article 26 states that where two or more controllers can determine their respective responsibilities for compliance with the obligations under the GDPR, however, the organization can split the responsibilities amongst the controllers. Furthermore, Controllers will be responsible for exercising the rights of the data subject in accordance with Article 13 and 14.[74] Firstly, Article 13 requires that information to be provided where personal data are collected from the data subject. Secondly, Article 14 requires that information to be provided where personal data has not been obtained from the data subject.

Nonetheless, Controllers could be located in jurisdictions (other countries) outside of the EU. Article 27 provides that representatives of controllers can be established, while not being located in the EU or any of its member states.[75] However, this does not apply to large scale processing of special categories of data as referred to in Article 9(1) or processing of personal data relating to criminal convictions and offences referred to in Article 10. In other words, Article 9 restricts certain data that can reveal the race, ethnic origin, political or religious belief. It also applies to membership of a trade union and revealing biometric data or sex and sexual orientation of a person.[76] The restrictions are all consistent with the fundamental rights set out in the European Charter of Human rights 2000.

---

[72] Regulation 2016/679, Article 24.

[73] Ibid, Article 25.

[74] Ibid, Article 13–14.

[75] Ibid, Article 27.

[76] Ibid, Article 9.

3.4.1 Processor

A Controller may appoint a Processor who is required to comply with the GDPR.[77] There is a binding obligation on the Controller when appointing a Processor, which must be done in writing.[78] This legislative step also sets the EU apart from other jurisdictions discussed throughout this book. The agreement must state the following, that the Processor:

- only act on the controller's documented instructions;
- impose confidentiality obligations on all personnel who process the relevant data;
- must ensure the security of the personal data that it processes;
- abide by the rules regarding appointment of sub-processors;
- implement measures to assist the controller in complying with the rights of data subjects;
- assist the controller in obtaining approval from Data Protection Assessor where required;
- at the controller's election, either return or destroy the personal data at the end of the relationship (except as required by EU or Member State law); and
- provide the controller with all information necessary to demonstrate compliance with the GDPR.[79]

The Processor can only act on written instructions of the Controller. A Processor may appoint a sub-processor but only on the approval of the Controller. Article 28(2)(4) states that a Processor must not appoint a Sub-Processor without the prior written consent of the Controller.[80] Upon agreement of the controller of a Sub-Processor (s), those Sub-Processors must be appointed on the same terms as are set out in the contract between the Controller and the Processor.[81] This approach ensures there is consistency and continuity in the application of the GDPR between the Controller, Processor and Sub-Processor (s). Article 28(3)(h) requires that, in the event that a Processor believes that the Controller's instructions conflict with the requirements of the GDPR or other EU or Member State laws, the Processor must immediately inform the Controller.[82] A Processor must keep records of its processing activities performed on behalf of the Controller. This includes the details of the Processor; the categories of processing activities performed; information regarding Cross-Border Data Transfers; and a general description of the security measures implemented.[83]

---

[77] Ibid, Article 28(1)(3).

[78] Ibid, Article 28(1)(3).

[79] Ibid.

[80] Ibid.

[81] Ibid.

[82] Ibid.

[83] Ibid, Article 30(2).

The Processor has further responsibility and must implement appropriate security measures to protect personal data against accidental or unlawful destruction or loss, alteration and unauthorized disclosure. Some of the measures to be taken include encryption, reviews, testing and back-ups.[84] Furthermore, the liability for non-compliance with the GDPR is high, as it allows a data subject to claim directly against the Processor. Where non-compliance has been detected the Processor will be liable personally for any damage or loss.[85] However, it remains to be seen how this will operate in practice. This forces the Processor to ensure they have taken all the appropriate steps to ensure compliance with the GDPR. It also places a form of co-responsibility on both the Controller and Processor. This co-responsibility viewed positively provides that throughout the collection, use and management of personal data, someone is accountable. However, a problem could arise where their respective responsibilities are not clearly defined and have been blurred.

3.4.2 Data Protection Officer

Data Protection Officers[86] (DPO) add a further layer to the process, and must be appointed by all public authorities, where the activities of the controller or processor require monitoring of data subjects on a large scale. Article 39 sets out specific tasks the DPO must undertake, such as informing and advising the Controller or Processor of their obligations under GDPR.[87] This is a critical role in providing support to the Controller and Processor. The role provides advice on data impact assessments to ensure the national supervisory authority is appropriately informed. The role is an important contact point whereby the supervisory authority can contact this known individual within an organization. The structure within an organization allows for a robust and systematic approach to the implementation of the GDPR. The structure also reinforces the notion that the organization must be able to regulate themselves, with minimal oversight from the regulator, but, be accountable to a regulator. This is reinforced by Article 38, whereby the Controller has responsibility for ensuring the officer is appropriately involved and supported. The DPO is subject to the same level of confidentiality as the Controller and Processor, when processing data. Even to this point, one can see how the GDPR has established a multi-layered approach to capture most stages of data processing, collection, storage and use.

[84] Ibid, Article 28(1), (3)(e), (4), 32.

[85] Ibid, Article 82(1)(2).

[86] Ibid, Article 37.

[87] Ibid.

3.5 Right to Be Forgotten

The right to be forgotten has become an important part of the overall regulatory toolbox in protecting a data subject's personal data. The right enables a person to have their information removed from the Internet. Jeffery Rosen, in referring to the Vice President of the European Commission from 2010 to 2014, believes that regulators across the EU are of the opinion that all citizens face the difficulty of escaping their past.[88] This is even more evident now that the Internet records everything and forgets nothing. [89] When Commissioner Reding announced the new right to be forgotten, she noted the particular risk to teenagers who might reveal compromising information that they would later come to regret. Commissioner Reding then articulated the core provision of the right to be forgotten provides an individual who no longer wants his personal data to be processed or stored by a data controller, and if there is no legitimate reason for keeping it, to request that the data be removed from the system.[90]

Moreover, technological changes also bring about new regulatory challenges. The Internet, cloud computing, and mobile devices allow each of us to access our data everywhere and at any time. Our personal data races from Munich to Miami and to Hong Kong in fractions of a second. In this new data world, we all leave digital traces every moment, everywhere.[91] Do people care about how their personal data is protected? Do our rules need to be strengthened to give people more confidence and to make it easier for businesses to operate on, and in, Europe's digital single market? The simple answer is yes – when it comes to European citizens. In Europe, people do care, with 72% of Europeans saying that they are concerned about how companies use their personal data.[92] From this concern, one can see how today, the right to be forgotten appears to be firmly entrenched in EU law.

However, the right to be forgotten does not come without its critics. It has been referred to as a political slogan.[93] Similar to other political slogans, people can see in it what they want. The debate would sound quite different if the slogan were actually something more descriptive, for example, the right to delete.[94] The right to be forgotten is viewed as a re-branding of long-standing data protection principles, in

---

[88] Rosen J *The Right to Be Forgotten* Stanford Law Review Online 64, (2012) http://www.stanfordlawreview.org/online/privacy-paradox/right-to-beforgotten, accessed 5 May 2018. Viviane Reding, Vice President, Eur. Comm'n, The EU Data Protection Reform 2012: Making Europe the Standard Setter for Modern Data Protection Rules in the Digital Age 5 (Jan. 22, 2012), *available at* http://europa.eu/rapid/pressReleasesAction.do?reference=SPEECH/12/26&format=PDF, accessed 10 May 2018.

[89] Ibid.

[90] Ibid.

[91] Ibid.

[92] Ibid.

[93] Fleischer P, *Foggy Thinking About the Right to Oblivion*, Privacy? (2011), http://peterfleischer.blogspot.com/2011/03/foggy-thinking-about-right-to-oblivion.html, accessed 5 May 2018.

[94] Ibid.

particular the right to rectify one's own personal data, the right to oppose processing of one's personal data in the absence of legitimate purposes, or the principle of data minimization.[95] On the other hand, many believe the right is not new and is simply an attempt to apply long-standing data protection principles to the new worlds of the Internet and modern technologies.[96] Nonetheless, the right to be forgotten has been explicitly written into the GDPR. Article 17(1) states:

> "[t]he data subject shall have the right to obtain from the controller the erasure of personal data concerning him or her without undue delay and the controller shall have the obligation to erase personal data without undue delay.[97]

However, the above is only applicable where one of the following grounds apply. That is, the personal data is no longer necessary in relation to the purposes for which it was collected or otherwise processed. Secondly, the data subject withdraws their consent on which the processing is based according to point (a) of Article 6(1), or point (a) of Article 9(2).[98] In addition, the right to be forgotten will only be applicable where there is no other legal ground for the processing, or, where the data subject objects to the processing pursuant to Article 21(1)[99] Furthermore, this will also apply where there is no overriding legitimate ground (s) for the processing, or the data subject objects to the processing pursuant to Article 21(2).[100] The remaining points where the personal data have been unlawfully processed, or the personal data has to be erased for compliance with a legal obligation across the Union or Member State law to which the controller is subject, will also apply. The right will also be applicable where the personal data has been collected in relation to the offer of information society services referred to in Article 8(1).[101]

Importantly, there is no other legislation, from the other jurisdictions discussed throughout this book that have incorporated such a comprehensive provision, to codify the right to be forgotten. The right is further underpinned by Recital 65 and 66 of the GDPR.[102] According to Recital 66, the right to be forgotten has been included to strengthen the right to erasure in the online environment, the right to erasure should also be extended in such a way that a controller who has made the personal data public should be obliged to inform the controller (s) that are processing such personal data to erase any links to, or copies, or replications of the personal data.[103] In doing so, the controller is required to take reasonable steps, by considering the available technology and the means available to the controller, including

[95] Ibid.

[96] Ibid.

[97] Regulation 2016/679, Article 17.

[98] Ibid, Article 6, 8, 9, 21.

[99] Ibid.

[100] Ibid.

[101] Ibid.

[102] Recital 65–66.

[103] Ibid, Recital 66.

technical measures, to inform the controllers of the data subject's request.[104] Arguably, this provision strengthens the concepts of privacy by design and default.

Moreover, in accordance with Recital 65, the right to rectification and erasure provides a data subject with the ability to have their personal data concerning him or her rectified and a 'right to be forgotten' where the retention of such data infringes on the GDPR, in addition to Union or Member State law.[105] In particular Recital 65 provides that a data subject should have the right to have his or her personal data erased. Furthermore, a data subject has the right for their personal data not to be processed, where the personal data is no longer necessary in relation to the purposes for which it was collected or otherwise processed. The right to be forgotten should also apply where a data subject has withdrawn his or her consent or objects to the processing of personal data concerning him or her, or where the processing of his or her personal data does not otherwise comply with this Regulation.[106] That right is relevant in particular where the data subject has given his or her consent as a child and is not fully aware of the risks involved by the processing, and later wants to remove such personal data, especially on the internet.[107] The data subject should be able to exercise that right notwithstanding the fact that he or she is no longer a child.

However, Recital 65 goes onto provide exceptions to the right to be forgotten. Recital 65 states that the further retention of the personal data should be lawful where it is necessary, for exercising the right of freedom of expression and information, for compliance with a legal obligation.[108] Another exception applies where the performance of a task carried out in the public interest. The public interest [principle] is very broad and could apply to health and security, amongst others.[109] A further exception applies where the exercise of official authority vested in the controller is on the grounds of public interest in the area of public health.[110]

The right to be forgotten has a close connection with the right to withdraw consent (discussed below). In order to exercise the right to be forgotten, one condition is that the data subject withdraws consent on which the processing is lawful and there is no other legal ground for the processing.[111] Nevertheless, the right is not specific and there is no black or white application of the law. Rather, it is argued that, as the right evolves along with technology, it will be assessed on a case by case basis.

---

[104] Ibid.

[105] Ibid, Recital 65.

[106] Ibid.

[107] Ibid.

[108] Ibid.

[109] Ibid.

[110] Ibid, applicable for archiving purposes in the public interest, scientific or historical research purposes or statistical purposes, or for the establishment, exercise or defence of legal claim.

[111] Regulation 2016/679, Article 17 (1)(b).

3.6 Agency [Regulator] – Authority

There is a multi-layered approach to regulatory oversight between the EU and member states. Each member state is responsible for establishing a Supervisory Authority (SA) otherwise referred to as Data Protection Authority.[112] The SA has a critical role at the national level to ensure the GDPR is applied and implemented consistently with other member states. There is provision for more than one SA to be established in a member state, and this might be required for the larger more populated states. An SA is a key contact point between the member state and EU Commission. One of the most important roles the SA has is identifying a lead controller or processor is carrying out the cross-border processing of personal data. This is to ensure there is a seamless transfer and transaction of personal data across the internal borders of the EU member states. Article 4(23) of the GDPR defines 'cross-border processing' as either the:

- processing of personal data which takes place in the context of the activities of establishments in more than one Member State of a controller or processor in the Union where the controller or processor is established in more than one Member State; or the
- processing of personal data which takes place in the context of the activities of a single establishment of a controller or processor in the Union but which substantially affects or is likely to substantially affect data subjects in more than one Member State.[113]

Thus, where an organization has establishments in France and Germany, and the processing of personal data takes place in the context of their activities, then this will constitute cross-border processing. The CJEU affirmed this position in *S.R.O. v Nemzeti Adatvédelmi és Információszabadság Hatóság*.[114] The court held that each member state is to designate one or more public authority to be responsible for monitoring the application within its territory of the national provisions adopted by the member states on the basis of the previous Directive 95/46.[115] The court went onto say that:

> where the supervisory authority of a Member State, to which complaints have been submitted in accordance with Article 28(4) of the Directive, reaches the conclusion that the law applicable to the processing of the personal data concerned is not the law of that Member State, but the law of another Member State, Article 28(1), (3) and (6) of that Directive must be interpreted as meaning that that supervisory authority will be able to exercise the effective powers of intervention conferred on it in accordance with Article 28(3) of that directive only within the territory of its own Member State. Accordingly, it cannot impose penalties

[112] Regulation 2016/679, Article 51. Recitals (117) to (123). Article 29 Working Party Guidelines on the Lead Supervisory Authority, WP 244.

[113] Ibid, Article 4(23).

[114] Case 230, *S.R.O. v Nemzeti Adatvédelmi és Információszabadság Hatóság*, The Court of Justice, 2015.

[115] Ibid.

> on the basis of the law of that Member State on the controller with respect to the processing of those data who is not established in that territory, but should, in accordance with Article 28(6) of that Directive, request the supervisory authority within the Member State whose law is applicable to act.[116]

The case subtly highlighted the need for the SA not only to collaborate with their counterparts in other member states, but also, take a greater role in better understanding how foreign data controllers operate. Doing so, will ensure compliance with the GDPR and minimize the unauthorized use of data. The SA is also required to interpret the effects on a case by case basis.[117] They need to take into account the context of the processing, the type of data, the purpose of the processing and factors such as whether the processing there is likely to cause, damage, loss or distress to individuals. The role of the SA also extends to determining the effect of an individuals' health, well-being, peace of mind, economic status, or particularly the personal data of children where the personal data can be intrusive.[118]

In addition to the above, the EU has appointed a European Data Protection Supervisor (EDPS),[119] who is an independent data protection authority. The role of the EDPS is to monitor the protection of personal data that has been processed by EU institutions.[120] The EDPS has a unique role to intervene where possible before the Court of Justice of the European Union to provide expert advice in relation to data protection law. The EDPS is also charged with cooperating with member state national supervisory authorities, in collaborating on improving consistency in protecting personal data and information, more broadly.

The European Data Protection Board (EDPB) has also been established to assist with the international transfer of personal data to third countries. Article 69 provides autonomy to the EDPB to act independently amongst others, but not limited to monitor the application of the GPR and advise the Commission on any issue related to the protection of personal data within the Union.[121] Its main role is to determine disputes between national supervisory authorities. The Board also provides advice and guidance as well as approving whole of EU codes and certifications.[122]

---

[116] Ibid, para 60.

[117] Ibid.

[118] Article 29 Working Party Guidelines on the Lead Supervisory Authority, 1.1.1.

[119] European Data Protection Supervisor, https://edps.europa.eu/about-edps_en, accessed 5 December 2017.

[120] Ibid.

[121] Regulation 2016/679, Articles 70 and 71.

[122] Ibid.

3.7 Public and Private

The GDPR applies to both public and private organizations. In other words, it applies to the European institutions, including member state institutions along with the private sector. There is no exemption for the public sector. Article 4 specifies and includes public authorities in the definitions of controllers and processors.[123] Furthermore, Article 37, requires all public authorities to identify a data protection officer. The only exemption to this is the courts.[124] This sets the GDPR apart from most other jurisdictions that have predominantly limited the scope of data protection law to the private sector or specific elements of the private sector.

3.8 Consent

The concept of consent has begun to pervade most of the legal framework arising from the GDPR. Consent provides a data subject with greater control and ownership of their personal data that is defined by the law. Firstly, Article 7 of the GDPR requires that consent to be freely given.[125] Article 4.11, states that 'consent' of the data subject means any freely given, specific, informed and unambiguous indication of the data subject's wishes by which he or she, by a statement or by a clear affirmative action, signifies agreement to the processing of personal data relating to him or her.[126] This takes away any ambiguity surrounding what an agreement might constitute. Secondly, Recital 32 requires consent should be given by a clear affirmative act establishing a freely given, specific, informed and unambiguous indication of the data subject's agreement to the processing of personal data.[127] In other words, by the data subject ticking a box that they have visited the internet website, is enough to constitute consent. It is a new process introduced by the GDPR and enables consent to be tracked, by the data subject. This places responsibility on the organization to have a system in place on their website, which will require a tick box of some description that has the user, for example 'accept all cookies, accept first party cookies or reject cookies.[128] Therefore, the GDPR has a list of purposes for how personal data is to be used, it calls for active consent. This will enable the user to better understand how their data is being managed, used and processed. Nonetheless, the EDPS has expressed concern that any tracking of consent, must also only be

[123] Ibid, Article 4.

[124] Ibid, Article 37.

[125] Ibid, Article 7.

[126] Ibid, Article 4.11.

[127] Recital 32.

[128] EPrivacy Regulation, European Data Protection Supervisor, Opinion on the Proposal for a Regulation on Privacy and Electronic Communications (ePrivacy Regulation), Article 10 and Recital 23.

undertaken with the consent of the individual.[129] However, where processing is based on consent, the controller shall be able to demonstrate that the data subject has consented to processing of his or her personal data. If the data subject's consent is given in the context of a written declaration which also concerns other matters, the request for consent shall be presented in a manner which is clearly distinguishable from the other matters.[130] It must be in an intelligible and easily accessible form, using clear and plain language. Any part of such a declaration which constitutes an infringement of this Regulation shall not be binding.

Thirdly, the data subject shall have the right to withdraw his or her consent at any time. The withdrawal of consent shall not affect the lawfulness of processing based on consent before its withdrawal. Prior to giving consent, the data subject shall be informed thereof. It shall be as easy to withdraw as to give consent. Fourthly, when assessing whether consent is freely given, utmost account shall be taken of whether, inter alia, the performance of a contract. This also includes the provision of a service, which is conditional on consent to the processing of personal data that is not necessary for the performance of that contract. Consent is not provided if the individual has no genuine or free choice or is unable to refuse or withdraw consent at any time.[131]

The GDPR provides some guidance on how consent should operate. That is, consent should be given by a clear affirmative act.[132] It will be important for an organization to establish systems and processes to monitor and record whether actual consent has been provided or not. For instance, the GDPR suggests that this could include ticking a box when visiting an internet website, choosing technical settings for information society services. In addition, consent could come in the form of another statement or conduct which clearly indicates the data subject's acceptance of the proposed processing of his or her personal data.[133] Silence, pre-ticked boxes or inactivity should not therefore constitute consent. Consent should cover all processing activities carried out for the same purpose or purposes. When the processing has multiple purposes, consent should be given for all of them. If the data subject's consent is to be given following a request by electronic means, the request must be clear, concise and not unnecessarily disruptive to the use of the service for which it is provided. To process the data the individual 'has given consent to the processing of his or her personal data for one or more specific purposes'.[134] Furthermore Article 9 provides that 'explicit consent' is generally required to process 'special categories' of personal data. Businesses must inform individuals about this right to withdraw consent.[135]

---

[129] Ibid.

[130] Regulation 2016/679, Article.

[131] Ibid, Article 7.

[132] Ibid, Preamble at 32.

[133] Ibid.

[134] Regulation 2016/679, Article 6(1)(a).

[135] Regulation 2016/679, Article 7(3).

3.8.1 *Children's Consent*

The idea of enabling children to provide their personal data is problematic. They have become one of the most vulnerable groups in society because the Internet has allowed this group access to information, like never before. Hence, some safeguards have been established to ensure that children of a certain age, usually under the control of parents, are unable to provide actual consent unless there is a form of supervision. This is not new, children must acquire the consent from parents in many other areas of the law, such as medical etc. Although, a notable difference from the other jurisdictions laws discussed in this book, is the EU's specific reference to children. There are also specific requirements in relation to children's consent. A person under the age of 16 who wishes to use online services, can only provide consent through one of the child's parents.[136] Children 16 years or older may give consent for processing data related to themselves. However, member states may introduce domestic laws to lower this age to not less than 13 years.[137]

3.9 Extra-Territorial Reach

The GDPR applies to processing of personal data, whether undertaken within or outside the Union.[138] Article 3 applies to the processing of personal data in the context of the activities of an establishment of a controller or a processor in the Union, regardless of whether the processing takes place in the Union or not.[139] The GDPR applies to the processing of personal data of data subjects who are located in the Union by a controller or processor that is not necessarily established in the Union, or where the processing activities are related to:

(a) offering of goods or services, irrespective of whether a payment of the data subject is required, to such data subjects in the Union; or
(b) monitoring of their behavior as far as their behavior takes place within the Union.[140]

The GDPR also applies to the processing of personal data by a controller not established in the Union, but in a place where Member State law applies by virtue of public international law. In *S.R.O. v Nemzeti Adatvédelmi és Információszabadság Hatóság* the court ruled that:

[136] Regulation 2016/679, Article 8.

[137] Ibid.

[138] Regulation 2016/679, Article 3.

[139] Ibid.

[140] Ibid.

> the establishment constitutes where the organization exercises a real and effective activity, by having stable arrangements in place.[141]

The Commission has responsibility for deciding whether a third countries' regulatory framework provides similar protections to that of the GDPR, before data can be transferred to third countries. In circumstances where a third country's laws differ significantly to that of the EU, the data may still be transferred.[142] The data controller, through binding corporate rules, can establish an agreement to facilitate the transfer or data.[143] In addition, where a code of practice has been developed and approved or an approved certification has been issued, and the appropriate safeguards met, the personal data can be transferred outside of the EU.[144]

In summary, the reach of the GDPR is quite extensive and has extended its reach from the previous Directive 95/46/EC. This has been reinforced by scholars who have stated that the most important finding from the case mentioned above, *Google Spain,* has been the territorial scope provided by EU data protection laws. EU data protection law will apply in the case of search engine operations through a branch or a divisional office within any Member State of the EU, even if the main company originated and is based outside of the EU.[145] Therefore, in practical terms, the GDPR can be interpreted as not only a regional but also an international data protection law.[146] For example, when registrars and registries established outside the EU provide their domain name registration services to natural persons in the EU, the GDPR can apply. Moreover, both the member states national Data Protection Authority and the European Data Protection Board have a role in monitoring the application and implementation of the GDPR.

3.10 Retention

Article 5(1)(e) of the GDPR[147] states that data kept in a form which permits identification of data subjects for no longer than is necessary for the purposes for which the personal data are processed; personal data may be stored for longer periods insofar as the personal data will be processed solely for archiving purposes in the public interest, scientific or historical research purposes or statistical purposes in accordance with Article 89(1)[148] subject to implementation of the appropriate

---

[141] Case C-230, *S.R.O. v Nemzeti Adatvédelmi és Információszabadság Hatóság,* 1 October 2015.

[142] Regulation 2016/679, Article 46.

[143] Ibid.

[144] Ibid.

[145] Bu-Pasha S *Cross-border issues under the EU data protection law with regards to personal data protection*, Taylor & Francis, (2017) pp. 213–228.

[146] Ibid.

[147] Regulation 2016/679, Article 5.

[148] Regulation 2016/679, Article 89(1).

technical and organizational measures required by this Regulation in order to safeguard the rights and freedoms of the data subject ('storage limitation'). What this means to any organization is that personal data should not be retained for a period than is necessary (storage limitation).

The storage limitation is likely to challenge organizations whose core business centers on data collection and storage. Organizations will need to balance the requirements to comply with the GDPR and the business needs of the organization. The cyber security risks are well demonstrated, and storing large amounts of data for lengthy periods only increases the risk of that data being exposed to security breaches (see Chap. 16). The retention of data will fall to the data controller who will have to ensure that the period is kept to a strict minimum.[149] The storage limitation principle will require an organization to delete personal data.[150] The only limited exception applies for archiving purposes in the public interest, scientific or historical research purposes or statistical purposes.[151] Even though there is no specific time limit provided, the right to be forgotten and restrictions of profiling such as the use of Big data[152] is likely to assist in managing any lengthy storage of personal data.

The European Court of Human Rights in *Affaire Aycaguer c. France (Requête no. 8806/12) Arret, Strasbourg*[153] confirmed that there had been a breach of the right to a person's private life as a result of an order to provide a biological sample to be included in the national DNA data base of France.[154] The Court ruled that the:

> national regulations on the storage of DNA profiles did not provide the data subjects with sufficient protection, owing to its duration and the fact that the data could not be deleted. [155]

The national regulations therefore failed to strike a fair balance between the competing public and private interests. Furthermore, DNA profiles constitute personal data, and depending on the jurisdiction, this data would be defined as sensitive data.

3.11 Principles and Codes

In leading the way in the development and implementation of data protection laws, the GDPR establishes seven core principles. These include:

---

[149] Regulation 2016/679, Recital 39.

[150] Ibid. Failure to comply could result in fines as high as 4% of annual worldwide turnover or €20million – whichever is the greater.

[151] Regulation 2016/679, Articles 5 (1)(e) and 89(1).

[152] Regulation 2016/679, Articles 17, 18, 21, 22.

[153] *Affaire Aycaguer c. France (Requête no. 8806/12) Arret,* Strasbourg, 22 September 2017.

[154] Ibid.

[155] Ibid.

- Accountability, which aims to guarantee the enforcement of the GDPR principles.[156]
- Accuracy, because of the personal, social and commercial risks of producing in accurate data.[157]
- Data minimization,[158] which is subject to limited exceptions requires an organization should only process the personal data that it actually needs to process. Organization need to be conscious by establishing sound best practice management systems and processes to ensure there is not an over collection and retention of data that could inevitably used illegally.
- Fair, lawful and transparent processing, so as the data subject understands that their data will be used for a particular purpose.[159]
- Limitation requires that personal data collected for one purpose should not be used for a new, or incompatible purpose.[160]
- Retention periods as discussed above ensure that data is not retained for unnecessary timeframe for which is was collected. That is data collected, used and retained in 2018 may no longer be relevant in 2022, and therefore the organization needs to have a process in place to remove and delete the data.[161]
- Security is obviously a key feature to protect data. The GDPR requires that controllers be responsible for ensuring that personal data are kept secure from external threats.[162]

While it is out of scope to analyze and discuss each of these principles, it is argued that they are very important to establishing trust and certainty for data subjects when using the Internet. These principles are very different to the other countries, but achieve a similar result by maintaining integrity in the law. The principles also assist by ensuring compliance is achieved. They are all important principles that must be followed, although the one of the more important principles is accountability. Article 5 requires the data controller to be responsible for, and demonstrate their compliance with the law. This requires the controller to establish organizational processes, implement those processes and undertake periodic reviews. Arguably they go a long way to meeting, strengthening and expanding on the OECD principles.

---

[156] Regulation 2016/679, Article 5(2), the controller is responsible for compliance with the Data Protection Principles.

[157] Regulation 2016/679, Article 5(1)(d).

[158] Regulation 2016/679, Article 5(1)(c).

[159] Regulation 2016/679, Article 5(1)(a).

[160] Regulation 2016/679, Article 5(1)(b).

[161] Regulation 2016/679, Article 5(1)(e).

[162] Regulation 2016/679, Articles 5(1)(f), 24(1), 25(1)–(2), 28, 39, and 32, data must be processed in a manner that ensures appropriate security of those data.

Article 40, requires member states to establish a code of conduct to ensure the GPR is appropriately implemented.[163] The guidance provided by the GDPR ensure member states will be consistent and include:

- Fair and transparent processing.
- The legitimate interests pursued by controllers in specific contexts.
- The collection of personal data.
- The pseudonymisation of personal data.
- The information provided to the public and to data subjects.
- The exercise of the rights of data subjects.
- Information provided to and the protection of children and the manner in which the consent on the holders of parental responsibility over children is to be obtained.
- General data protection obligations of data controllers, including privacy by design and measures to ensure security of processing.
- Notification of personal data breaches to supervisory authorities and communication of such personal data breaches to data subjects.
- Transfer of personal data to third countries or international organizations.
- Out-of-court proceedings and other dispute resolution procedures for resolving disputes between controllers and data subjects with regard to the processing, without prejudice to the rights of data subjects.[164]

Pseudonymisation[165] is a newly introduced concept by the GDPR. Article 4 defines the concept to mean, the processing of personal data in such a manner that the personal data can no longer be attributed to a specific data subject without the use of additional information, provided that such additional information is kept separately and is subject to technical and organizational measures to ensure that the personal data are not attributed to an identified or identifiable natural person.[166] However, the concept of anonymization has not been defined, but rather, Recital 26 states the principles of data protection should therefore not apply to anonymous information, namely information which does not relate to an identified or identifiable natural person. It is considered personal data rendered anonymous in such a manner that the data subject is not or no longer identifiable.[167] Leslie Stevens believes that pseudonymisation is a privacy-enhancing technique which removes

[163] Ibid, Article 40.

[164] Regulation 2016/679, Article 40.

[165] Regulation 2016/679, Article 4, pseudonymisation' means the processing of personal data in such a manner that the personal data can no longer be attributed to a specific data subject without the use of additional information, provided that such additional information is kept separately and is subject to technical and organizational measures to ensure that the personal data are not attributed to an identified or identifiable natural person.

[166] Ibid.

[167] Ibid, Recital 26.

direct identifiers by pseudonyms. [168] Pseudonyms as indirect identifiers, is data that can be used to identify a person.[169]

The Code reinforced the principles set out in Article 5, and therefore, the EU is able to construct an enforcement framework that businesses are responsible to. The Code also enables greater regulatory oversight by establishing a co-regulatory framework between government and the private sector. Article 24, establishes the controller's responsibilities in relation to processing personal data and promotes the idea that a Code is in place to ensure compliance.[170] In addition, Article 28 and Recital 81, provide that a processor's conduct be undertaken according to an approved Code.[171] While Article 32 requires adherence to an approved Code or an approved certification mechanism for the purposes of processing data, and require that the controller and the processor shall implement appropriate technical and organizational measures to ensure a level of security.[172] Arguably, codes of practice further strengthen the legal and policy framework for managing the practical aspects of handling and managing personal data.

3.12 Cross Border Transfer

The transfer of data across international borders is not new and will only grow. To assist in facilitating this process the EU through Article 44 provides that:

> 'Any transfer of personal data which are undergoing processing or are intended for processing after transfer to a third country or to an international organization shall take place only if, subject to the other provisions of this Regulation, the conditions laid down in this Chapter are complied with by the controller and processor, including for onward transfers of personal data from the third country or an international organization to another third country or to another international organization. All provisions in this Chapter shall be applied in order to ensure that the level of protection of natural persons guaranteed by this Regulation is not undermined'.[173]

A Code has a further function apart from the above, and will also assist in facilitating the transfer of data across international borders. Article 45 allows the Commission, to assess the adequacy of protection for the transfer of data.[174] The transfer to a third country can be undertaken provided standard data protection clauses adopted by the EU Commission are applied.[175] To date the only countries to

[168] Stevens L, *The Proposed Data Protection Regulation and its Potential Impact on Social Sciences Research in the UK*, European Data Protection Law Review Vol.1, (2015) pp. 97–112.

[169] Ibid.

[170] Regulation 2016/679, Article 24.

[171] Regulation 2016/679, Article 28, Recital 81.

[172] Regulation 2016/679, Article 32.

[173] Regulation 2016/679, Article 44.

[174] Regulation 2016/679, Article 45.

[175] Regulation 2016/679, Article 46.

obtain recognition of adequacy include Andorra, Argentine, Canada, Faroe Islands, Guernsey, Israel, Isle of Man, Japan, Jersey, New Zealand, Switzerland, Uruguay and the United States.

It can be argued that the EU is stating to third countries that where they offer a level of protection that is considered adequate, but may not necessarily mean the protections are consistent with the EU, that data can be transferred across international borders. These clauses can become part of an agreement, and could for instance be included in the accreditation agreement between ICANN (Internet Corporation for Assigned Names and Numbers) and the registrars, in order to ensure the legality of third country transfers. However, it should be noted that the existing model clauses have been subject to criticism and that their validity is to be tried by the Court of Justice of the EU. Their future validity, at least in their current form, is somewhat uncertain. Furthermore, Article 47 ensures there are binding corporate rules in place that can be enforced.[176] The corporate rules create a self-regulatory framework whereby an organization must comply with its own rules. This is a positive sign because it places greater responsibility on industry to regulate themselves, rather than the traditional position of government doing this role on behalf of the community. However, and as this book highlights there is a long way to go before any form, or even substantive self-regulation exists that is effective. The EU Commission has responsibility for determining whether a third country or international organization ensures an adequate level of protection.[177]

The cross border transfer of personal data can be exposed to many variables. Thus, the EU are of the view that an organization must establish strong processes and systems, to ensure any personal data collected, stored and used is accurate. Any data that is inaccurate must be deleted.[178] Recital 30 requires that every reasonable step should be taken to ensure that personal data which are inaccurate are rectified or deleted.[179] While it may be a broad concept to take ‘reasonable’ steps, the threshold could be quite narrow, depending on the situation. This is a further area of research.

Article 20 of the GDPR establishes a new right to data portability.[180] The right to portability has two elements. Firstly, the right of data subjects to receive the personal data that they have provided to a controller.[181] Secondly, the right to receive personal data in a structured, commonly used and machine-readable format, and to

[176] Regulation 2016/679, Article 63.

[177] Regulation 2016/679, Article 45, a transfer of personal data to a third country or an international organisation may take place where the Commission has decided that the third country, a territory or one or more specified sectors within that third country, or the international organisation in question ensures an adequate level of protection. Such a transfer shall not require any specific authorisation.

[178] Regulation 2016/679, Article 5 and 16, Recital 39.

[179] Recital 30.

[180] Regulation 2016/679, Article 20.

[181] Ibid.

transmit the data to another data controller.[182] In addition, data portability also allows for the direct transmission of personal data from one data controller to another. The right is supposed to restrict a data controller form possessing data, and create competition between data controllers. The inclusion of this right, has from a policy perspective allowed the EU to continue to transpose the single market concept across all areas of the economy, and to ensure there will be a Digital single market. The EU has further argued that this data portability right also represents an opportunity to "re-balance" the relationship between data subjects and data controllers, through the affirmation of individuals'.[183] It is also argued that it creates a level of competition. Nonetheless, as the GDPR is very much in its infancy, this is another important area to watch.

The EDPB[184] is required to provide the Commission with an opinion assessing the adequacy of a country or organization's level of data protection.[185] The EDPB will be responsible for setting guidance to controllers, processors and business to comply and determine disputes between national supervisory authorities. The emphasis by the Commission is determining whether the third country has satisfactory safeguards to protect the transfer. For example, it would not be unreasonable to expect that the Commission consider the transfer of data to Singapore or Australia, two countries that have mature privacy laws. In the case where the Commission considers the transfer of data will not have adequate protection, that transfer can still take place, provided the data controller established approved binding corporate rules. In addition, the controller has an agreement in place outlining the standard data protection clauses, and approved certification has been obtained.[186]

Article 48 and 49 of the GDPR also refer to the transfer of data in circumstances where a foreign tribunal or administrative body has ordered the transfer that is not permitted, or where appropriate safeguards have not been established.[187] In *Maximilian Schrems v Data Protection Commissioner*[188] the court was required to determine whether the transfer of personal data had adequate protections – from Facebook within the EU (Ireland) to servers located in the United States. Mr. Schrems lodged a complaint with the Irish Data Protection Commissioner saying that in light of the revelations made in 2013 by Edward Snowden concerning the activities of the United States intelligence services – the National Security Agency ('NSA'), the law and practice of the United States did not offer adequate protection against surveillance by public authorities of the data transferred to that country.[189]

[182] Ibid.

[183] Article 29 Data Protection Working Party (Revised 5 Apr 2017) Guidelines on the right to data portability, pp. 3–6.

[184] Regulation 2016/679, Article 68, formerly established by Article 29 Working party, Directive 95/46/EC.

[185] Regulation 2016/679, Article 75 (1)(s).

[186] Regulation 2016/679, Article 46.

[187] Regulation 2016/679, Article 48–49.

[188] C-362/14 *Maximilian Schrems v Data Protection Commissioner*, The Court of Justice 2015.

[189] Ibid.

The Irish authority rejected the complaint taking the view that an investigation into the matters raised by Mr. Schrems was unfounded and that there was no evidence that Mr. Schrems' personal data had been accessed by the NSA.[190] Further, the Commissioner held that the allegations raised by Mr. Schrems would have to be determined in accordance with Safe Harbor Decision 2000/520[191], and the Commission had found that the United States ensured an adequate level of protection.

The court struck down the transatlantic US-EU Safe Harbor agreement that had been in place for 15 years.[192] This agreement enabled businesses to transfer data from Europe to the United States (US). Effectively, businesses in the US could self-certify that they would comply with EU data protection standards in order to allow for the transfer of European data to the United States. The CJEU found that European data was not sufficiently protected in the United States.[193] The CJEU ruled that:

> a member states' supervisory body is independent and have the power to confirm compliance with the data protection laws.[194]

It must be noted that a national supervisory body can verify the level of protection afforded to the transfer of data to a third country. Importantly, any assessment must be according to European standards, and not that of a third country, unless there was an exemption.[195] Thus, there is a commitment to ensure that EU standards are not compromised, where that data is being transferred to a third country.

For Asia, the heightened importance and impact of the GDPR to national data has resulted in a limited number of countries seeking EU assessments. In early 2017 the EU's Communication Exchanging and Protection Personalized data in a globalized world attempted to reach an adequacy decision with Japan and other countries.[196] This was as a result of the *Schrems* case, which requires any adequacy decision to balance the needs for the free flow of personal data to countries, but must be undertaken only where that country has equivalent data protection rules to the EU.

---

[190] Ibid.

[191] Safe Harbor Decision 2000/520 – NEED FULL REF.

[192] C-362/14 *Maximilian Schrems v Data Protection Commissioner*, The Court of Justice 2015.

[193] Ibid.

[194] Ibid.

[195] Ibid, European Broadcasting Union, the *Schrems* case was one of the catalysts for legal reform in the EU. The centrepiece of this reform is the General Data Protection Regulation which is currently being finalized in so-called trilogue meetings between the European institutions. The Court of Justice, in its ground-breaking Grand Chamber judgment, follows Advocate General Bot's opinion delivered only a fortnight earlier on 23 September. Two aspects are especially noteworthy, https://www.ebu.ch/files/live/sites/ebu/files/News/2015/10/Case%20note%20Schrems.pdf, accessed 6 may 2018.

[196] European Commission 'Commission proposes high level of privacy rules for all electronic communications and updates data protection rules for EU institutions', 10 January 2017, http://europa.eu/rapid/press-release_IP-17-16_en.htm, accessed 10 January 2018.

3.13 Breach

The GDPR has narrowed the gap in regards to breaches of the Regulation, by placing a time limit on reporting any breach. A controller must inform the supervisory authority within 72 h of the breach, after becoming aware of it.[197] Importantly, the reporting must include the nature of the personal data breach, approximate number of data subjects and personal data records. The reporting must also provide the name and contact details of the data protection officer, the impact of the breach and the measures taken to be taken by the controller to address the breach.[198]

The emphasis placed on reporting the type of breach arguably has several benefits. Firstly, ensuring compliance with the GDPR. Secondly, the critical elements to the breach identifying the number and personal information concerned. Thirdly, the requirement for the controller to outline what measures will be taken to address such a breach, ensures that two things. The first is that similar breach should not occur again, and requiring a self-regulatory process for the controller and organization to review and improve its internal self-regulatory systems and processes. Article 83 (2) details a list of infringements relating to:

- the nature, gravity and duration of the infringement;
- the intentional or negligent character of the infringement;
- any action taken by the controller or processor to mitigate the damage;
- the degree of responsibility of the controller or processor and organizational measures implemented by them;[199]
- any previous infringements;
- the degree of cooperation with the supervisory authority;
- the categories of personal data affected;
- the manner in which the infringement became known;
- where controller or processor has been issued warnings or reprimanded according to article 58;
- adherence to approved codes of conduct pursuant to Article 40 or approved certification mechanisms pursuant to Article 42[200]; and
- any financial benefits gained, or losses avoided.[201]

Along with the breach, the extent of the possible infringements identified, while not exhaustive ensures that the organization, controller and processor meet the obligations of the GDPR. This reinforces the concept that apart from focusing on right pertaining to data management, the EU is pushing a co-regulatory and self-regulatory framework.

---

[197] Regulation 2016/679, Article 33 and 55.

[198] Ibid.

[199] Regulation 2016/679, pursuant to Articles 25 and 32.

[200] Codes as part of Article 40 and 42.

[201] Regulation 2016/679, Article 83(2).

The potential issue for the EU rules governing jurisdiction of claims based on the alleged infringement of privacy rights, and the rules that determine which court, or courts, are entitled to decide the substance in cross-border dispute. However there are potential concerns regarding whether, and in what circumstances, the ongoing application of the rules of the Brussels I Regulation[202] regarding parallel proceedings and prorogation of jurisdiction could actually be regarded as prejudicial to Article 79(2) GDPR.[203] Article 79 provides the right to an effective judicial remedy against a controller or processor. It states:

1. Without prejudice to any available administrative or non-judicial remedy, including the right to lodge a complaint with a supervisory authority pursuant to Article 77[204], each data subject shall have the right to an effective judicial remedy where he or she considers that his or her rights under this Regulation have been infringed as a result of the processing of his or her personal data in non-compliance with this Regulation.
2. Proceedings against a controller or a processor shall be brought before the courts of the Member State where the controller or processor has an establishment. Alternatively, such proceedings may be brought before the courts of the Member State where the data subject has his or her habitual residence, unless the controller or processor is a public authority of a Member State acting in the exercise of its public powers.[205]

However, Franzine points out that, to date, nothing seems to prevent a general rule such as Article 25[206] of the Brussels I Regulation from being relied upon to uphold an agreement that, while preserving the operation of Article 79(2) GDPR, would allow a data subject to also bring proceedings in the courts of one more Member States, other than those provided for under the GDPR.[207] Article 79 of the GDPR does not specify whether jurisdiction is exclusive.[208] Furthermore, Article 79 does not clarify the grounds that can be derogated from under an agreement between

---

[202] Brussels I Regulation No 1215/2012 on jurisdiction and the recognition and enforcement of judgments in civil and commercial matters, Official Journal of the European Union L 351, clarifies the jurisdictional regime: the rules which courts of European Union Member States use to determine if they have jurisdiction in cases with links to more than one country in the European Union.

[203] Franzina, P (2016) *Jurisdiction regarding Claims for the Infringement of Privacy Rights under the General Data Protection Regulation*, in Alberto de Franceschi (2016) *European contract law and the digital single market: the implications of the digital revolution*, Cambridge, Antwerp, Portland, Intersentia, p 81–103.

[204] Regulation 2016/679, Article 77.

[205] Regulation 2016/679, Article 79.

[206] Regulation 2016/679, Article 25.

[207] Franzina, P (2016) *Jurisdiction regarding Claims for the Infringement of Privacy Rights under the General Data Protection Regulation*, in Alberto de Franceschi (2016) *European contract law and the digital single market: the implications of the digital revolution,* Cambridge, Antwerp, Portland, Intersentia, p 108.

[208] Regulation 2016/679, Article 79.

parties.[209] Franzine argues that the general rules of the Brussels I Regulation on explicit prorogation of jurisdiction should be deemed to prejudice the operation of Article 79(2) of the GDPR whenever the application would restrict the right of the data subject to sue a controller before the courts specified under the latter provision.[210]

3.14 Cyber Security

The EU has for some time been undertaking a review of its broader approach to cyber security. Personal data in cyber security and cybercrime has become evident, and entities can profit extensively from these activities (see Chap. 15). The EU Cyber Security Strategy[211] highlights five strategic priorities, and include:

- Achieving cyber resilience;
- Drastically reducing cybercrime;
- Developing cyber-defence policy and capabilities related to the Common Security and Defence Policy);
- Develop the industrial and technological resources for cybersecurity; and
- Establish a coherent international cyberspace policy for the European Union and promote core EU values.[212]

The important point regarding the establishment of a coherent international cyberspace policy for the European Union, arguably is a key component of EU core values. In other words, the balance between the economic and social policy of cyberspace, cybercrime, cyber security, privacy rights and data protection form an integral part of the overall all policy direction of the EU. They are all intertwined, complex and from time to time in competition and conflict with each other.

3.15 Conclusion

The 2018 GDPR and its 1995 predecessor have, arguably, set the scene and road-map for data protection and privacy throughout the world. It is our view that the EU has dragged other regions and countries to the table in establishing similar laws to

[209] Ibid.

[210] Franzina, P (2016) *Jurisdiction regarding Claims for the Infringement of Privacy Rights under the General Data Protection Regulation*, in Alberto de Franceschi (2016) *European contract law and the digital single market: the implications of the digital revolution,* Cambridge, Antwerp, Portland, Intersentia.

[211] Cybersecurity Strategy of the European Union: An Open, Safe and Secure Cyberspace, Joint Communication to the European parliament, The Council, The European Economic and Social Committee and the Committee of the Regions, (2013), pp. 2–5.

[212] Ibid.

protect privacy and personal data over the Internet. The GDPR has raised the benchmark for data protection and privacy. In other words, the previous 1995 Directive on data protection, was just that – a Directive. The GDPR is a Regulation, and therefore is a higher level of law within the EU framework. It is our view, that the EU continuing to elevate data protection and privacy within its legal framework, only reinforces the importance and complexity in this area of law and policy. Should this continue, it may result in the EU establishing a dedicated Treaty or Convention, to further strengthen data protection and subsequently privacy law.

It is asserted that the GDPR could be viewed as potentially overreaching and elevating the right to privacy over other economic needs. It could be conceived that the GDPR may impinge or dilute corporate profitability in trying to comply with its every detail, that some of its provisions invite different constructions. Additionally, EU courts diverge over the privacy protection versus economic/public interest in having access to GDPR information. For example, some EU member state courts, such as in the Netherlands, have held, in the prelude to the GDPR, that privacy concerns should be focused on commercial interests, rather than the public's right to know, or, a person seeking to have their personal data removed, deleted from the Internet. The balance between the various legal and policy issues that compete and conflict with each other could become even more complex as technology continues to develop and evolve.

Member states are obliged to implement EU law and are all signatories to the European Convention on Human Rights and Fundamental Freedoms 1950. They are also signatories to the 2000 Charter of Fundamental Rights. Importantly, the 2000 Charter is binding on member states. This Chapter has highlighted the multi-layered approach taken by EU in this area of law. Today, the GDPR requires organizations to establish data controllers and processors, amongst. This multilayered approach sets the EU apart from its counterparts – Australia and Singapore. It also establishes itself very separately to the other countries discussed throughout the book. The rights based approach, and the introduction of the right to be forgotten has been widely accepted by the courts and other countries have begun to follow suit.

The GDPR has introduced a number of new legal concepts such as the right to be forgotten, data portability, anonymization and pseudonymisation, which all contribute to strengthening the risk-based regulatory approach to data protection and privacy over the Internet. Along with the broad definition of personal data, the concept of consent is in no doubt playing an important role in the data protection law of the EU. However, it remains to be seen as to whether the current status of personal data, based on the minimal jurisprudence at present, does not force these and other concepts to be reviewed as technology expands. Furthermore, the right to object has also emerged as another important concept and principle because it not only assists in protecting peoples' online privacy, it also extends to restricting direct marketing from the use of personal data to contact data subjects. In other words, Article 21(2) restricts of the GDPR, to some extent anti-competitive behavior by organization through direct marketing.

Throughout the EU, every organization will be required to comply with the GDPR, which requires entities to review their internal operating procedures to ensure long term compliance.[213] In other words, the GDPR applies to EU institutions, national governments and the private sector, unlike other jurisdictions where their respective laws (if they exist) may only apply to the private sector. Importantly, the GDPR has extra-territorial reach in respect of EU data subjects outside the EU. This forces organisations that collect and hold personal data on eu citizens, for example in Australia, are required to also comply with the GPDR. There is no doubt that as the EU continues work in this area, the balance between human rights and the single market has the potential to become increasingly more complex. No doubt, EU policy makers, the private sector, legal profession, legal scholars and industry experts will review its implementation over the next decade. Finally, as the book will highlight, the EU model is at one end of the spectrum, while Singapore and Australia's model (legal frameworks) are at the other end. That is, Singapore's model takes a business friendly approach, whereas, Australia's framework sits somewhere between the two. The EU framework also sets itself apart from the models that currently exist in India, Indonesia, Malaysia, Japan and Thailand.

References

Albrecht J, (2016) *How the GDPR Will Change the World, European Data Protection Law Review*, Volume 2, Issue 3, pp. 287–289

Bu-Pasha S, (2017) *Cross-border issues under the EU data protection law with regards to personal data protection*, Taylor & Francis, pp. 213–2 28

Franzina, P (2016) *Jurisdiction regarding Claims for the Infringement of Privacy Rights under the General Data Protection Regulation*, in Alberto de Franceschi (2016) *European contract law and the digital single market: the implications of the digital revolution*, Cambridge, Antwerp, Portland, Intersentia, p 81–103

Greenleaf, G (2014) *Asian Data Privacy Laws: Trade and Human Rights Perspective*, Oxford University Press, pp. 35–42

Greenleaf, G (2017) *Asia Data Privacy Laws – Trade and Human Rights Perspectives*, University New South Wales.

Greenleaf G, (2018) *The Legal and Business Risks of Inconsistencies and Gaps in Coverage in Asia Data Protection Laws* Session II Materials, *Asian Business Law Institute (ABLI) Data Privacy Forum*, Singapore

Hintze M, LaFever G (2018) *Meeting Upcoming GDPR Requirements While Maximizing the Full Value of Data Analytics, Balancing the Interests of Regulators, Data Controllers and Data Subjects, Unlock Big Data Value while Complying with the GDPR,* http://files8.design-editor.com/93/9339158/UploadedFiles/1B4F2EF8-BC8D-A12D-C9B1-7DF644A29C1F.pdf, accessed 24 April 2018.

Pormeister K, (2017) *Genetic data and the research exemption: is the GDPR going too far?, International Data Privacy Law*, Volume 7, Issue 2, pp. 137–146

Stevens L, (2015) *The Proposed Data Protection Regulation and its Potential Impact on Social Sciences Research in the UK*, European Data Protection Law Review Vol.1, pp. 97–112

[213] IT Governance Privacy Team, EU *General Data Protection Regulation (GDPR): An Implementation and Compliance Guide*, https://www.itgovernance.co.uk/download/EU_GDPR_Implementation_and_Compliance_Guide_sample.pdf, accessed 5 May 2018.

Chapter 4
Singapore

Abstract Singapore has, within 50 years, come from the third world to the first world and established a reputation as a trade, investment and legal hub for Asia. Singapore has recognized that data, including personal data, is an increasingly important resource in the digital age (Chesterman, S *Data Protection Law in Singapore Privacy and Sovereignty in an Interconnected World*, Second Edition Academic Publishing (2018). This Chapter will draw on the important work of Simon Chesterman, David Alfred, Jansen Aw, Warren Chik, Lanx Goh, Hannah Lim, Abu Bakar Munir, Daniel Seng, Bryan Tan, David Tan, Steve Tan and Tan Cheng Han). As these scholars have observed, the discerning and intelligent application of personal data will allow us to unlock the benefits of our Smart Nation initiatives and enhance the lives of Singaporeans (Ibid). For businesses in Singapore, the ability to harness personal data will be a significant competitive advantage in the digital economy. As the impetus to collect, apply and share personal data grows, Singapore's policies must strike a careful balance between enabling the use of data for innovative business purposes and addressing legitimate societal concerns over privacy and safeguards (Ibid). Even though it is widely understood that the Singapore model has looked to the EU, UK, Canada, Hong Kong, Australia and New Zealand, as well as the OECD Guidelines on the Protection of Privacy and Transborder Flow of Personal Data, and the APEC Privacy Framework, it is our view that Singapore's data protection framework is stand alone (Personal data Protection Commission Singapore, https://www.pdpc.gov.sg/Legislation-and-Guidelines/Personal-Data-Protection-Act-Overview, accessed 2 December 2018). It is also our view that Singapore's approach to data protection law, being a business and consumer friendly model, sits at one end of the spectrum while the EU's model sits at the other end. In other words, and as highlighted in the previous Chapter, the EU certainly takes a higher level focus on protecting the privacy of their citizens than that of any other country discussed in this book.

Singapore have adopted the data protection principles and concepts of the OECD. It enacted the *Personal Data Protection Act 2012* (No. 26 of 2012) on 15 October 2012. Consistent with the other Chapters throughout this book, Chap. 4 will only discuss the key concepts and elements attributed to the Personal Data Protection Act 2012. This discussion will include the definition of personal data, controller, and evaluate whether the laws of Singapore apply to both the public and

© Springer Nature Singapore Pte Ltd. 2019

R. Walters et al., *Data Protection Law*,
https://doi.org/10.1007/978-981-13-8110-2_4

private sectors. In addition, this Chapter will evaluate the concepts of consent and collection, accuracy, retention, data transferred to a foreign country, enforcement, extraterritorial—reach and the regulator. Moreover, this Chapter will also discuss Singapore's Do Not Call Registry, and other concepts, such as the loss or damage of personal data, the right to be forgotten, and briefly outline the supporting cyber security laws that have been established in the Island state.

4.1 Introduction

Singapore has a remarkable story. One of the first countries throughout South East Asia to establish data protection laws. Singapore rose from the third to the first world in less than 50 years. Just how remarkable Singapore is, cannot be underestimated. They have no natural resources. It is an Island state that adopted the common law system of the United Kingdom. Their leadership throughout South East Asia and beyond in areas of the law and trade has seen the country rise as a world leader in business and services. Singapore has created a business friendly environment and is strategically placed in the Asia Pacific region.

The *Personal Data Protection Act 2012* (No. 26 of 2012) strengthens Singapore's business and trade competitiveness and current position as a trusted, world-class hub for trade and investment. This approach is subtly different to the European Union that have placed a greater emphasis on privacy as a human right.[1] The PDPA, provides the minimum standard for the protection of personal data across Singapore society.[2] The legislation demonstrates the commitment by government to protect the collection, use and disclosure of personal data. It also comprises nine main obligations which organizations must comply with when undertaking activities relating to the collection, use or disclosure of personal data. In the course of meeting these obligations, organizations are required to develop and implement policies and practices that are necessary for the organization to comply with the PDPA.[3] The PDPA recognizes the balance between the need to protect individuals' personal data and the need of organizations to collect, use, transfer or disclose personal data.

Section 3 of the PDPA states that the purpose of the legislation is to govern the collection, use and disclosure of personal data by organizations in a manner that recognizes both the right of individuals to protect their personal data and the need of organizations to collect, use or disclose personal data.[4] In addition, personal data cannot be collected and used for purposes that a reasonable person would consider inappropriate in the circumstances. Furthermore, the PDPA has been developed to

---

[1] Report of the Committee on the Future Economy Pioneers of the next generation, https://www.gov.sg/~/media/cfe/downloads/cfe%20report.pdf?la=en, accessed 15 May 2017.

[2] Chesterman, S *Data Protection Law in Singapore, Privacy and Sovereignty in an Interconnected* World, Academy Publishing, 2014, pp. 208–218.

[3] Personal Data Protection Act 2012, The principles include consent; purpose; reasonableness; access; correction; accuracy; protection; limited retention and limited transfer.

[4] Personal Data Protection Act 2012, section 3.

strengthen and develop trust in data protection by Singapore, and reflects the broader policy objectives of the legislation and the state. Trust and certainty are two very important principles in data protection law. Without, business and the broader community would be less inclined to engage in the use of technology that collects and uses personal data. Locally, the PDPA seeks to maintain the trust of individuals in organizations that manage data.[5]

Arguably, Singapore has looked to Australia, the EU, United Kingdom, Canada, Hong Kong, the OECD Guidelines on the Protection of Privacy and Transborder Flow of Personal Data, and the APEC Framework to develop its 2012 law.[6] The PDPA is supported by a number of industry sector specific guidelines for the health, real estate, education and telecommunications sectors.[7] Graham Greenleaf believes that Singapore has implemented all of the OECD concepts and principles related to data protection (see Chap. 16).[8] However, Singapore law has a number of exceptions which do not form part of the discussion in this Chapter. Notably and consistent with the state's policy direction, Singapore does not recognize a standalone right to privacy.[9] The right can be found in the common law and legislation,[10] even though, as highlighted in Chap. 1, since Singapore was founded as an independent state, it has made no apology for having to encroach on people's private life.

Nonetheless, two trends have emerged from Singapore's data protection laws. Simon Chesterman highlights that the first trend in relation to personal data was not based on a privacy or the rights of data subjects. Rather, it arose from the commercial realities of globalization and the integration of information economies—and the need for Singapore to retain its business friendly environment.[11] The second trend is that the changing data processing practices are forcing a reconsideration of basic premises of privacy laws and data protection in Singapore. In other words, there is a perceived need to move focus from limiting the collection of data, and allow the market to operate—to regulating its use.[12]

---

[5] Wong YongQuan, B *Data privacy law in Singapore: the Personal Data Protection Act 2012* International Data Privacy Law, Vol. 7, No. 4 (2017).

[6] Chesterman, S *Data Protection Law in Singapore Privacy and Sovereignty in an Interconnected World*, Second Edition Academic Publishing (2018).

[7] Personal Data Protection Commission, https://www.pdpc.gov.sg/Legislation-and-Guidelines/Guidelines/Sector-Specific-Advisory-Guidelines, accessed 4 March 2018.

[8] Graham Greenleaf, G *Singapore's Personal Data Protection Act 2012: Scope and Principles with so Many Exemptions, it is only a 'Known Unknown'* Privacy Laws & Business International Report, Issue 120, December (2012), p, 1–7.

[9] Chesterman, S *Data Protection Law in Singapore Privacy and Sovereignty in an Interconnected World*, Second Edition Academic Publishing (2018).

[10] *Protection form Harassment Act 2014, Copy Right Act 1987.*

[11] Chesterman S (2012) *After Privacy: The Rise of Facebook, the Fall of WikiLeaks, and Singapore's Personal Data Protection Act 2012*, Singapore Journal of Legal Studies, p. 392.

[12] Ibid.

4.2 Definition Personal Data

Personal data is defined as data, whether true or not, about an individual who can be identified from that data; or from that data and other information to which the organization has or is likely to have access.[13] Personal data that can be used to identify a person or otherwise includes the:

- Full name of a person;
- NRIC Number or FIN (Foreign Identification Number);
- Passport number and mobile telephone number;
- Facial image of an individual (e.g. in a photograph or video recording);
- Voice of an individual (e.g. in a voice recording);
- Fingerprint;
- Iris image; and
- DNA profile.[14]

The general definition of personal data includes many elements similar to other countries who have also defined this data as sensitive (data). Interestingly, the advancement of technology today has enabled a lot more identifying information to be collated, stored and used. For instance, only 50 years ago there was very little or no discussion in relation to iris imagery or DNA profiling. Nevertheless, Singapore has excluded personal data that is contained in a record that has been in existence for at least 100 years, or personal data that relates to a person who has been deceased for more than 10 years.[15]

On 5 February 2018, the Minister for Health was asked by the Singaporean Parliament, as a result of the introduction of the National Electronic Health Record (NEHR), how the Ministry would safeguard the confidentiality of its records. This was because of the potential that data leaks can affect a patient's employability and any future career prospects.[16] Furthermore, the question was asked as to what security measures would be put in place to ensure that not every employee in a clinic has access to personal records; and what measures are put in place to prevent data breaches.[17] As to background and functions, the NEHR, are responsible for collecting and consolidating data from different health institutions across Singapore. The information collected, collated and retained includes admission history, discharge summary, test results, radiology results, medication history, any surgery or procedures, allergies or adverse drug reactions. Most, if not all this data contains personal information that is defined by the PDPA. The NEHR has a role to centralize all health records from all hospitals, general practitioners and other health care

[13] Personal Data Protection Act 2012, section 2.

[14] Personal Data Protection Act 2012.

[15] Ibid.

[16] Singapore Parliament, Thirteenth Parliament of Singapore First Session, Paper, No. 58, 5 February 2018.

[17] Ibid.

providers. Apart from the challenges with uptake by the private sector, the Minister for Health stated in response to questions about security measures it would put into place, that it would employ the proposed legislation to centralize all health records in Singapore.[18] It recognized that patients can only realize the full potential of the NEHR if the data is comprehensive. Additionally, the Department of Health added that, for NEHR data to be comprehensive, every provider and healthcare professional needs to contribute relevant data to it.[19] It added that a series of audits and checks would be undertaken to confirm that the data and information will not be used in a manner that compromises the individual's privacy. It was also noted that data subjects, as part of the proposed changes in 2019, will also have access to the system to identify who and what organization has access to and has accessed personal health records. The challenges from such an initiative may not be fully realized in the initial stages. That is, unless the citizens of Singapore consider this ease of access to personal data, being a problem.

Nevertheless, there have been concerns raised in relation to the level of personal data protection axross Singapore. In *Re Full House Communications Pte Ltd*[20] the court highlighted the concerns aa data subjects personal data was secure protected. On 1 March 2015, the Complainant and his mother had attended the Furniture Fair and had purchased items which entitled the Complainant to participate in the Respondent's lucky draw. To participate in the lucky draw, a participant was required to register his personal details on the laptop provided by the Respondent at the redemption counter, including the individual's name, identity card number, occupation, contact number, e-mail address and residential address.[21] The form would then be printed out and dropped into a box for the lucky draw. While entering the personal details of his mother in the computerized form, the Complainant had concerns about the level of protection of the personal data that was provided by the Respondent.[22] As a result the Commission undertook an investigation. The Commission noted that the Respondent's responses to the Commission were as follow:

(a) The Respondent acknowledged that the auto-fill function had been enabled for all the fields in the form for the convenience of customers;
(b) the Respondent maintained that the personal data entry into the laptops had been in the presence of its staff, and they would watch the customers and ensure that no one would be able to take photos of the personal information displayed on the laptops;

---

[18] Channel News Asia, 5 February 2018.

[19] Ibid.

[20] [2016] SGPDPC 8.

[21] Ibid.

[22] Section 24 of the Personal Data Protection Act 2012 states that an organization is obliged to protect personal data in its possession or control by making reasonable security arrangements to prevent unauthorized access, collection, use, disclosure, copying, modification, disposal or similar risks. Additionally, "personal data" as referred to in section 2 refers to data, whether true or not, about an individual who can be identified: (a) from that data; or (b) from that data and other information to which the organization has or is likely to have access.

(c) the forms were not accessible to the Internet; and
(d) subsequent to receiving the Commission's notification of this matter, the Respondent had taken remedial actions during the ongoing Furniture Fair.[23]

In summary, the case highlighted that, where a person's full name can be located in the drop down box, it would enable identification of that data subject. The importance of this, is that the Commission recognizes that there are other systems within technology that can disclose an individual's identity. Furthermore, some of the information obtained and disclosed falls within the definition of personal data under the PDPA.

A year later in 2017, the Commission in Singapore have had to grapple with disclosing and identifying personal information, when an individual is communicating through WhatsApp. In *Re Executive Coach International Pte. Ltd*[24] the Commission had to decide whether the personal history of an individual was disclosed to participants in a WhatsApp Group. At issue was whether the organization is responsible for Mr. L's disclosure of the personal data, and if the organization is liable for Mr. L's disclosure, whether the organization is in breach of ss 13 and 20 of the PDPA for the said disclosure.[25]

The facts were that the organization provided life and executive coaching services to individual and corporate clients. The Complainant was a former employee of the organization. She was the personal assistant to (Mr L), a director of the organization. The Complainant left the organization on unamicable terms. The WhatsApp Group, comprising of the organization's employees and volunteers, was created on 22 August 2013. The Complainant and Mr. L were both participants in this WhatsApp Group.[26] At the material time on 7 April 2015, there were a number of other participants in this WhatsApp Group.[27] On 7 April 2015, Mr. L disclosed highly sensitive information about the Complainant's personal history, namely her past drug problem and issues with infidelity in her relationship, to the participants in the WhatsApp Group.[28] The organization did not dispute that the personal history of the Complainant was personal data. The disclosure of the personal data was made by Mr. L following allegations that she was undermining the organization's authority by persuading the employees and volunteers of the organization to leave the organization.[29] The Complainant claimed that the Personal Data was disclosed by her to Mr. L in the context of Mr. L being the Complainant's employer, teacher and coach. On 11 May 2015, the Commission notified the organization of the complaint and requested the organization to co-operate and assist in investigations. In the course of the investigation, the organization presented to the Commission that;

---

[23] Ibid.

[24] [2017] SGPDPC 3.

[25] Ibid. The Complainant and Organization disagreed on the exact number of participants in the WhatsApp Group on 7 April 2015. The Complainant claimed that the WhatsApp Group contained 117 participants. The Organization claimed that there were only 58 participants and that a group could only accommodate a maximum of 100 participants.

[26] Ibid.

[27] Ibid.

[28] Ibid.

[29] Ibid.

(a) Mr. L disclosed the Personal Data in his personal capacity and not as an employee of the Organization; and (b) the Personal Data was only known to Mr. L and not the Organization, and that the Organization did not authorize Mr. L to disclose the Personal Data.
(b) the Personal Data was only known to Mr. L and not the Organization, and that the Organization did not authorize Mr. L to disclose the Personal Data.[30]

It was concluded that the personal data disclosed involved sensitive data of the individual's personal history, and in this instance, there is no question, and it is not disputed, that such information falls within the definition of "personal data" under the PDPA.[31] The nature of the personal data, including the fact that the Complainant was identified in the WhatsApp Group, put it beyond doubt that the information was information "about an individual who can be identified from that data".[32] Under s 53(1) of the PDPA, any act(s) done or conduct engaged in by an employee in the course of his employment shall be treated for the purposes of the PDPA as done or engaged in by his employer as well as him, whether or not it was done or engaged in with the employer's knowledge or approval. This matter alone highlights the complexity in having to respond to claims about the misuse of personal data that has been defined under national law. The Commission found that there were significant breaches of the PDPA, notably sections 13 and 20, even though the breach was within the organizations employment group chat.[33]

It was also concluded that the content of individuals' communications, such as email messages and text messages, in and of themselves may not be considered personal data, unless they contain information about an individual that can identify the individual.[34] Arguably, this expanded recognition of personal data highlights that the current definition is unlikely to be settled, and could be expanded in the future. This confirms that the disclosure of personal data, whether general or sensitive, can occur easily and under many different fora.

Personal data that has been anonymized ceases to be personal data, for the purposes of the PDPA.[35] This is because personal data refers to data about an individual. The treatment of anonymized data under the Singapore data protection regime is similar, but there are difference to the EU data protection regime.[36] Wong YongQuan highlights that the differences between the two regimes began to develop when accounting for the distinction between anonymized data and pseudonymized[37]

[30] Ibid.

[31] Ibid.

[32] Ibid.

[33] Ibid.

[34] Ibid.

[35] Wong YongQuan, B *Data privacy law in Singapore: the Personal Data Protection Act 2012* International Data Privacy Law, Vol. 7, No. 4 (2017).

[36] Ibid.

[37] Ibid, Pseudonymisation, defined in the GDPR, means: the processing of personal data in such a manner that the personal data can no longer be attributed to a specific data subject without the use of additional information, provided that such additional information is kept separately and is sub-

data.[38] In Singapore, the PDPC appears to regard pseudonymized data as anonymized data, which falls outside of the definition of personal data in the PDPA.[39] YongQuan goes onto to say that, strictly applied the definition of personal data in the PDPA cannot support the equation of pseudonymized data with anonymized data.[40] Rather, on the contrary, pseudonymized data remains personal data.

4.3 Controller

The controller has become a central part of data protection law. The controller provides a point of contact within an organization, who is responsible for the collection and use of personal data. However, Singapore unlike the EU do not specify that a controller is to be established within an organisation. Rather, Singapore only refer to a designated individual, however, they play a similar role to the controller under EU law. Section 11 of the PDPA requires an organization to designate an individual to be responsible for ensuring its compliance with the legislation.[41] In addition, section 11(3) provides that an organization is to designate one or more persons to be responsible for ensuring that the organization complies with the PDPA.[42] Apart from the seriousness the Singaporean government places on data protection, the resulting affect is to place a greater regulatory burden and accountability on organizations to comply with the legislation. In relation to responding to access and correction requests, at least one aspect of the business contact information of this designated individual should be a mailing address (for example an office address) or an electronic mailing (email) address. Importantly, the legal responsibility for complying with the PDPA remains with the organization and does not "pass" to the individual designated by the

---

ject to technical and organizational measures to ensure that the personal data are not attributed to an identified or identifiable natural person. Pseudonymization differs from anonymization in that pseudonymization is reversible by the organization possessing the pseudonymized data, whereas anonymization is irreversible. Pseudonymized data can be re-associated with particular individuals with the aid of other information held by the organization.

[38] Personal Data Protection Commission (PDPC), Singapore, Guide to Basic Anonymisation Techniques, January 2018.

[39] Ibid, the Selected Topics Guidelines, the PDPC clarifies that anonymized personal data is not personal data and then lists pseudonymization as an 'anonymisation technique'. Crucially, the PDPC also provides an example which suggests that pseudonymized data is regarded as anonymized data as long as the organization concerned puts in place controls to prevent the specific department using the pseudonymized data from reidentifying individuals from the pseudonymized data. Pseudonymisation is also referred to as coding. Pseudonyms can be irreversible, where the original values are properly disposed and the pseudonymisation was done in a non-repeatable fashion, or reversible (by the owner of the original data), where the original values are securely kept but can be retrieved and linked back to the pseudonym, should the need arises.

[40] Ibid.

[41] Personal Data Protection Act 2012, section 11.

[42] Personal Data Protection Act 2012, section 11.

organization.[43] Furthermore, section 13 prohibits organizations from collecting, using or disclosing an individual's personal data unless the individual gives, or is deemed to have given, his consent.[44] Thus, the designated individual has an added layer of responsibility for managing personal data they receive on behalf of the organization.

4.4 Public and Private

The application of data protection law in Singapore is quite narrow. Similar to its northern neighbour Malaysia, public agencies are exempt from the legislation,[45] and any organization can be included in the gazette as a 'public agency'.[46] The PDPA has been described as a regime that is bifurcated, with a comprehensive legislative regime for the private sector and a separate set of rules for the public sector.[47] This could place businesses and organizations in situations of uncertainty, and something that would have a further cost impact, due to the administrative work required to determine what an organization is dealing with. Furthermore, Simon Chesterman argues that, an organization or a business will have no idea whether that business or organization is acting on behalf of the government.[48] This Catch-22 positon is unique to Singapore, as there is no recourse when personal data is provided under the false assumption of protection.[49] Over the next decade it will be interesting to see whether Singapore change their policy position on this point of law.

The most notable exclusion of the PDA is the way in which Singapore has enacted legislation that is inferior to other current, past and future legislation.[50] The PDPA appears to step outside the boundary of the tradition common law and equity. Moreover, the existing provisions that have effect on data protection would operate instead of the PDPA and not in addition to it.[51] Notably, the private sector exemption for an individual acting in a personal or domestic capacity exists, however, there is no exemption for non-commercial information. In other words, business contract information can be exempted from the PDPA except in accordance with section 4(5)

[43] Personal Data Protection Act 2012, section 11(6).

[44] Personal Data Protection Act 2012, section 13.

[45] Chesterman, S *Data Protection Law in Singapore, Privacy and Sovereignty in an Interconnected* World, Academy Publishing, (2014).

[46] Personal Data Protection Act 2012, sections 2 and 4.

[47] Wong YongQuan, B *Data privacy law in Singapore: the Personal Data Protection Act 2012* International Data Privacy Law, Vol. 7, No. 4 (2017).

[48] Ibid.

[49] Chesterman, S *Data Protection Law in Singapore Privacy and Sovereignty in an Interconnected World*, Second Edition Academic Publishing (2018).

[50] Wong YongQuan, B *Data privacy law in Singapore: the Personal Data Protection Act 2012* International Data Privacy Law, Vol. 7, No. 4 (2017).

[51] Ibid.

of the PDPA. While there is the potential for this information to be abused, it is considered that the Do Not Call provisions and anti-spam laws go some way to minimize any abuse.[52]

Arguably, one of the reasons for adopting this approach is for Singapore to maintain its status as welcoming business and trade across all sectors into the state. The assumption is that, if Singapore was to establish laws that in some way restrict or push away organizations and business, the multiplying effect would likely have significant economic consequences for the small island state and its people, particularly its business community.

4.5 Consent and Collection

Similar to the other countries and jurisdictions discussed throughout this book, the collection of personal data can only be undertaken, provided the data subject has provided consent. Consent can be obtained in a number of different ways. As good practice, an organization should obtain consent that is in writing or recorded in a manner that is accessible for future reference, for example, if the organization is required to prove that it had obtained consent.[53] An organization may also obtain consent verbally although it may correspondingly be more difficult for an organization to prove that it had obtained consent. For these types of situations, at a minimum, the organization should document, in some way, the consent that was provided, for example, by noting that oral consent was provided by an individual for a certain purpose.[54] The importance of consent cannot be underestimated as it cuts across most areas of the PDPA. In practice, it has been determined that an individual has not given consent unless that person has been notified of the purposes for which the data will be collected. In circumstances where the organization fails to inform the individual of the purposes for which data will be collected, used and disclosed, any consent provided does not amount to an actual consent.[55] Thus, to reinforce this point, the person must be informed and have provided consent for the use or disclosure of the data that has been collected in relation to them. However, the PDPA does not prescribe the precise mechanisms by which organizations should obtain consent.[56] Nevertheless, the PDPC notes that it is good practice to 'obtain consent that is in writing or recorded in a manner that is accessible for future reference'.[57]

---

[52] Chesterman, S *Data Protection Law in Singapore Privacy and Sovereignty in an Interconnected World*, Second Edition Academic Publishing (2018).

[53] Ibid.

[54] Ibid.

[55] Personal Data Protection Act 2012, Division 1.

[56] Wong YongQuan, B *Data privacy law in Singapore: the Personal Data Protection Act 2012* International Data Privacy Law, Vol. 7, No. 4 (2017).

[57] Personal Data Protection Commission, Advisory Guidelines on Key Concepts in the Personal Data Protection Act at para 12.5.

Section 14(1) of the PDPA states how an individual provides consent. Moreover, sections 13–17 of the PDPA deal with a number of issues relating to the Consent Obligation.[58] Importantly, the PDPA does not affect existing legal or regulatory requirements that organizations have to comply with. Organizations may collect, use and disclose (as the case may be) personal data without the individuals' consent if required or authorized to do so under the PDPA or other written law, although those organizations may need to comply with other requirements of the Data Protection Provisions which are not inconsistent with its obligations under other written law.[59] In particular, an individual has not provided consent, unless that individual has been notified of the purposes for which his personal data will be collected, used or disclosed.[60] If an organization fails to inform the individual of the purposes for which the personal data will be collected, used and disclosed, any consent given by that individual would not amount to consent under section 14(1). Further details on the organization's obligation to notify the individual are explained in the section on the "Notification Obligation".[61]

The PDPA provides that personal data can be collected, used and disclosed without consent.[62] Circumstances where personal data may be collected, used or disclosed without consent was emphasized in *Jump Rope (Singapore).*[63] The PDPC stated that, in exceptional circumstances, it may be reasonable for an organization to disclose personal data of an individual without consent.[64] This would apply when the disclosure of personal data of an individual who has been dismissed, blacklisted or undergoing disciplinary proceedings for the purpose of warning others.[65] However, the PDPC said that the organization must comply with the neighbouring obligations.

The PDPA neither defines collection, use and disclosure, as specific terms. They are subjective terms to enable broad interpretation so as technology evolves collection, use and disclosure of data is likely to also evolve. In addition, section 13(b) provides that the consent of the individual is not required in circumstances where the collection, use or disclosure of personal data is statutorily mandated or authorized.[66] Generally, collection refers to any act or set of acts through which an orga-

[58] Personal Data Protection Act 2012, division 1, section 13–17.

[59] Personal Data Protection Commission, Advisory Guidelines on Key Concepts in the Personal Data Protection Act, (2017).

[60] Ibid.

[61] Ibid.

[62] Personal Data Protection Act 2012, section 15 and 17. In accordance with Second Schedule (collection), Third Schedule (use) and Fourth Schedule (disclosure).

[63] [2016] SGPDPC 21.

[64] Ibid.

[65] Yip, M *Personal Data Protection Act 2012: Understanding the consent obligation*, Singapore Management University (2017).

[66] Personal Data Protection Act 2012, section 13.

nization obtains control over or possession of personal data.[67] Secondly, use refers to any act or set of acts undertaken by an organization to use the data.[68] Notwithstanding the above, these exceptions are generally characterized by necessity, reasonableness and/or fairness.[69] Many Yip argues that some of the exemptions appear to be very wide, for instance, collection *necessary* for "evaluative purposes" and where the personal data is publicly available.[70] Even so, the interpretation of these exemption are likely to be left to the courts to make a final determination of when and how such exemptions will apply. No doubt, any exemption will be provided on a case by case basis.

Organizations are required to notify the data subject of the purposes of their collection, use or disclosure of personal data. Transparency in relation to the collection, use or disclosure of personal data comprises two distinct duties.[71] Firstly, the organization is collecting, using or disclosing personal data about an individual pursuant to actual consent from that individual.[72] Secondly, the organization is required to inform the individual of the purposes of the collection, use, or disclosure.[73] It must be noted that the use of data can come in many different forms, such as making available to a third party, displaying the date in a publication or on a website, or using it to simply know who and what the person does for a living.

The exceptions have been clarified by necessity, reasonableness or fairness. Therefore, the PDPA has established that acceptable uses of personal data do not require an individual's consent. However, the broader issue is the extent to which consent from an individual can be applied. In *Universal Travel Corp Pte Ltd*[74] four passengers required confirmation of cancellation of their flight so as they could apply for an insurance claims. The agency disclosed to all four passengers the entire list that contained the individual's personal data. It was held that:

> the passengers could not be deemed to have consented to the disclosure. There was no urgency that required paragraph 1(a) of the Fourth Schedule to be invoked, and the information could have been provided to the individual's separately without disclosing who the information of the other people.[75]

Section 18 of the PDPA becomes important because it limits the purposes for which an organization may collect, use, or disclose personal data (Purpose

---

[67] Personal Data Protection Act 2012, section 20, Notification Obligation.

[68] Ibid.

[69] Yip, M *Personal Data Protection Act 2012: Understanding the consent obligation*, Singapore Management University (2017). The PDPA acknowledges that certain forms of socially, morally or legally acceptable uses of personal data do not require the individual's consent.

[70] Ibid.

[71] Wong YongQuan, B *Data privacy law in Singapore: the Personal Data Protection Act 2012* International Data Privacy Law, Vol. 7, No. 4 (2017).

[72] Ibid.

[73] Ibid.

[74] *Universal Travel Corp Pte Ltd* [2016] SGPDPC 4.

[75] Ibid.

Limitation Obligation).[76] Section 18 states that an organization may collect, use or disclose personal data about an individual only for purposes that a reasonable person would consider appropriate in the circumstances. Benjamin Wong YongQuan notes that under the Purpose Limitation Obligation principle, if the organization has notified the individual of the purposes of the collection, use or disclosure pursuant to section 20, then the organization may only collect, use or disclose the personal data for those purposes.[77]

The Personal Data Protection Commission (PDPC) also emphasized the independence and importance of the neighbouring obligations, the notification obligation and the reasonableness of purpose obligation under section 18.[78] The PDPC's approach underscores that the PDPA's protection framework is not based solely on consent. In particular, the PDPC provided that the reasonableness obligation is "an important aspect of the PDPA as it is effective in addressing excesses in the collection, use or disclosure of personal data" under a broadly-worded consent clause. Even if an organization has not breached the consent obligation, it may be guilty of breaching the neighbouring obligations.

Section 21(1) of the PDPA allows an individual to request access to their personal data and information about the ways in which that personal data has been or may have been used or disclosed by the organization within a year before the date of the individual's request.[79] An individual can submit a request to gain access to personal data about him or her, and to some or all personal data and information about the ways the personal data has been used.[80] However, there are limitations to such a request that include, but are not limited to, an organization only providing such personal data if it is feasible for it to do so.[81] This subjective and broad approach does provide an organization significant flexibility. In addition, an organization is not required to provide access to information which is no longer within its possession or under its control.

Section 21(3), provides circumstances in which an organization 'must not' provide personal data or other information.[82] A provision such as is this, is important for, and applies to, the protection of physical or mental health, or, reveals the identity of an individual who has provided personal data about another individual. Therefore, no data or information is to be released that is in the national interest.[83] That is, where that information could be a security threat to the state and its citizens.

---

[76] Wong YongQuan, B *Data privacy law in Singapore: the Personal Data Protection Act 2012* International Data Privacy Law, Vol. 7, No. 4 (2017).

[77] Ibid.

[78] Personal Data Protection Act 2012, section 18.

[79] Personal Data Protection Act 2012, section 21.

[80] Personal Data Protection Act 2012, section 21.

[81] Personal Data Protection Act 2012, section 21(2), Third, Fourth and Fifth Schedule provides a list of exemptions. Use of Data without consent, Disclosure of data without consent and Exemption from Access.

[82] Personal Data Protection Act 2012, section 21(3).

[83] Ibid.

In summary, the provision provides the data subject with a greater degree of control over their personal data that is under the control of organizations.[84] It allows the data subject to ascertain whether personal data about themselves is being held by an organization, and to make corrections when such personal data is erroneous.

In another matter *Re Asia Renal Care (Katong) PTE LTD and Another*[85] the Commission had to resolve the issue of whether disclosure of personal data was deemed to constitute consent when personal data was used to respond to the Complainant's complaint. There were two organizations involved. The first issue was whether the Complainant consented to, or could be deemed to have consented to the organization collecting and using his personal data for the purposes of sending him the letter in question, pursuant to the Consent Obligation. The second issue was whether the first organization is permitted to disclose the Complainant's personal data to the second organization pursuant to the Consent Obligation.[86]

The facts of the case were that the Complainant was a dialysis patient for a number of years at a clinic operated by the First Organization. The second Organization was the majority shareholder of the First Organization.[87] On 8 June 2015, the managing director and operations manager of the second Organization delivered a letter to the Complainant. The letter concerned the complaints that the Complainant had with the service he had received at the clinic, but it also addressed the Complainant's combative behaviour towards the staff, nurses, doctors and other patients at the clinic.[88] However, the second Organization was of the view that such behaviour was disruptive to the operations of the clinic, it raised the possibility of terminating the clinic's services to the Complainant in the letter.[89] The Complainant's name and residential address was set out as the addressee at the top of the letter. The Complainant lodged a complaint with the Commission, alleging that there was an unauthorized collection and use of his personal data by the Second Organization without his consent.[90] The Complainant further alleged that the First Organization ought not to have disclosed his personal data to the Second Organization.

The Commission found the:

> Complainant had, before his receipt of the letter in question, already been raising complaints about the service at the clinic directly to the Second Organization and, on occasion, to both Organizations simultaneously.[91]

---

[84] Wong YongQuan, B *Data privacy law in Singapore: the Personal Data Protection Act 2012* International Data Privacy Law, Vol. 7, No. 4 (2017).

[85] 1 February [2016].

[86] Ibid.

[87] Ibid.

[88] Ibid.

[89] Ibid.

[90] Ibid.

[91] Ibid.

Additionally, while corresponding with the second Organization, the Complainant had provided his contact details to the second Organization directly. It was noted that:

> By such actions, the Complainant must be taken to have consented or be deemed to have consented to the Second Organization's use of the Complainant's personal data for the purposes of engaging with him over the issues he raised and on the subject matter of his complaints, for instance his service experience at the clinic.[92]

In conclusion, the Commission found that the First and Second Organizations did not breach their Consent Obligations under the Personal Data Protection Act 2012. In the Singaporean context, Man Yip is of the opinion that consent should be assessed by considering the other interests that are worth protecting.[93] Yip believes that having an overemphasis on consent in personal data protection law would undoubtedly lead to higher compliance costs for businesses and slower transaction rates.[94] These consequences would affect both organizations as well as individuals. Beyond purely economic consequences, organizations may require an individual's personal data for legitimate and/or reasonable activities.[95]

It is our view the point made by Yip can also be seen in other countries and jurisdictions that do not follow the EU model of putting the data subject's rights (privacy) at the forefront of data protection law. Furthermore, this highlights the complex nature of data protection and privacy law, whereby the nation state have to balance the economic (business and innovation) needs with protecting citizen's personal data (a level of privacy) as defined by the law. Arguably, it is inevitable that nation states have the sovereign right to make that decision in the best interest of that state. On the other side, it can be argued that the current framework around consent is inadequate, because, in practice, individuals are not fully informed of what they are consenting too.

4.6 Accuracy

The management of personal data must be accurate. If not, the potential legal ramifications can be vast and varied such as wrongfully identifying the data subject. Section 23 of the PDPA[96] requires an organization to make a reasonable effort to ensure that personal data collected by or on behalf of the organization is accurate and complete, if the personal data:

(a) is likely to be used by the organization to make a decision that affects the individual to whom the personal data relates; or
(b) is likely to be disclosed by the organization to another organization.[97]

---

92 Ibid.

93 Yip, M *Personal Data Protection Act 2012: Understanding the consent obligation*, Singapore Management University (2017).

94 Ibid.

95 Ibid.

96 Personal Data Protection Act 2012, section 23.

97 Ibid.

This obligation is an important part of the overall process for the collection, storage, use and dissemination of personal data and information. The consequences to the individual for an organization not providing accurate data is enormous and could have ramifications that go far beyond the individuals, personal and professional standing. To ensure that personal data is accurate and complete, an organization must make a reasonable effort to ensure that (a) it accurately records personal data which it collects (whether directly from the individual concerned or through another organization); (b) personal data it collects includes all relevant parts thereof (so that it is complete); (c) it has taken the appropriate (reasonable) steps in the circumstances to ensure the accuracy and correctness of the personal data; and (d) it has considered whether it is necessary to update the information.[98]

However, the obligation only extends to the organization making a "reasonable" effort, by take into account the following:

(a) the nature of the data and its significance to the individual concerned (e.g. whether the data relates to an important aspect of the individual such as his health);
(b) the purpose for which the data is collected, used or disclosed;
(c) the reliability of the data (e.g. whether it was obtained from a reliable source or through reliable means);
(d) the currency of the data (that is, whether the data is recent or was first collected some time ago); and
(e) the impact on the individual concerned if the personal data is inaccurate or incomplete.[99]

Arguably, the problem lies in the obligation for the organization to confirm whether the information is accurate. There is currently no regulatory requirement for this check to be undertaken and it is for the organization to perform its own risk assessment. Thus, the onus is for the organization to use and demonstrate every reasonable effort to ensure the accuracy and completeness of such personal data that is likely to be used to make a decision that will affect the individual.

4.7 Retention

The collection and use of data and information will result in organizations retaining that data and information. The use of IT systems to undertake this function is a relatively new concept when compared to the traditional use of paper and hard copy filing systems. For instance, an employer of an organization has been able to retain the personal data and information on individuals for a defined period. In Singapore, section 25 does not specify a maximum period of time for retaining data

[98] Personal Data Protection Act 2012.

[99] Ibid.

collected.[100] The flexible approach applied to this provision allows an entity to retain relevant personal data and information for what they deem is a 'reasonable' time that is no longer necessary for legal or business purposes.

The Retention Limitation Obligation prevents organizations from retaining personal data in perpetuity where it does not have legal or business reasons to do so.[101] It is widely recognized that each organization will have its own business reasons for retaining data. The Retention Limitation Obligation does not specify a fixed duration of time for which an organization can retain personal data. Thus, the timeframe for retaining data will be determined along the lines that require the organization to retain that data.

Arguably, where it is demonstrated that data has been retained for 10 years, and that it has not been utilized in the past 5 years, it could be determined that the retention of that data is no longer reasonable. Moreover, it is argued that it is a subjective test, which will be evaluated on a case by case basis. Some guiding principles handed down by the PDPC to assist organizations include how much effort and resources the organization would need to expend in order to use or access the personal data, and whether any third parties have been given access to that personal data. [102]

4.8 Data Transferred to a Foreign Country

An organization is not able to transfer any personal data to a country or territory outside Singapore when there is comparable protection in place.[103] The Commission may, on the application of any organization, request an exemption of the transfer of personal data by that organization. However, this must be in writing.[104] The Commission may also specify certain conditions as to when the transfer of data can be exempted. The Commission has the power to, at any time, add to, vary or revoke any condition imposed under this section.

Moreover, regulation 9 of the *Personal Data Protection Regulations 2014* plays a vital role in regulating the transfer of personal data outside of the island state.[105] Regulation 9 states:

> For the purposes of section 26 of the Act, a transferring organization must, before transferring an individual's personal data to a country or territory outside Singapore take appropriate steps to ensure that the transferring organization will comply with Parts III to VI of the Act, in respect of the transferred personal data while it remains in the possession or under the control of the transferring organization; and take appropriate steps to ascertain whether,

[100] Personal Data Protection Act 2012, section 25.

[101] Advisory Guidelines on the PDPA, 2017.

[102] Ibid.

[103] Personal Data Protection Act 2012, section 26 (1).

[104] Personal Data Protection Act 2012, section 26 (2).

[105] Personal Data Protection Regulations 2014, Regulation 9.

and to ensure that, the recipient of the personal data in that country or territory outside Singapore (if any) is bound by legally enforceable obligations (in accordance with regulation 10) to provide to the transferred personal data a standard of protection that is at least comparable to the protection under the Act.[106]

Furthermore, the transferred personal data, while it remains the under the control of the transferring organization, the personal data is data in transit, and could become publicly available in Singapore. Importantly, the transferring organization is taken to have satisfied the requirements of paragraph (1)(b) in respect of an individual's personal data which it transfers to a recipient in a country or territory outside Singapore if:

(a) subject to paragraph (4), the individual consents to the transfer of the personal data to that recipient in that country or territory;
(b) the transfer of the personal data to the recipient is necessary for the performance of a contract between the individual and the transferring organization, or to do anything at the individual's request with a view to the individual entering into a contract with the transferring organization;
(c) the transfer of the personal data to the recipient is necessary for the conclusion or performance of a contract between the transferring organization and a third party which is entered into at the individual's request;
(d) the transfer of the personal data to the recipient is necessary for the conclusion or performance of a contract between the transferring organization and a third party if a reasonable person would consider the contract to be in the individual's interest;
(e) the transfer of the personal data to the recipient is necessary for the personal data to be used under paragraph 1(a), (b) or (d) of the Third Schedule to the Act or disclosed under paragraph 1(a), (b), (c), (e) or (o) of the Fourth Schedule to the Act, and the transferring organization has taken reasonable steps to ensure that the personal data so transferred will not be used or disclosed by the recipient for any other purpose;
(f) the personal data is data in transit; or
(g) the personal data is publicly available in Singapore.[107]

Consent has an important role to the data subject maintaining a level of control over their personal data, before it leaves the state.[108] As highlighted by Chesterman,

---

[106] Ibid.

[107] Personal Data Protection Regulations 2014, regulation 9.

[108] Ibid. In other words, an individual is not taken to have consented to the transfer of the individual's personal data to a country or territory outside Singapore if the individual was not, before giving his consent, given a reasonable summary in writing of the extent to which the personal data to be transferred to that country or territory will be protected by a standard comparable to protection under the Act; the transferring organization required the individual to consent to the transfer as a condition of providing a product or service, unless the transfer was reasonably necessary to provide the product or service to the individual; or transferring organization obtained or attempted to obtain the individual's consent for the transfer by providing false or misleading information about the transfer, or by using other deceptive or misleading practices. There is nothing in the law that

Alfred and Goh[109] overall the international transfer of personal data and information in accordance with section 26 of the PDPA, is consistent with the framework that has been adopted by the OECD and APEC (see Chap. 16). However, in July 2017, the Singapore Personal Data Protection Commission commenced public consultation in relation to the PDPA. One of the proposed changes could see the concept of consent being modified. How, and in what form the proposed modification will look like has not been finalised. It will likely make it easier for businesses to make use of the personal data they collect (e.g. transfer), however, at the time of writing the book it was unlikely that any of the safeguards surrounding consent would be diminished. Organizations would need to conduct a risk and impact assessment to mitigate against any risk.[110] The resulting effect would be to reduce the burden to organization in the first instance, from the inconvenience of consent. On the other hand, an organization would be afforded greater responsibility by having to assess the risk and demonstrate their accountability in protecting the data subjects' personal data as defined by the law. The proposal would also expand the exceptions to consent, and create 'Legal or Business Purpose' consent. How this would operate is yet to be concluded. Nonetheless, it is proposed that consent would not apply where it is not desirable or appropriate, and where the benefits to the boarder public outweigh and adverse risk.

4.9 Enforcement

The PDPA provides for enforcement to be taken by PDPC.[111] The relevant considerations include the:

(a) number of third parties to whom the disclosure has been made;
(b) period of disclosure;
(c) the amount of personal data disclosed;
(d) level of sensitivity of the disclosed personal data; and
(e) the impact of disclosure upon the individual.[112]

An additional to the above consideration is required as to whether the disclosure was caused by willful or systemic failures of the organization. In *Chua Yong Boon Justin*,[113] the PDPC imposed a $500 fine on the breaching party, on account of the fact that the breach was willful. At issue was where the Respondent was a registered

---

prevents an individual from withdrawing any consent given for the transfer of the personal data to a country or territory outside Singapore.

[109] Chesterman, S *Data Protection Law in Singapore Privacy and Sovereignty in an Interconnected World*, Second Edition Academic Publishing (2018), p. 382.

[110] Personal Data Protection Commissioner, Guide to Data Protection Impact Assessments (2017).

[111] Personal Data Protection Act 2012, section 39.

[112] Ibid.

[113] *Chua Yong Boon Justin* [2016] SGPDPC 13, 19(c).

salesperson who obtained personal data of the Complainant and his wife in the course of his real estate agency work and hence in the course of carrying on his business.[114] The Personal Data Protection Commission held the Respondent:

> was not allowed to claim that the subsequent disclosure of the personal data was made in a "personal or domestic capacity", which would have allowed him to dispense with the need to obtain consent under Section 4(1)(a) of the PDPA.[115]

Furthermore, the Commission held that the Respondent:

> continued to hold such personal data in the course of his business, and needed to comply with his Consent Obligation when disclosing the personal data.[116]

However, the PDPC set the fine amount at "the lower end of the spectrum" in view of the fact that the disclosure was limited, one-off, and did not cause a harmful impact on the individual.[117] In 2016, the PDPC, issued Advisory Guidelines on the Enforcement of the Data Protection provisions of the PDPA. The Guidelines underpin the Act.[118] The Guidelines themselves are not binding, and are designed to provide guidance to the Commission on the interpretation of the PDPA's provisions relating to the enforcement of the PDPA.[119]

Rather than describe a complaints mechanism within the PDPA. The PDPC has established a complaints and review process. Again this process is also underpinned by the powers afforded to the PDPC in accordance with the PDPA[120] and the Guidelines. While a complaint may be made, there are several options available to the Commissioner to resolve the issue.[121] That is, where the Commission believes that any complaint by an individual against an organization may be more appropriately resolved by mediation, the Commission may, with the consent of the complainant and the organization, refer the matter for mediation.

---

[114] Ibid.

[115] Ibid.

[116] Ibid.

[117] Ibid, 21.

[118] Personal Data Protection Act 2012, section 49(1).

[119] Personal Data Protection Commission, Advisory Guidelines on the Enforcement of the Data Protection 2016. The Guidelines are advisory in nature and are not legally binding on the Commission, or any other party. They do not modify or supplement in any way the legal effect and interpretation of any laws cited including, but not limited to, the PDPA and any subsidiary legislation (such as regulations and rules) issued under the PDPA. Accordingly, these Guidelines shall not be construed to limit or restrict the Commission's administration and enforcement of the PDPA. The provisions of the PDPA and any regulations or rules issued thereunder will prevail over these Guidelines in the event of any inconsistency.

[120] Personal Data Protection Act 2012, section 27. he Commission's powers in relation to alternative dispute resolution do not include deciding on disputes between a complainant and an organization or ordering an organization to compensate a complainant who suffers a loss as a result of a contravention of any of the Data Protection Provisions by the organization. The PDPA provides that individuals who suffer loss or damage as a direct result of a contravention of Part IV, V or VI of the PDPA may commence civil proceedings in the courts against the organization.

[121] Ibid.

4.9.1 Notification of Breach

In July 2017, the Singapore Personal Data Protection Commission commenced public consultations in relation to the PDPA. The changes had not been finalized at the time of writing this book (note this to be updated upon publication—if changes). Nonetheless, the notable inclusions would see Singapore continue to follow other jurisdictions and establish a mandatory data breach notification mechanism. The proposal would require an organization to notify a person (data subject) within 72 h of the breach—related to them. Organizations will also be required to notify the Commission of the breach. The proposal will extend to all breaches, no matter whether the breach is actual or perceived.[122] Moreover, the enforcement mechanisms in Singapore's PDPA are significantly stronger than its regional Asian counterparts—in both the variety of penalties and their relative seriousness, including fines of up to SG$1 m.[123]

4.9.2 Data Protection Impact Assessments

The PDPC introduced guidelines on Data Protection Impact Assessments, in 2017, to strengthen the management of personal data on systems (public facing websites, cloud storage platforms, customer relationship management) and processes in accordance with the PDPA.[124] The key tasks in a DPIA include:

- identifying the personal data handled by the system or process, as well as the reasons for collecting the personal data;
- identifying how the personal data flows through the system or process;
- identifying data protection risks by analysing the personal data handled and its data flows against PDPA requirements or data protection best practices;
- addressing the identified risks by amending the system or process design, or introducing new organisation policies; and
- checking to ensure that identified risks are adequately addressed before the system or process is in effect or implemented.[125]

The impact assessments are also considered a key component in enhancing the concept of data protection by design, whereby organizations are encouraged to consider the protection of personal data from the earliest possible design stage. The assessment promotes the idea that organizations are then encouraged to consider the risks associated with managing personal data throughout the entire cycle from collection to long term storage.

---

[122] Personal Data Protection Commissioner, Guide to Data Protection Impact Assessments (2017).

[123] Greenleaf, G *Data Privacy in Asia*, in Chesterman, S *Data Protection Law in Singapore Privacy and Sovereignty in an Interconnected World*, Second Edition, Academic Publishing (2018).

[124] Personal Data Protection Act 2012, section 11 and 12.

[125] Ibid.

4.10 Extraterritorial – Reach

Singapore's personal data protection laws, to date, do not reach beyond the state. The legislation only applies to the collection, disclosure, and use within Singapore. However, other supporting cyber security legislation does have extraterritorial reach. It remains to be seen whether the ongoing interconnectedness of the digital economy, and whether in the future the PDPA will be amended to take on a greater role and reflect that of the European Union model, as businesses expand their operations to, within and outside Singapore.

4.11 Agency [Regulator], Principles and Codes

The Cyber Security Agency (CSA) is the national agency overseeing cybersecurity strategy, operation, education, outreach, and ecosystem development.[126] The CSA assists in implementing the national strategy for cyber security and the so called four pillars to assists government, business and community.[127] This includes strengthening the resilience of the critical infrastructures; and mobilizing businesses and the community to make cyberspace safer and protecting personal data. The CSA recognizes the challenges faced by the SME (small and medium sized) sector, from the lack of awareness to cost impacts on implementing and maintaining secure IT infrastructure.[128] The CSA will build on such efforts and continue to work with industry partners and associations to encourage SMEs to adopt measures to improve and enhance their cybersecurity.[129] The CSA, along with the Singapore Business Federation, continue collaboration and the Employee Cybersecurity Kit was launched to assist the SME sector.

The PDPC has responsibility as the Regulator for the PDPA, and in 2016, released Advisory Guidelines on Enforcement of the Data Protection Provisions (Guidelines). The Guidelines have been made pursuant to section 49(1) of the PDPA.[130] The Guidelines were issued as a result of a number breach and enforcement actions having been undertaken. The Guidelines make clear when to decide whether to exercise its powers to enforce the Data Protection Provisions. Moreover, the PDPC takes into account two main objectives; (1) the resolution of an individual's complaint; and (2) ensuring that organizations comply with the data protection provisions. The factors that would prompt the PDPC to conduct an investigation into an organization's failure to comply with its data protection obligations, including, amongst others, the following:

---

[126] Cyber Security Agency https://www.csa.gov.sg, accessed 20 December 2018.

[127] Ibid.

[128] Ibid.

[129] Ibid.

[130] Personal data Protection Act 2012, section 49.

- whether the organization's conduct indicates a systemic failure to comply with the PDPA or establish and maintain the necessary policies and procedures to ensure compliance;
- the number of individuals who are, or may be, affected by the organization's conduct;
- the impact of the organization's conduct on the individual who may be affected, for example, whether the individual may have suffered loss, injury or other damage as a result of the organization's contravention of the PDPA or whether they have been exposed to a significant risk of the same; and
- whether the organization has been approached by the individual to seek a resolution; and e. the public interest in the PDPC conducting an investigation.[131]

Organizations need to take seriously the need to protect an individual's personal data and, where there have been lapses, take immediate and corrective steps to remedy the breach, which will include cooperating with any investigation. The Guidelines make it clear, the decision whether to impose a financial penalty on a defaulting organization will depend on a number of factors, in particular "the seriousness and impact of the organization's breach and the immediacy and effectiveness of corrective actions" to address the breach.

In January 2018, Singapore's privacy regulator awarded SG$6000.00 in damages against a telemarketer for selling personal data, without consent. The personal data consisted of individual's names, their NRIC numbers, mobile phone number and annual income ranges.[132] The PDPC highlighted that profiteering from the sales of personal data by organizations at the expense of consumers and individuals is the very kind of activity which the PDPA seeks to curb, and hence, must be severely dealt with.[133] In achieving the objectives of the PDPA, Singapore has adopted similar principles that reflect those of other jurisdictions and include the: (a) Consent; (b) Purpose Limitation; (c) Notification; (d) Access and Correction; (e) Accuracy; (f) Protection; (g) Retention Limitation; (h) Transfer Limitation; and (i) Openness Obligation.[134]

Moreover, and in delivering on the above principles, and strengthening the co-regulatory model, the development of Codes of Practice are strongly encouraged on a sectorial basis. However, they are not specified by the Act. For instance, a sector association may establish a sector code of practice. Any code of practice, provided it meets the core principles of the data protection laws also meet section 11 whereby an organization shall consider what a reasonable person would consider appropriate in the circumstances. Associations, such as the banking and health sectors, have adopted codes of practice so as to foster a better understanding of what legal obligations apply when dealing with personal data. However, an association developed code of practice is not binding on its members, but rather used as a guide. The question arises whether Singapore should enable codes of practice to be enforced?

---

[131] Advisory Guidelines on Enforcement of the Data Protection Provisions, section 15.3.

[132] Tham, I *Privacy watchdog fines first 'data monger'*, Singapore Strait Time, 29 January 2018, A6.

[133] Ibid.

[134] Personal Data Protection Act 2012, sections 13–26.

The Commission can appoint a Commissioner and Deputy Commissioners.[135] The Commission has the power to undertake reviews of complaints made in relation to an organization's handling of personal data. The Commission can request access to documentation and other information to ensure and organization is complying with the laws. They are able to take copies of any document produced, and serve notice on individuals to ensure related documents are produced. Furthermore, the Commission has broad powers and can enter most premises to gain access to information, documents and equipment or articles relevant to an investigation. This is an extensive power, because the Commission does not require a search warrant to enter a premises. Nonetheless, a warrant can be obtained by the Court to effect a search.

Section 29(1) of the PDPA provides that the Commission may, if it is satisfied that an organization is not complying with the law, direct the organization to undertake certain activities to ensure compliance.[136] A direction can also include ceasing to collect, use or disclose personal data; destroying personal data and paying any penalty imposed.

The PDPC also has the authority to resolve disputes. Section 27(1) of the PDPA, where any complaint by an individual against an organization may be more appropriately resolved by mediation.[137] The PDPC can with the consent of the complainant and the organization, refer the matter for mediation. Furthermore, in accordance with section 27(2) of the PDPA,[138] the PDPC may direct a complainant or an organization or both to attempt to resolve the complaint of the individual in the way directed by the PDPC. In certain circumstances the PDPC may commence an investigation in relation to an organization's compliance with sections 21 or 22 of the PDPA. These situations may include, but are not limited to where the organization does not comply with the Enforcement Regulations during a review.[139]

4.12 Do Not Call Registry

Part IX of the PDPA provides for the 'Do Not Call Registry' to restrict and limit marketing messages being disseminated by marketing agencies to anyone. The Registry will initially comprise three (3) separate registers kept and maintained by the PDPC[140] and covers telephone calls, text messages and faxes. Individuals can register their Singapore telephone number(s) on one or more Do Not Call Registers, depending on their preferences in relation to receiving marketing messages through telephone calls, text messages or faxes. Australia has a similar registry, and both go some way to regulating or restricting competition practices from the use of personal data and information (see Chap. 5).

[135] Personal Data Protection Act 2012, sections 7 and 8.

[136] Personal Data Protection Act 2012, section 29 (1).

[137] Personal Data Protection Act 2012, section 27(1).

[138] Personal Data Protection Act 2012, section 27(2).

[139] Advisory Guidelines on Enforcement of the Data Protection Provisions, section 14.

[140] Personal Data Protection Act 2012, section 39.

Simon Chesterman highlights that the Do Not Call Registry, while a slightly odd fit within the *PDPA*, is suggested by having its own interpretive clause.[141] Chesterman, goes onto say that the obligations created are additional to those in the *Spam Control Act 2007*, on the basis that whereas that Act puts conditions on the sending of unsolicited commercial messages in bulk, the Do Not Call Registry determines whether a specified message may be delivered to a specific telephone number.[142] A specified message for this purpose is defined by reference to the content, presentation, or linked information; a message is covered by the provision if, having regard to that information, "it would be concluded" that one of the purposes of the message is to advertise or otherwise offer to supply goods or services, an interest in land, or a business or investment opportunity.[143] In a similar way to Australia, the Singapore Registry, provides a mechanism for individuals to opt-out from receiving specified messages. However, this only applies to a Singapore telephone number.

The purpose of section 37 of the PDPA is to ensure that the content of the message; the presentational aspects of the message; the content that can be obtained using the numbers, URLs or contact information (if any) mentioned in the message; and if the telephone number from which the message is made is disclosed to the recipient.[144] The content (if any) can be obtained by calling that number, it would be concluded that the purpose, or one of the purposes, of the message is to:

(i) offer to supply goods or services;
(ii) advertise or promote goods or services;
(iii) advertise or promote a supplier, or prospective supplier, of goods or services;
(iv) offer to supply land or an interest in land;
(v) advertise or promote land or an interest in land;
(vi) advertise or promote a supplier, or prospective supplier, of land or an interest in land;
(vii) offer to provide a business opportunity or an investment opportunity;
(viii) advertise or promote a business opportunity or an investment opportunity;
(ix) advertise or promote a provider, or prospective provider, of a business opportunity or an investment opportunity; or
(x) any other prescribed purpose related to obtaining or providing information.[145]

[141] Chesterman, S *After Privacy: The Rise of Facebook, the Fall of WikiLeaks, and Singapore's Personal Data Protection Act 2012*, Singapore Journal of Legal Studies, pp. 391–415, 2012, p 413. Messages may not be sent to a Singapore telephone number that has been entered on a Do Not Call Register. The Do Not Call provisions apply to a "person" who sends such messages, who is under an obligation to check the Do Not Call Registry within a period to be prescribed.

[142] Ibid.

[143] Ibid.

[144] Personal Data Protection Act 2012, section 37.

[145] Personal Data Protection Act 2012, section 37.

Notwithstanding the above framework, Warren Chick notes the Registry had not come without criticism.[146] Chick highlights how the Exemption Order (EO), which is applicable to tax and fax messages, was released on 26 December 2013.[147] However, there was backlash against the EO which was unexpected. Chick argues that there was initially no distinction in the treatment of all three methods of communication within the Act or when the register was first opened for signature. It was only after the register was opened, and after the PDPC garnered overwhelming response that the exemption order was made; apparently also only after concomitant and ongoing consultations with businesses only.[148] The exemption reinforces the need to balance business and individual interests, an approach that the Government promised during the preparation stages of the PDPA and that is explicitly stated in section 3 of the Act. Such an approach necessitates the constant adjustment of the allocation of rights, even after the Act is enacted; and that is one of the main tasks that the PDPC was set up to perform. In fact, it is predicted that the balance will largely be recalibrated in favour of businesses in time to come, given the economy-centric policies of the Singapore government over privacy interests.[149] Finally, Chik argues that the parameters of the EO must be further defined and confined, through the "ongoing relationship" requirement, in such a way that it does not weaken the overall "opt-out" regime for text-based messages under the DNC regime.[150] At the time of writing, the authors note that the EO, based on the 2013 Fact Sheet, appears to not have been updated to address the concerns raised by Chik.

4.13 Loss or Damage

A data subject who incurs a loss or damage directly as a result of a contravention of any provision in Part IV, V or VI pf the PDPA by an organization shall have a right of action for relief in civil proceedings in a court.[151] If the Commission has made a decision under the Act in respect of a contravention specified in subsection (1). Where there is no action accruing under subsection (1) action may be brought in respect of that contravention until after the decision has become final as a result of there being no further right of appeal.[152] The court may grant to the plaintiff relief by way of injunction or declaration, damages or some other relief as the court thinks fit.

[146] Chik, W *Thee Singapore Do Not Call Register and the Text and Fax Exemption Order* Singapore Management University (2014).

[147] Ibid.

[148] Ibid.

[149] Ibid.

[150] Ibid.

[151] Personal Data Protection Act 2012, section 32 (1).

[152] Ibid.

4.14 Right to Be Forgotten

The right to be forgotten does not expressly exist in Singapore. The right to be forgotten or otherwise referred to as the right to erasure is very different to the right to be left alone. The concepts achieve different things. The right to be left alone, as provided for by the Do Not Call Registry, allows the data subject the option of not being contacted by certain industry sectors. Section 16 provides for the withdrawal of consent 'on giving reasonable notice to the organization, an individual may at any time withdraw any consent given, or deemed to have been given under this Act, in respect of the collection, use or disclosure by those organizations of personal information about the individual'.[153] Additionally, where an individual withdraws consent to the collection, use or disclosure of personal data, the organization shall cease collecting, using or disclosing the personal data. Furthermore, section 22 allows an individual to request an organization to correct an error or omission in the personal data about the individual that is in the possession or under the control of the organization. However, the laws do not allow for a person to request that their personal data to be erased or deleted.

4.15 Supporting Cyber Security Laws

In 2017, The Singaporean Parliament introduced the Computer Misuse and Cybersecurity (Amendment) Bill (CMCA) in 2017.[154] In 1993 the Computer Misuse Act was introduced to strengthen the capability and of the IT sector by criminalizing unauthorized access or modification of computer material. The new laws criminalize the act of dealing in personal information where criminals have use personal information obtained illegally from a computer hacking to commit or facilitate the commission of crimes (e.g. identity fraud).[155] For instance, criminals may trade hacked credit card information even though they themselves may not have been responsible for hacking the credit card information. Not only does this have implications to individuals and businesses, it is also a major concern to the banking and finance sector, especially in an age of heightened scrutiny of money laundering and money used to fund terrorist activities.

Of notable interest and when compared to the PDPA, the CMCA now has extraterritorial application where it can be determined that there is a "serious harm" in Singapore. Serious harm would include illness, injury or death of individuals in Singapore, as well as disruptions to essential services in Singapore. That's is, where there has been a major disruption of breach of security to the Singaporean state, its government and institutions. However, it is asserted that measuring the level of

[153] Personal Data Protection Act 2012, section 16.

[154] Computer Misuse and Cybersecurity Act 2017.

[155] Computer Misuse and Cybersecurity Act 2017, section 3–8.

harm to a data subject is different to an entity, and is not settled. Furthermore, criminals operating in the cyber world have the ability to conduct multiple unauthorized acts. Thus, the legislation allows charges to be amalgamated into a single charge. It remains to be seen as to how this provision will be used.

Since 2016, Singapore has been developing the Cyber Security Act. The Bill[156] was scheduled to enter the Singaporean Parliament in 2017. However, due to further public consultation, the Bill has been postponed to 2018. The legislation aims to ensure that operators take proactive steps to secure critical information infrastructure, and report incidents of abuse and infringements. The Act will prevent, manage and respond to cybersecurity threats and incidents, to regulate owners of critical information infrastructure.[157] The aim is to establish a framework for the sharing of cybersecurity information, to regulate cybersecurity service providers. It will also empower the Cyber Security Agency (CSA) to manage cyber incidents and raise the standards of cybersecurity providers in Singapore.

The proposal confers power on the Commissioner of Cybersecurity to investigate threats and incidents to ensure that essential services including telecommunications, transport, healthcare, banking and energy are not disrupted. The Bill aims to harmonize the requirements to protect critical information infrastructure across the public and private sectors. Banking and privacy rules that forbid the sharing of confidential information will be superseded by the Cybersecurity Bill.[158] The bill appoints Assistant Commissioners that will be appointed from individual sectors such as banking and energy. This will provide greater collaboration, oversight, and better coordinate sector regulators varying legislative powers.

There will be an increased focus on prevention from further cyber attacks, whereby essential services will be required to report and notify the Commissioner of any such attacks. Furthermore, it will enable the Commissioner to identify and designate new systems during times of national emergency. Finally, the proposal will establish a licensing framework for cyber security vendors, providing services to investigate work that involves hacking and forensic examination, and non-investigative work such as managed security operations. Investigative cyber-security hackers must also apply for an individual licence.

Furthermore, and in addition to the above, the PDPA is not the only source of law that relates to data protection. For instance, the Banking sector has legislation that describes how customer information will be handled by Banks[159] in Singapore. The *Human Biomedical Research Act 2015* governs the human biomedical research, which provides that no human biomedical research can be conducted if the appropriate consent of a person for participation as a research subject.[160]

---

[156] Cyber Security Bill, https://www.csa.gov.sg/~/media/csa/cybersecurity_bill/draft_cybersecurity_bill_2017.ashx?la=en, accessed 4 December 2017.

[157] Ibid.

[158] Second Reading Speech On Cybersecurity Bill 2018 By Dr. Yaacob Ibrahim, Minister For Communications And Information, During Parliamentary Sitting On 5 February 2018.

[159] Banking Act 2008.

[160] Human Biomedical Research Act 2015.

Finally, there is ongoing work and reviews being undertaken in Singapore. The most current include the Public Consultation for Managing Unsolicited Messages and the Provision of Guidance to Support Innovation in the Digital Economy April–June 2018.[161] In addition, Public Consultation for Proposed Advisory Guidelines on the PDPA for NRIC Numbers.[162]

4.16 Conclusion

Singapore is considered the Switzerland of South East Asia. Arguably, they are, and have been for a number of decades a central pillar for trade and the rule of law throughout the South East Asia. Singapore have one of the highest reputations around the world for following and implementing the rule of law, and providing an environment whereby the public sector sets the benchmark for the lack of corruption.[163]

Nonetheless, and on the backdrop of the above, the PDPA, not only supports this view, but also is relatively young when compared to its Australian and EU counterparts. Moreover, the co-regulatory approach taken by Singapore has similarities to the privacy and personal information regulatory framework in Australia, the EU and other countries that arguably have a mature regulatory framework. Compared with the EU, the Singapore approach is not as rigid. Under the GDPR personal data can only be collected for specified, explicit and legitimate purposes and not further processed in a manner that is incompatible with those purpose.

Moreover, the right to erasure (forgotten) is another area of great difference between Singapore law and other countries, particularly the EU. The GDPR more or less can compel an organization to delete—remove personal data. Even though the PDPA requires data to be destroyed, there is a lot of flexibility as to how this can be undertaken. Furthermore, Singapore has progressed very quickly to reinforce to business and the community, the seriousness of data protection, which is demonstrated by the number of decisions and fines issued by the Commissioner in 2017 and 2018. However, and as already highlighted, privacy in Singapore has traditionally not been a priority for Singaporean's or their government. The importance of privacy to the community may well change as people become more aware of the impact this is having to their personal privacy. Conversely, Singaporean's have had to live in an environment where government reach consistently encroaches on people's privacy, and it will be interesting to see whether they have the same level of acceptance for the private sector, in years to come.

---

[161] Personal Data Protection Commission Singapore, https://www.pdpc.gov.sg/Legislation-and-Guidelines/Public-Consultations, accessed 2 August 2018.

[162] Ibid.

[163] Chesterman, S *Data Protection Law in Singapore Privacy and Sovereignty in an Interconnected World*, Second Edition Academic Publishing (2018), p. 465.

Simon Chesterman suggests the ongoing competing models of privacy and business broadly reflect the tensions in the theoretical approaches to privacy, but the impetus for reform is not always a desire to protect privacy as such.[164] In Singapore, at least, reform is not being driven by the desire to defend the rights of data subjects, rather, it is based primarily on economic considerations.[165] Arguably, it is our view that Singapore has established a business friendly model. It is further argued, this approach serves Singapore well, however, the challenge in the future will be how states react to different legal requirements in other countries. Will Singapore have the opportunity to influence and set the future direction of data protection law? Or, will they be required to fall in behind the EU or other countries?

Most notably, when compared to other jurisdictions, the PDPA is limited in scope and does not apply to all personal data processing activities. The PDPA does not apply to the public sector or any organization acting as an agent of a public sector. Another limitation is where business contact information is excluded from the operation of the PDPA, along with data intermediaries. However, data intermediaries are required to ensure they delete personal data when it is no longer needed. Arguably, these limitations can be viewed by other jurisdictions as not being compatible in the international arena. For Singapore the policy direction ensures minimal impact to business. The number of exemptions provided to the concept of consent stands out as potentially diluting its effectiveness. For instance, deemed consent is considered to have been provided voluntarily as soon as a data subject has handed over their personal data. Thus, it will be interesting to see what direction Singapore takes in the future, as the GDPR and other regional countries develop their data protection laws.

It could be argued that Singapore has certainly been a leader across Asia and ASEAN countries in developing a data protection regime. Even though, it was initially developed with the economy front and center, there is greater awareness of the personal impact that technology is having on individual's personal data and privacy within Singapore. This can be demonstrated when in 2018, health records were hacked, and the Prime Minister's along with another estimated 1.5 million people's health records being illegally accessed.[166] In conclusion, the ability for a data subject to obtain compensation from the misuse of personal data, defined by laws that has resulted in humiliation, is not subject to direct regulation. However, this does not mean that data subject could not seek compensation through the courts.

[164] Ibid.

[165] Chesterman, S *After Privacy: The Rise of Facebook, the Fall of WikiLeaks, and Singapore's Personal Data Protection Act 2012*, Singapore Journal of Legal Studies, pp. 391–415, 2012, p 414.

[166] Singapore says hackers stole 1.5 m health records, https://www.theaustralian.com.au/national-affairs/health/singapore-says-hackers-stole-15m-health-records/news-story/c372cc1f4136a0b-93316f0a1a15ffcfe, accessed 2 August 2018.

References

Chesterman, S (2014) *Data Protection Law in Singapore, Privacy and Sovereignty in an Interconnected* World, Academy Publishing, pp. 208–218

Chesterman, S (2018) *Data Protection Law in Singapore Privacy and Sovereignty in an Interconnected World*, Second Edition Academic Publishing

Chik, W (2014) *Thee Singapore Do Not Call Register and the Text and Fax Exemption Order* Singapore Management University

Greenleaf, G (2012) Singapore's Personal Data Protection Act 2012: Scope and Principles (with so Many Exemptions, it is only a 'Known Unknown') Privacy Laws & Business International Report, Issue 120, p, 1–7

Greenleaf, G (2018) *Data Privacy in Asia*, in Chesterman, S *Data Protection Law in Singapore Privacy and Sovereignty in an Interconnected World*, Second Edition, Academic Publishing

Wong YongQuan, B (2017) *Data privacy law in Singapore: the Personal Data Protection Act 2012* International Data Privacy Law, Vol. 7, No. 4

Yip, M (2017) *Personal Data Protection Act 2012: Understanding the consent obligation*, Singapore Management University

Chapter 5
Australia

Abstract This Chapter provides an outline of the current privacy laws in Australia. Privacy regulation and law in Australia is multilayered and includes regulation by government (primary legislation, regulations and codes), and industry self-regulation. The *Privacy Act 1988* (Cth) is the principal legislation that regulates privacy, personal data and personal information. Australia sets itself apart from the EU and Singapore by naming its laws as 'privacy', rather than data protection. Yet, the principles and concepts enshrined in the law have similarities to those jurisdictions. Due to the complex nature of data protection and privacy law, Australia has established the Australian Privacy Principles, which play an important role in underpinning the primary Act. It will be argued that Australia's legal framework is the third model discussed in this book, that in our view sits between Singapore and the EU. The Australian privacy framework balances business, government policy and human rights, it does not, in the same way as Singapore, create a business friendly model – that is stand alone. In addition, Australia's model having been originally established in the 1980s took a very different focus, with an emphasis on the credit industry. Furthermore, in support of the legislative framework, Australia has relied on the common law to determine how privacy is determined. This Chapter does not deal with any privacy laws that have been established by either of the Australian states or territories, because the scope of this book is only to compare the national or in the case of the EU, supranational laws. This Chapter highlights how the Privacy Act defines personal information and regulates the collection, use and disclosure of personal information about individuals. The Australian legislation is technical and cumbersome hence to avoid the over analysis of the technical aspects of the law, this Chapter as has been the case with other Chapters in this book will only focus on key terms, concepts and principles. This includes, but not limited to whether the law applies to both public and private sectors; the definition of personal information, and its extra-territorial reach. In addition, the Chapter will also identify how Australia has dealt with establishing an agency [regulator], principles and codes of practice, along with the concept of consent, collection of personal information, and quality of information – (accuracy). Finally, this Chapter will discuss retention of personal information, notification of a breach in the law, and determine whether the right to be forgotten has been considered. It also examines the principles data portability, loss or damage and enforcement, impact assessments and, additional legislation and standards in regards to cybersecurity.

© Springer Nature Singapore Pte Ltd. 2019

R. Walters et al., *Data Protection Law*,
https://doi.org/10.1007/978-981-13-8110-2_5

5.1 Introduction

Australia is a Federation that is comprised of a Federal Government and five states governments including, New South Wales, Queensland, South Australia, Tasmania and Victoria, and two Territories; the Northern Territory and the Australian Capital Territory.[1] Even though there is no explicit reference to privacy or data protection in the constitution, the Commonwealth has the power to make laws in relation to external affairs.[2] This power enables the Commonwealth to give effect to its international obligations under a bona fide treaty such as the ratification and adoption of the International Covenant on the Civil and Political Rights 1966.[3]

The *Privacy Act 1988* (the Act) has been designed by the Australian Government to balance the protection of the privacy of individuals with that of the interests of entities in carrying out their functions or activities.[4] The Act also promotes the responsible and transparent handling of personal information by entities while facilitating an efficient reporting system and free flow of information. However, the Act does not apply to commercial data, except credit data. Credit reporting was included into the Act in the early 1990s following public concern regarding the credit industry's introduction of credit reporting, which essentially allows for the disclosure of personal data and information on individuals seeking to obtain credit.[5] This is a notable difference to the other laws discussed in this book, which do not include credit data.

Six years following the implementation of the 1988 Act, the Australian Privacy Charter was released. The Charter reinforced Australia's values to protecting the privacy rights of its citizens.[6] While not enforceable, the Charter identifies key concepts from the OECD principles that include, but not limited to justification, consent, accountability, observance, openess, communication, space, physical, collection, security and disclosure. The Charter also reinforced the point that

---

[1] *Data-matching Program (Assistance and Tax) Act 1990 (Cth)* regulates the commonwealth government data-matching using tax file numbers. The Tax File Number Guidelines 2011 issued under the Privacy Act also regulate the collection, storage, use, disclosure, security and disposal of individual's tax file numbers by public agencies and private organizations. The States and Territories of Australia have each established their own laws, they include and not limited to: Australian Capital Territory – *Information Privacy Act 2014*. New South Wales, *Privacy and Personal Information Protection Act 1998* and *Health Records and Information Privacy Act 2002*. Northern Territory, *Information Act*, in force at 12 April 2017. Queensland, *Information Privacy Act 2009*. South Australia, South Australia has issued an administrative instruction requiring its government agencies to generally comply with a set of Information. Tasmania, *Personal Information and Protection Act 2004*. Victoria, *Privacy and data Protection 2014* and *Freedom of Information Act 1992*.

[2] *Australian Constitution* section 51(xxix).

[3] *Victoria v Commonwealth* (1996) 187 CLR 416 at 487.

[4] Privacy Act 1988, section 2A.

[5] History of the Privacy Act 1988, Office of Information Commissioner, https://www.oaic.gov.au/about-us/who-we-are/history-of-the-privacy-act, accessed 20 December 2018.

[6] Australian Privacy Charter Council *"The Australian Privacy Charter"* (1995) PrivLawPRpr 31; 1995, 2(3) Privacy Law & Policy Reporter 44.

Australia, being a free and democratic society respects the autonomy of individuals, and limits the power of both State and private organizations to intrude on that autonomy. It went onto say that privacy is a core value that underpins human dignity and is associated with the freedom of association and freedom of speech.

The privacy laws of Australia are unique. Apart from facilitating the protection of personal information, the laws also establish a credit reporting system. The aim is to protect personal information by emphasizing the need for information collectors to be open, fair and accountable in the use of information.[7] These laws have been specifically established so as to balance the needs between the credit agency and the needs of the individual seeking to obtain credit. The Australian credit reporting system also helps ensure that credit providers are able to comply with their responsible lending obligations under the *National Consumer Credit Protection Act 2009.*[8] The Act places obligations on the type or personal information that credit providers can disclose, the organizations that can handle the information and the purpose for which the information is to be handled. In other words, a person who makes an application to a credit provider, the provider is able to use that personal data to verify the person's credit rating or risk.

Importantly, schedule one of the Act outlines the Australian Privacy Principles (APP).[9] The APPs underpin the Act by providing organizations with a governance framework to transparently manage personal data. Organizations are required to implement practices, procedures and systems to comply with the APPs and a registered APP code. The APPs along with the Act apply to any public or private organizations,[10] to which they must comply. In *Re TYGJ and Information Commissioner*[11], the Australian Administrative Tribunal (AAT) stated:

> The *Privacy Act* is an Act to protect the privacy of individuals. Section 13 sets out the circumstances in which an act or practice is an interference with a person's privacy. Of relevance in this case is s 13(a) which provides an act or practice is an interference with the privacy of an individual if the act or practice: (a) in the case of an act or practice engaged in by an agency (whether or not the agency is also a file number recipient, credit reporting agency or credit provider) – breaches an Information Privacy Principle in relation to personal information that relates to an individual.[12]

---

[7] Credit reporting Code of Conduct issued by the Privacy Commissioner under the Privacy Act, September 1991 and including all amendments as at March 1996, Privacy Commissioner, Human Rights and Equal Opportunity Commission, 1996. ISBN 064224846 X.

[8] Privacy Act 1988, Part IIIA.

[9] Office of information Commissioner, in December 2000, the *Privacy Amendment (Private Sector) Act 2000* extended coverage of the Privacy Act to some private sector organisations. The amendments commenced on 21 December 2001. These amendments introduced 10 National Privacy Principles (NPPs) into the Privacy Act, which set out standards in relation to private sector organisations collecting, using and disclosing, keeping secure, providing access to, and correcting personal information. The *Privacy Amendment (Enhancing Privacy Protection) Act 2012*, which commenced on 12 March 2014, introduced many significant changes to the Privacy Act, including: the Australian Privacy Principles, which replaced the IPPs and the NPPs, to regulate the handling of personal information by Australian and Norfolk Island Government agencies and some private sector organisations.

[10] Privacy Act 1988, section 6.

[11] *Re TYGJ and Information Commissioner [2017] AATA 1560.*

[12] *Re TYGJ and Information Commissioner [2017] AATA 1560*, 10–18.

The AAT reinforced the purpose and objective of the Act is to protect the privacy of Australian citizens, and that the Privacy Principles also form an important part of the overall regulatory framework for the management of people's personal information.

The APPs apply to all Australian and Norfolk Island Government agencies, all private sector and not-for-profit organisations.[13] The APPs also apply to all private health service providers and some small businesses (otherwise known as an 'APP entities') must handle, use and manage personal information. The principle objective of the APPs is to ensure:

- the open and transparent management of personal information including having a privacy policy;
- an individual having the option of transacting anonymously or using a pseudonym where practicable;
- the collection of solicited personal information and receipt of unsolicited personal information including giving notice about collection;
- that how personal information can be used and disclosed (including overseas);
- maintaining the quality of personal information;
- personal information is kept secure; and
- the right for individuals to access and correct their personal information is maintained.[14]

In addition to the above, there are also separate APPs that deal with the use and disclosure of personal information for the purpose of direct marketing,[15] cross-border disclosure of personal information[16] and the adoption, use and disclosure of government related identifiers.[17] APP 7 provides that an organization must not use or disclose personal information it holds for the purpose of direct marketing unless an exception applies. APP 7 may also apply to an agency in the circumstances set out in 7A. Direct marketing involves the use or disclosure of personal information to communicate directly with an individual to promote goods and services. An organization must, on request, provide its source for an individual's personal information, unless it is impracticable or unreasonable to do so. This is an important point because, to some extent the basis for restricting the use of personal information (data) by an organization to create or establish dominance in the market is directly related to competition law (see Chap. 14). The principles also place more stringent obligations on an APP entity when they handle sensitive information – as highlighted and defined above.

Nonetheless, the courts have had, and continue to play a critical role in determining the common law of privacy throughout Australia and its states and territories. In

[13] Privacy Act 1988, schedule 3, The Australian Privacy Principles.

[14] Australia Privacy Principles.

[15] Australia Privacy Principle, 7.

[16] Australia Privacy Principle, 8.

[17] Australia Privacy Principle, 9.

Plenty v Dillon[18] the Australian High Court has argued that there is a public interest in protecting and enforcing the private freedoms, rights and interests of individuals. The Australian High Court in *Plenty* stated that:

> if the courts of common law do not uphold the rights of individuals by granting effective remedies, they invite anarchy, for nothing breeds social disorder as quickly as the sense of injustice which is apt to be generated by the unlawful invasion of a person's rights.[19]

Thus, all cannot be understood or applied unless the crucial variables such as privacy versus public interest are understood. The facts were that Plenty, the owner and occupier of a small farm, expressly forbade Dillon, a constable, from entering his land. Dillon wanted to enter the land to serve some legal documents on Plenty. A further point is of importance which has been touched upon above, namely the entry to property with consent. Gaudron and McHugh JJ stated that:

> A person who enters the property of another must justify that entry by showing that he or she either entered with the consent of the occupier or otherwise had lawful authority to enter the premises.[20]

The important point is that the policy behind the law is to protect the possession of property and the privacy and security of its occupier. A person who enters the property of another must justify that entry by showing that he or she either entered with the consent of the occupier or otherwise had lawful authority to enter the premises. While this case has nothing to do with modern technology, the Internet or personal information or personal data, the issue of consent in privacy, within the common law dates back a long way, and is also associated with trespass cases as they relate to privacy of personal property. Thus, as has been demonstrated throughout the book, thus far, consent has become a fundamentally important concept in relation to data protection and privacy law more generally (see country Chapters).

The use of video material by the public and organizations has increased significantly over the past decade. However, with this use there has been issues in regards to privacy. A decade following the Plenty case, the Australian High Court in *Australian Broadcasting Corporation v Lenah Game Meats Pty Ltd*[21] had to determine whether the actions taken to publish a film was unconscionable, and whether the right to privacy existed as a result of this action. The respondent was successful in obtaining an injunction against the appellants from publishing a film displaying possums being stunned and killed at an abattoir. The film had been obtained from a third party while trespassing.[22] The Court found that it was not unconscionable for the appellants to publish the film and a corporation did not have a right to privacy. Gleeson J stated that:

---

[18] *Plenty v Dillon* (1991) 171 CLR 635, 655.

[19] Ibid.

[20] Ibid.

[21] *Australian Broadcasting Corporation v Lenah Game Meats Pty Ltd* (2001) 208 CLR 199, 42.

[22] Ibid.

> certain kinds of information about a person, such as information relating to health, personal relationships, or finances, may be easy to identify as private; as may certain kinds of activity, which a reasonable person, applying contemporary standards of morals and behaviour, would understand to be meant to be unobserved. The requirement that disclosure or observation of information or conduct would be highly offensive to a reasonable person of ordinary sensibilities is in many circumstances a useful practical test of what is private. [23]

Gleeson CJ argued that the equitable action for breach of confidence may be the most suitable legal action for protecting people's private information, stating:

> [E]quity may impose obligations of confidentiality even though there is no imparting of information in circumstances of trust and confidence. And the principle of good faith upon which equity acts to protect information imparted in confidence may also be invoked to 'restrain the publication of confidential information improperly or surreptitiously obtained'. The nature of the information must be such that it is capable of being regarded as confidential. A photographic image, illegally or improperly or surreptitiously obtained, where what is depicted is private, may constitute confidential information … If the activities filmed were private, then the law of breach of confidence is adequate to cover the case … There would be an obligation of confidence upon the persons who obtained [images and sounds of private activities], and upon those into whose possession they came, if they knew, or ought to have known, the manner in which they were obtained …The law should be more astute than in the past to identify and protect interests of a kind which fall within the concept of privacy. For reasons already given, I regard the law of breach of confidence as providing a remedy, in a case such as the present, if the nature of the information obtained by the trespasser is such as to permit the information to be regarded as confidential.[24]

However, it must be noted that the private information referred to by the court and Gleeson J, was not necessarily the personal (private) information defined by the *Privacy Act 1988*. Nevertheless, Gummow and Hayne JJ, with whom Gaudron J agreed, considered a broader range of privacy invasions and left open the direction that the future development of the law protecting privacy may take:

> In the present appeal Lenah encountered … difficulty in formulating with acceptable specificity the ingredients of any general wrong of unjustified invasion of privacy. Rather than a search to identify the ingredients of a generally expressed wrong, the better course, is to look to the development and adaptation of recognized forms of action to meet new situations and circumstances … Lenah's reliance upon an emergent tort of invasion of privacy is misplaced. Whatever development may take place in that field will be to the benefit of natural, not artificial, persons. It may be that development is best achieved by looking across the range of already established legal and equitable wrongs. On the other hand, in some respects these may be seen as representing species of a genus, being a principle protecting the interests of the individual in leading, to some reasonable extent, a secluded and private life, in the words of the *Restatement*, 'free from the prying eyes, ears and publications of others'. Nothing said in these reasons should be understood as foreclosing any such debate or as indicating any particular outcome.[25]

This case has been influential in the development of privacy in Australia, the court restricted its development at common law. However, privacy continues to evolve and in the past 4 years the courts in Australia have had to consider how to

---

[23] Ibid.

[24] Ibid, [34], [39], [40], [55].

[25] Ibid, 109–134.

deal with remedies associated with the breach of privacy, even though it may not necessarily involve deciding on whether the personal information defined by the law has constituted the breach.

The production and use of explicit video material has also been considered by the courts, particularly in relation to the possible infringement on a person's privacy. In *Giller v Procopets*[26] the issue of invasion of privacy was again considered. However, this matter involved the production of an explicit video that was made public, which was based on equity and a breach of confidence. For just over 3 years, Giller and Procopets lived in a de facto relationship during which time Procopets assaulted Giller on several occasions. After the couple ceased cohabiting in July 1993, Procopets visited Giller and their twin sons from time to time and occasionally stayed overnight and assisted Giller with the children. Their sexual relationship continued until December 1996. In November 1996 Procopets used a hidden camera to secretly record the sexual activity between himself and Giller. Giller became aware of the camera and acquiesced to its use.[27] A month later, when their relationship deteriorated further, Procopets threatened to show, and then attempted to show, videos depicting their sexual activity to Giller's family and friends, and told her employer he had a video of her engaged in sexual activity with a client. Procopets videotape sexual activity between himself and Giller was initially without her knowledge, but subsequently with her knowledge and acquiescence.[28] In the Supreme Court of Victoria, Giller claimed damages for breach of confidence based on the showing of the sexually explicit videos.[29] The trial judge found that the sexual relationship between the parties was confidential and Procopets had breached that confidence. The court held that Giller could not recover damages for breach of confidence because, Giller had not sought an injunction; and secondly, Australian law did not permit an award of damages for breach of confidence for mental distress falling short of a psychiatric injury.[30] On appeal in 2008 the Victorian Court of Appeal:

> unanimously upheld the action for breach of confidence and, by majority, awarded Giller the sum of $40,000, including $10,000 for aggravated damages.[31] The aggravated damages were awarded because the court was satisfied that Procopets had deliberately breached his duty of confidence so as to humiliate, embarrass and distress Giller. The fact that Giller had not sought an injunction to restrain Pocopets from showing or distributing the video did not deprive the court of its power to award damages because [t]hat power exists so long as a court has jurisdiction to grant an injunction.[32]

[26] *Giller v Procopets* [2008] VSCA 236; (2008) 24 VR 1.

[27] Ibid.

[28] Ibid.

[29] Ibid.

[30] Ibid.

[31] Ibid.

[32] Ibid.

The court noted that in other words, "it is both necessary and sufficient that an injunction could have been brought".[33] Professor Butler argued that the Victorian Court of Appeal in *Giller* held that a generalized tort of unjustified invasion of privacy should not be recognized where there was an existing cause of action that could be developed and adapted to meet new circumstances.[34] Such a position accords with the need to preserve coherency in the law.[35] Arguably, the Victorian Court of Appeal's decision in *Giller* has provided a foundation for the protection of an individual's privacy,[36] across Australia, from an equitable perspective (confidence), rather than a full tort.

Furthermore, in 2013, the court in the Australian state of Queensland had to decide whether an invasion of privacy arose from an action for breach of confidence resulting in the intentional infliction of emotional distress, from the use of pictures on a website hosted and registered by Yahoo [37] In *Doe v Yahoo!7 Pty Ltd*[38], Mr. Anderson, on behalf of Yahoo!, did not take any point that the facts were not deposed to in an affidavit by the Plaintiffs. The Plaintiffs allege that individuals can become registered users of this service and can post photographs and comments through the use of a profile on the Yahoo! website. On 24 November 2009 an article was written about Jane Doe in The Gladstone Observer newspaper (The Observer).[39] The article contained a photograph of Ms. Doe and noted she had an autoimmune disease. In another internet-based article from 'The Observer', another photograph of Ms. Doe appeared. The Plaintiff alleges Mr. Pagett created a profile in the name—Jane Doe‖ on the Yahoo! website sometime in March 2010. It is alleged that the photograph used on the profile was from 'The Observer' article.[40] The profile (which changed over time) was used by Mr. Pagett to post objectionable comments on various articles on the Yahoo! website.[41] The court concluded that the:

> photograph and personal information, which was published while losing its confidentiality upon publication would be a breach of confidence where consent was not provided. Because this is an area of developing law in Australia, the court held that it would be inappropriate to strike out any action on this basis. It may be argued that the misuse of the photograph and information about Doe constituted a breach of confidence in that the information was converted into something which was offensive without her consent. Furthermore, in relation to a breach of privacy, the court concluded that being a developing area of law, it would be

---

[33] Ibid.

[34] Ibid.

[35] Butler, D *The Dawn of the Age of the Drones: An Australian Privacy Law Perspective*, University of NSW Law Journal 434 (2014).

[36] Rivette, M *Litigating privacy cases in the wake of Giller v Procopets*, Media and Arts Law Review (2010).

[37] *Doe v Yahoo!7 Pty Ltd* [2013] QDC 181.

[38] Ibid.

[39] Ibid.

[40] Ibid.

[41] Ibid, 16–22.

wring for the court to strike out any action. It was noted that a breach of privacy as a cause of action had been recognized internationally in Canada, New Zealand and United States.[42]

More recently in 2015, the court in Western Australia had to consider whether there was a breach of privacy from the dissemination of explicit images following a break down in a relationship. In *Wilson v Ferguson*, the court considered a claim alleging breach of confidence, from the production of explicit personal material. The issue raised was how an Australian court exercising equitable jurisdiction should respond to the publication by a jilted ex-lover, to a broad audience via the internet, of explicit images of a former partner which had been confidentially shared between the sexual partners during their relationship.[43] In this case the court was satisfied that:

> such a publication occurred in breach of an equitable obligation of confidence owed by the defendant to the plaintiff.[44]

The appropriate relief for the breach of that obligation in the present circumstances is the grant of an injunction prohibiting further publication of the images and an award of equitable compensation. The equitable compensation should include an award to compensate the plaintiff, so far as money can, for the humiliation, anxiety and distress which has resulted from the defendant's publication of the images, in breach of the obligation of confidence he owed to her. The court stated it represents a development in the equitable doctrine in Australia.[45] The court referred to:

> only one other decision, an unreported judgment of the Victorian County Court, in which equitable compensation has been awarded for non-economic loss, occasioned by a breach of confidence.[46]

The court was not able to locate any other Australian cases in which such an award has been made. However, prospective developments in the equitable doctrine of breach of confidence to protect privacy values were contemplated by at least some members of the High Court in *Lenah Game Meats*.[47] The court went further by stating that the:

> defendant shall not, either directly or indirectly, publish in any form any photographs or videos of the plaintiff engaging in sexual activities or in which the plaintiff appears naked or partially naked (including with breasts exposed) other than: (a) as may be required by law; (b) to professional advisers for the purpose of obtaining professional advice; (c) with leave of this Court; or (d) with the express written consent of the plaintiff. [48]

The court ordered the defendant to pay to the plaintiff equitable compensation in the amount of $48,404.00. 3.

---

[42] Ibid.

[43] Ibid.

[44] *Wilson v Ferguson* [2015] WASC 15.

[45] Ibid, at 76.

[46] Ibid.

[47] Ibid.

[48] Ibid.

Currently in Australia, there is a distinction between the disclosure of personal data such as health and finance records to a third party or unauthorized access, or intrusion to an individual's privacy. They are considered separate issues of privacy and can constitute different practices. Intrusion is closely associated with trespass, which has been attributed to the traditional notion of privacy to personal property and not the Internet. On the other side, disclosure can come in different forms such as media and photographs being disseminated over Facebook is based on breach of confidence, and is forming part of the legal and policy discourse in relation to privacy over the Internet. Yet, a tort for breach of privacy in Australia has not been widely accepted for the privacy breaches over the Internet. Australia, by taking a common law approach to determining the extent to which privacy is breached has, to date, applied equity to a breach of confidence. Arguably, Australia is still grappling with the concept of privacy at common law, particularly over the Internet, even though the Privacy Act has been in operation for 30 years.

Notably, to date, the Australian High Court has had little to say in relation to these and other elements of privacy, rather leaving it to the lower level jurisdictional courts. This is in part because there have been very few, if any, appeals to the federal courts on this issue. Moreover, the common law places privacy at the forefront in Australia, but there has been little to no reference to protecting people's personal data (information) has been decided upon.

Another important development in privacy along with data protection law has been the introduction of the principles of anonymity and pseudonymity. As highlighted in Chap. 3, anonymity and pseudonymity[49] enable individuals to exercise greater control over their personal information and decide how much personal information will be shared or revealed to others. APP 2.10 provides that a person can deal anonymously or pseudonymously with an APP entity.[50] That is, an individual has the option not to be identified and left alone. These principles serve to protect a person's privacy so as they cannot be located or contacted by a direct marketer or, for example, a former partner or family partners. They also serve to provide greater control and ownership over a person's individual personal information and data that has been defined by the law.

Notwithstanding the above, it can be argued that Australia has developed its modern day privacy laws that protect personal data, which are underpinned by the concepts and principles found within the data protection framework of the OECD[51]. Chapter 11 compares whether the other jurisdictions discussed in this book have also adopted these principles.

---

[49] Australian Privacy Principle 2.

[50] Australian Privacy Principle 2.10.

[51] Office of the Information Commissioner, The *Privacy Act 1988* (Privacy Act) was passed by the Australian Parliament at the end of 1988 and commenced in 1989. The Privacy Act gave effect to Australia's agreement to implement the *Organisation for Economic Cooperation and Development (OECD) Guidelines on the Protection of Privacy and Transborder Flows of Personal Data*, as well as to its obligations under Article 17 of the *International Covenant on Civil and Political Rights*, https://www.oaic.gov.au/about-us/who-we-are/history-of-the-privacy-act, accessed 20 December 2018.

5.2 Public and Private

Section 4 of the *Privacy Act 1988* states that the Act applies to the commonwealth, states and territories and extended territories of the Australia. By binding the Crown, the federal government and all federal government agencies must also comply with the Act.[52] The Act or APPs on the other hand does not apply to State and Territory governments. It is outside the scope of this book to discuss the extent of those laws. An APP entity includes government agencies, and private sector and not-for-profit organizations with an annual turnover of more than AU $3 million. For those small businesses with an annual turnover of less than $3 million dollars have a limited set of obligations. They are required to comply with the Act where that personal data is to be provided to a health service provider, trading in this data, a contractor of the commonwealth, reporting entity for money laundering, credit reporting, operator of residential tenancy database, employee registered under fair work legislation. In addition, a small business will need to comply with the Act where they conduct protection action ballots, or are businesses that are covered by the regulations. This is a significant difference between Australia's and the EU's laws. Australia, arguably takes a flexible approach, however, it is our view that this is an area that needs to be addressed. Australia needs to close the gap in this area of the law and ensure the laws apply to all entities. This is particularly important where smaller businesses rely on and use data as part of their business, in the future. Additionally, it also includes health service providers and those small businesses contracted to provide services to government or a credit reporting body. Applying the laws to both the public and private sector, arguable reinforces key societal principles related to data protection and privacy laws – that is, 'trust' and 'certainty'. It is argued that the principle of trust[53] has also become an important element of data protection law, to ensure that the community and industry are comfortable in using modern technology that consumes personal data and information.

5.3 Definition of Personal Information

The definition of personal information constitutes both general and sensitive personal identifiable information. Firstly, personal information constitutes whether a person can be reasonably identifiable from the information or opinion is true or not; and whether the information or opinion is recorded in a material form or not.[54] Section 6 of the Act does define what information can identify a person such as a person's full name, alias or previous name, date of birth, sex, current or last known address and driver's license. Important identifying information also includes current

[52] Privacy Act 1988, section 4.

[53] Hofman, D., Duranti, L., How, E *Trust in the Balance: Data Protection Laws as Tools for Privacy and Security in the Cloud Algorithms* MDPI (2017).

[54] Privacy Act 1988, section 6.

and last employer. Australia, rather than having a national identification card, like Singapore, has determined that further identifying information can be a person's Tax File Number (TFN).[55] However, this last identifying information is only relevant for a person that has acquired a TFN. Secondly, personal information also constitutes sensitive identifiable information.

Section 6 of the Act defines the data that is considered sensitive information.[56] The list includes an individual's:

- Racial or ethnic origin;
- Political opinions;
- Membership of a political association;
- Religious beliefs or affiliations;
- Philosophical beliefs;
- Membership of a professional or trade association;
- Membership of a trade union;
- Sexual orientation or practices;
- Criminal record;
- Health information about an individual;
- Genetic information (that is not otherwise health information);
- Biometric information that is to be used for the purpose of automated biometric verification or biometric identification; or
- Biometric templates.[57]

The collection of this sensitive data cannot be undertaken without the individual data subjects consent.[58] Biometrics is an interesting example because every time a person enters and or exits most international airports they must provide some biometric information. This includes, but not limited to facial recognition, body mapping or finger prints, as a person goes through a customs check. At the time of the facial scan or taking of finger prints, the biometric data is verified with the person's passport. However, this data is considered in the public interest and therefore can be collected and stored. In *Re TYGJ and Information Commissioner*[59], the AAT argued that:

> Section 6(1) also defined the expression "personal information": personal information means information or an opinion (including information or an opinion forming part of a database), whether true or not, and whether recorded in material form or not, about an individual whose identity is apparent, or can reasonably be ascertained, from the information or opinion.[60]

[55] Ibid.

[56] Ibid.

[57] Ibid.

[58] Australian Privacy Principle 3, the collection is reasonably necessary for an APP entity's functions or activity, or a listed exception applies.

[59] *Re TYGJ and Information Commissioner [2017] AATA 1560, 15–20.*

[60] Ibid.

The AAT went onto to say:

> A person, body or agency to whom personal information is disclosed under clause 1 of this Principle shall not use or disclose the information for a purpose other than the purpose for which the information was given to the person, body or agency.[61]

This purposive approach is yet another key concept that has begun to play an important role in the development and implementation of data protection and privacy law, throughout the world. Personal information is collected in Australia. If it is collected from an individual who is physically present in Australia or an external Territory, regardless of where the collecting entity is located or incorporated.[62] An example is the collection of personal information from an individual who is physically located in Australia or an external Territory, via a website that is hosted outside Australia. This applies even if the website is owned by a company that is located outside of Australia or that is not incorporated in Australia.

5.4 Consent and Collection

Consent can mean different things to different people. However, consent within data protection and privacy law has been codified, to ensure data subjects understand what and how the concept is applied. In Australia, consent can be expressly or inferred (implied),[63] written, verbal or silence.[64] The definition of consent constitutes an individual being adequately informed of the issues and obligations before giving consent (express or implied).[65] Consent must be current and specific, or voluntary and more importantly the person must have the capacity to understand and communicate that consent.[66] This protection ensures people who require assistance or specialist advice to provide consent, can do so. To date there is no court authority

---

[61] Ibid.

[62] Explanatory Memorandum, Privacy Amendment (Enhancing Privacy Protection) Bill 2012, p 218.

[63] *Giller v Procopets* (2008) 24 VR 1.

[64] Office of Information Commissioner, Australian Government: Key Concepts, https://www.oaic.gov.au/agencies-and-organisations/app-guidelines/chapter-b-key-concepts, accessed 12 November 2018.

[65] Privacy Act 1988, section 6.

[66] Ibid. In Direct Marketing, APP 7.15 The 'reasonably expect' test is an objective test that has regard to what a reasonable person, who is properly informed, would expect in the circumstances. This is a question of fact in each individual case. It is the responsibility of the organization to be able to justify its conduct. 7.16 Factors that may be important in deciding whether an individual has a reasonable expectation that their personal information will be used or disclosed for the purpose of direct marketing include where: the individual has consented to the use or disclosure of their personal information for that purpose (see discussion in paragraph 7.23 below and Chapter B (Key concepts) for further information about the elements of consent): the organization has notified the individual that one of the purposes for which it collects the personal information is for the purpose of direct marketing under APP 5.1 (see Chap. 5 (APP 5)) the organization made the individual aware that they could request not to receive direct marketing communications from the organization, and the individual does not make such a request (see paragraph 7.21).

as to how far this protection extends, and how it will be determined that a person has the capacity to provide consent.

Notwithstanding the above, there are exceptions to this. APP 7.2, 7.3, 7.4 allow an organization to use or disclose personal data in direct marketing, when the organization has collected the information of the person.[67] Express consent is given explicitly, either orally or in writing. This could be a handwritten signature, oral statement, or use of an electronic or voice signature. Generally, it cannot be assumed a person has provided consent on the basis they did not object in the first place to allow their data to be processed or transferred to a third party. Furthermore, it will be difficult for an APP entity to establish that an individual's silence can be taken as consent.

Organizations may use an opt-out mechanism to substantiate a person's implied consent. However, any opt out presented to a person needs to be articulated clearly and succinctly so as the individual understands what the implication might be. More importantly, consent does not exist where it is deemed to have been provided under duress, coercion or some kind of pressure. An individual must be aware of the implications of providing or withholding consent, for example, whether access to a service will be denied if consent is not given to collection of a specific item of personal information. Nonetheless, consent is effective when the individual has been fully informed of the risks and implications. In Australia, it is sufficient that the individual is advised and consents in broad terms.[68]

A data subject may withdraw their consent at any time, and this should be an easy and accessible process. Once an individual has withdrawn consent, an APP entity can no longer rely on that past consent for any future use or disclosure of the individual's personal information. Individuals should be made aware of the potential implications of withdrawing consent, such as no longer being able to access a service.[69]

An organization[70] cannot collect personal information unless that information directly relates to one or more of the organization's functions.[71] Personal information cannot be collected unless the person provides consent and the information is required under Australian law, a court or tribunal order. Collection of personal data can be undertaken by government agency such as health or immigration for enforcement purposes. For example, personal data can be collected to verify if a person has entered and stayed in the country illegally. These situations would be in the public and national interest, in the same way as requiring data to be collected for the purposes of communicable disease outbreak (health purposes).

---

[67] Australian Privacy Principles, 7.2, 7.3, 7.4.

[68] *Rogers v Whitaker* (1992) 175 CLR 479, 490.

[69] Office of Australian Information Commissioner, Australian Government, https://www.oaic.gov.au/agencies-and-organisations/app-guidelines/chapter-b-key-concepts, accessed 31 August 2018.

[70] Privacy Act 1988, section 6.

[71] Australian Privacy Principles 3.

5.4.1 Children

The Privacy Act does not specify an age after which individuals can make their own privacy decisions. An APP entity will need to determine on a case-by-case basis whether an individual under the age of 18 has the capacity to consent.[72] An individual under the age of 18, can have the capacity to consent when they have sufficient understanding and maturity to understand what is being proposed. In some circumstances, it may be appropriate for a parent or guardian to consent on behalf of a young person.[73] For instance, if the child is young or lacks the maturity or understanding to do so themselves. If it is not practicable or reasonable for an APP entity to assess the capacity of individuals under the age of 18, the entity may presume that an individual aged 15 or over has capacity to consent, unless there is something to suggest otherwise.[74] An individual aged under 15 is presumed not to have capacity to consent.

5.5 Extra-Territorial Reach

Notably, the APPs extend to organizations or small business operator that have an Australian link. An organization or small business[75] operator is linked to Australia when an Australian citizen or a person who has a continued presence in Australia and is not subject to a legal time limitation.[76] This is usually the 183 day rule that is applied for taxation purposes.[77]

---

[72] Office of Information Commissioner, Australian Government: Key Concepts, https://www.oaic.gov.au/agencies-and-organisations/app-guidelines/chapter-b-key-concepts, accessed 12 November 2018.

[73] Ibid.

[74] Ibid.

[75] The general law concept of 'carrying on business' has been said to 'generally involve conducting some form of commercial enterprise, systematically and regularly with a view to profit' or to embrace 'activities undertaken as a commercial enterprise in the nature of a going concern, that is, activities engaged in for the purpose of profit on a continuous and repetitive basis'. In determining whether the business is being carried on in Australia, the courts have focused on whether the activity is undertaken in Australia as part of the entity's business. That is, there is a need for some physical activity in Australia through human instrumentalities, being activity that itself forms part of the course of conducting business. However, as noted in Australian Securities and Investment Committee, the court stated that 'provided that there are acts within Australia which are part of the company's business, the company will be doing business in Australia although the bulk of its business is conducted elsewhere and it maintains no office in Australia'. *Gebo Investments (Labuan) Ltd v Signatory Investments Pty Ltd* [2005] NSWSC 544, 38. *Hope v Council of the City of Bathurst* (1980) 144 CLR 1, 8. *Gebo Investments (Labuan) Ltd v Signatory Investments Pty Ltd* [2005] NSWSC 544, 38. *Australian Securities and Investments Commission v ActiveSuper Pty Ltd* (No 1) [2012] FCA 1519, 47.

[76] Office of Australian information and Privacy Commissioner, Australian Government.

[77] Australian Taxation Office, Australian Government.

The transfer or disclosure of personal data from Australia to a third country[78] can only be undertaken when the organization has taken reasonable steps to show that the recipient will not breach the APPs.[79] The organization transferring the data outside of the Australia will be responsible for the retention and use of that data by the recipient. An APP entity[80] discloses personal information to an overseas recipient it will also need to comply with APP 6.[81] In other words, it must only disclose the personal information for the primary purpose for which it was collected, and that the person has consented to that disclosure.

The transfer of data across international borders is frequent and will only increase. APP 8.1 has determined that an overseas recipient is one that is not located in or on the Australian territory.[82] In addition, they are not the APP entity disclosing the personal data, and not the person to whom the data relates. However, APP 8 does not apply to an organization that receives the personal information in a third country which is the same entity.[83] On the one side, where an APP entity has an office located in Australia and Singapore, and personal information is being sent from the Australian office to the Singapore Office, APP 8 will not apply because the recipient is the same entity.[84] However, it must be noted that in relation to the case where an APP entity in Australia sends personal information to a 'related body corporate' located outside of Australia, [85] and the related body corporate is a different entity to

[78] Australia Privacy Principles 8, Privacy Act 1988, section 16C.

[79] Office of Australian Information Commissioner.

[80] Privacy Act 1988, section 6 defines an APP entity to be an individual, including a sole trader, a body corporate, a partnership, any other unincorporated association, or a trust unless it is a small business operator, registered political party, state or territory authority or a prescribed instrumentality of a state section 6C.

[81] Australia Privacy Principles 6, outlines when an APP entity may use or disclose personal information. An APP entity that holds personal information about an individual can only use or disclose the information for a particular purpose for which it was collected unless an exception applies. Exceptions include: the individual consented to a secondary use or disclosure; the individual would reasonably expect the secondary use or disclosure, and that is related to the primary purpose of collection or, in the case of sensitive information, directly related to the primary purpose; the secondary use or disclosure of the personal information is required or authorized by or under an Australian law or a court/tribunal order; a permitted general situation exists in relation to the secondary use or disclosure of the personal information by the APP entity; the APP entity is an organization and a permitted health situation exists in relation to the secondary use or disclosure of the personal information by the organization; the APP entity reasonably believes that the secondary use or disclosure is reasonably necessary for one or more enforcement related activities conducted by, or on behalf of, an enforcement body; the APP entity is an agency (other than an enforcement body) and discloses personal information that is biometric information or biometric templates to an enforcement body, and the disclosure is conducted in accordance with guidelines made by the Information Commissioner for the purposes of APP 6.

[82] Australia Privacy Principles 8 and s 16C create a framework for the cross-border disclosure of personal information. Where an entity discloses personal information to an overseas recipient, it is accountable for an act or practice of the overseas recipient that would breach the APPs (s 16C).

[83] Explanatory Memorandum, Privacy Amendment (Enhancing Privacy Protection) Bill 2012, p 83.

[84] Australia Privacy Principles, 8.

[85] Privacy Act 1988, section 6, Section 6(8) provides 'for the purposes of this Act, the question

the APP entity in Australia – APP 8 will apply. APP 8.1 provides that before an APP entity discloses personal information about an individual to an overseas recipient, the entity must take reasonable steps to ensure that the recipient does not breach the APPs in relation to that information.

Australia has adopted a limited approach to data localization, compared to other countries that are beginning to expand this area of their legal framework such Russia and China. Australia, at this stage, only requires that 'health' records of its citizens be stored locally within the territory of Australia. With the developments internationally, it remains to be seen whether Australia will, at some stage, expand its data localisation policy.

5.6 Regulator

The Australian Government has established the Office of the Australian Information Commissioner (OAIC – Commissioner).[86] The Commissioner has the power to monitor, audit, advise, educate, report and provide guidance to the community on all privacy related issues.[87] The Commissioner is also afforded the power to monitor and report on the adequacy of equipment and user safeguards.[88] In addition, the broad based powers of the Commissioner extend to assessing the Australian Privacy Principles to determine their adequacy. The Commissioner can issue enforcement undertakings, conduct investigations, review complaints, exempt certain documents from freedom of information. The Commissioner can also undertake enforcement proceeding in the Federal Court or Federal Circuit Court of Australia. Moreover, the Commissioner plays a key role in reviewing and determining enforcement for and of commonwealth government agencies in privacy related matters.[89] It is outside the scope of this book to discuss the extent of how these powers are applied.

In addition to the legislation the Commissioner has also issued guidelines[90] to promote an understanding and acceptance of the APPs. They play an important role by outlining the Commissioners' functions, and provide examples on how the APPs may apply. However, the guidelines are non-binding and therefore not considered to be a legislative instrument. Even so, the guidelines are a good tool for any organization to develop their internal policies, procedures and risk management systems for handling data. These become an important tool for organizations to develop self-regulation so as to strengthen the management of personal data regulation. However, it is out of scope of this book to discuss all the respective guidelines issues by the Commissioner.

---

whether bodies corporate are related to each other is determined in the manner in which that question is determined under the *Corporations Act 2001*.

[86] Privacy Act 1988, section 28.

[87] Ibid.

[88] Privacy Act 1988, sections 27–33.

[89] Privacy Act 1988, sections 35–70.

[90] Privacy Act, section 28.

The support of the regulatory framework allows for Codes to be developed. Section 26 states that the Code allows government to extend support to industry and the community to ensure responsible personal information and data management.[91] The objectives of the Code are to set out specific requirements that agencies must comply with; enhance privacy capability and accountability; and promote good privacy governance. Moreover, the Code has been established to build trust and confidence in managing personal data and information. At an operational level, the Code provides a degree of flexibility for the Commission and Commissioner to investigate an organization where there has been a breach of the privacy principles.[92] Furthermore, the Code mandates that an organization must establish a privacy management plan. The management plan must document and set out how the organization will comply with the APP 1.2.[93] It is argued that this multilayered approach, while it could be perceived as over regulation, the process and standards ensure that entities take a closer look at the practices employed to manage personal data and information. In an evolving area of law, government regulation is difficult enough to change at short notice, thus the Code and/or management plan allow entities to review and update their systems and processes, as required.

The Code can be developed by an APP entity, either on its own initiative or on request from the Information Commissioner, or by the Information Commissioner directly.[94] Section 26C describes what an APP Code is and requires that is a written code of practice about information privacy. The APP code must set out how one or more of the Australian Privacy Principles are to be applied or complied with; and specify the APP entities that are bound by the code, or a way of determining the APP entities that are bound by the code. In addition, the APP Code is to set out the period during which the code is in force (which must not start before the day the code is registered). In accordance with section 26G, the Commissioner may develop an APP Code, only where the Commissioner is satisfied that it is in public interest to do so.[95] The Commissioner has the power to approve and register the Code.[96] Upon registration the organization is bound by the Code and must not do an act, or engage in a practice, that breaches that Code. A breach of a registered Code will occur where that is 'an interference with the privacy of an individual'.[97]

[91] Privacy Act 1988, section 26.

[92] Privacy Act 1988, section 40.

[93] Australia Privacy Principles 1.2.

[94] Privacy Act 1988, section 26E and 26G. Section 26E requires that the request from the Commissioner must specify the period within which the request must be complied with; and set out the effect of section 26A. The period must run for at least 120 days from the date the request is made; and may be extended by the Commissioner. The request may specify one or more matters that the APP code must deal with; and specify the APP entities, or a class of APP entities, that should be bound by the code.

[95] Privacy Act 1988, section 26G. However, despite subsection 26C(3)(b), the APP code must not cover an act or practice that is exempt within the meaning of subsection 7B(1), (2) or (3).

[96] Privacy Act 1988, section 26H.

[97] Privacy Act 1988, section 13 (1)(b).

Section 35A of the Act allows the Commissioner to recognize an external dispute resolution (EDR) scheme.[98] Firstly, this provides a mechanism for the handling of privacy related complaints. EDR schemes constitutes the second tier of a three-tiered complaint process where an individual should first make a complaint in writing to a respondent entity and allow the entity a reasonable time to respond.[99] Secondly, an individual who is not satisfied with the response or outcome may complain to a recognised EDR scheme of which the entity is a member; and thirdly, an individual who is not satisfied with the outcome of the EDR process may complain to the OAIC.[100] The OAIC will consider whether to accept the complaint or decline to investigate as allowed under section 41 of the Act. The AAT in *Re TYGJ and Information Commissioner*[101] affirmed the complaints mechanism provided by the Act, stating:

> the *Privacy Act* provides that an individual may complain to the Commissioner about an act or practice that may be an interference with the privacy of the individual. The complaint must be in writing and specify the respondent to the complaint. Where the complaint is made about an act or practice of an agency which is a Department, the respondent to that complaint is the Secretary to the Department.[102]

The AAT went onto to say that in most instances:

> a person must first complain to the respondent about the act or practice before the Commissioner is obliged to investigate a complaint made. Even then, the Commissioner may decide not to investigate, or not to investigate further, an act or practice about which a complaint has been made.[103]

The process of making a complaint is straight forward, however, there are further requirements for the Commissioner to undertake before a formal investigation will proceed. It is outside the scope of this Chapter to discuss the entire complaints process.

5.7 Quality of Information – Accuracy

An organization collecting, retaining and using personal data must ensure that the information collected is accurate, up-to-date and complete.[104] APP 10.3 provides further guidance to organizations, and states that handling poor quality personal information can have significant privacy impacts for individuals.[105] Not only is there

[98] Privacy Act 1988, section 35.

[99] Ibid.

[100] Ibid.

[101] *Re TYGJ and Information Commissioner [2017] AATA 1560, 15–20.*

[102] Ibid.

[103] Ibid.

[104] Australian Privacy Principles, 10.1. 10.2.

[105] Ibid.

an impact to the individual, the organization's reputation as a data manager will be impacted from poor or inaccurate collection, storage and use of standards. An organization must ensure the quality of the personal data is accurate at the time of collection, and when the data is used or disclosed. This provides a two-step process to ensure there is adequate oversight for managing accurate information.[106]

The collection and use of personal data, at time, may be inaccurate. Therefore, APP 13, allows a person to request that their personal data be corrected.[107] The organization has a broader responsibility to the individual, after that personal data has been corrected. In other words, they must notify all other organizations to whom they have provided that data, with the correct updated data.[108] In the event an organization refuses to correct the personal data, they must inform the individual of the reasons so that the data subject fully understands the business decision for not doing so. Moreover, the organization must inform the person of the process they can use to make a complaint, where they disagree with the decision to refuse correcting their personal data.

The person involved can also request that a statement be attached to their record, stating that the information is not accurate, out of date, incomplete, irrelevant or misleading.[109] Any request or complaint must be attended to within a reasonable time period, approximately 30 days, and can be made to both the organization in question and to the Commissioner.[110] Note, that the APP operates in conjunction with Part V of the *Freedom of Information Act 1982,* which allows a person to request access to their personal data and information.

5.8 Retention

The retention or storage of data under the Act only pertains to credit reporting. Credit reporting information must be deleted following the retention period.[111] Section 20W describes the type of credit information and the period in which it is to be retained.[112] Generally, the period ranges between 2 to 7 years for credit, repayment and payment information, bankruptcy and insolvency. APP 11 requires organizations to de-identify or destroy personal data and information, when it is no longer required.[113] However, there is not specific time period stated and does not

[106] Ibid, 10.4, 10.5.

[107] Ibid, 13.

[108] Ibid, 13.2.

[109] Ibid, 13.3.

[110] Privacy Act 1988, sections 36 & 40.

[111] Privacy Act 1988, section 20 V.

[112] Ibid, section 20 W.

[113] Principle 11 also Rule 11.1(a) of Privacy (Tax File Number) Rule 2015 (Cth) under section 17 Privacy Act 1988, which requires Tax File Number ('TFN') recipients to take reasonable steps to safeguard TFN information. The Personally Controllable Electron Health Records Act 2012 (Cth) ('PCEHR Act') requires that the Electronic Health Record System must take steps to secure data processed by that system. The Code of Banking Practice (Australian Bankers Association Inc.,

apply where the information forms part of the Commonwealth government record that is required by a court. This de-identification or destruction process must not be confused with the right to request for data to be deleted. There is no such requirement and this issue does not form part of the right to be forgotten.

In addition, the *Telecommunications (Interception and Access) Act 1979* requires telecommunications companies to retain telecommunications data for 2 years. This includes data about phone calls, phone numbers of the people talking to each other and the length of time the conversation. In relation to email (s), this includes information of email addresses and the date in which it was sent. There is no requirement for the telecommunications company to retain data that discloses a person web-browsing information. However, this information can be used to identify serious criminal or national security investigation, including murder, counter-terrorism, counter-espionage, sexual assault and kidnapping cases. Graham Greenleaf argues that as a result of these laws being strengthened in 2015, there were a large number of critical submissions to the laws, with majority public support. Nonetheless, younger Australians were divided on whether such laws are necessary.[114]

5.9 Breach & Notification

A breach of the Act not only occurs against the Act but also the APPs have not been complied with. In 2017, the *Privacy Amendment (Notifiable Data Breaches) Act 2017* was introduced and came into effect in February 2018. The new legislation applies to both the public and private sectors with an annual turnover of $3 million.[115] The amendments bring Australia closer to the EU's framework. An organization will be required to notify the Information Privacy Commissioner where there has been an eligible breach. An eligible breach occurs when there has been unauthorized access to, unauthorized access of, or loss of personal information held by

---

2013) while voluntary is an industry scheme overseen by the Australian Bankers Association Inc. The Telecommunications Consumer Protections Code (Communications Alliance Ltd., 2012) (in particular Clause 6.9).

[114] Greenleaf G *Going against the flow: Australia enacts data retention law* Privacy Laws & Business International Report, 134 (2015) pp. 26–28. Lowy Institute 'Data retention scheme has majority support from Australians' 27 March 2015, http://www.lowyinterpreter.org/post/2015/03/27/Data-retention-scheme-has-majority-support-from-Australians.aspx, accessed 2 September 2018. With the threat of terrorism featuring prominently in public perceptions, the Government's new data retention laws have the support of a clear majority of Australians. When asked about 'legislation which will require Australian telecommunications companies to retain data about communications such as phone calls, emails and internet usage, but not their content', 63% of Australians say this is 'justified as part of the effort to combat terrorism and protect national security'. Only one-third (33%) say it 'goes too far in violating citizens' privacy and is therefore not justified'. Younger Australians (aged 18–29) are more divided on the need for data retention, with almost equal numbers supporting and opposing the legislation (50% say it is justified, while 47% say it goes too far in violating citizens' privacy).

[115] Explanatory Memorandum, Privacy Amendment (Notifiable Data Breaches) Act (2017).

an organization.[116] Additionally, an eligible breach also occurs when access, loss or disclosure will result in harm to a person, which the information relates. Even though serious harm is not defined, there are guiding principles that can assist in determining what constitute serious harm. For instance, the kind of information, sensitivity of the information, the security protections in place, the type of person or people who obtained the information and the nature of the harm. The law also introduced the requirement for an organization to undertake a risk assessment, when a breach has been detected.[117] This is a positive step by government forcing industry to identity and rectify the issue, and provide greater transparency. Risk assessments are something that is common in many industries. A risk assessment is to be taken within 30 days of the eligible breach.[118] The obligation extends to notifying the data subject whose personal information has been involved in the data breach. This is particularly important where that data is likely to cause serious harm. The obligation further extends to notifying what recommendations are to be made that outlines the steps to be taken to manage the breach. However, what is not clear, is how the level of harm can and is measured.

The benefit of this provision, is that it provides individuals with an opportunity to reduce the impact of data security breaches, for instance, cancelling credit cards or changing account passwords, and it can increase public confidence in the handling of consumer information.[119] As noted by Sara Smyth, critics counter that data breach notification laws negatively impact businesses. It reinforces the risk-based approach to privacy and data protection laws by emphasizing a self-regulatory initiative.[120] The current risk-based and self-regulatory approach appears to be appropriate. This is in part due to the fluid nature of the evolution of technology, and recognition that governments cannot regulate the entire life cycle of personal data and information. This framework also allows for the rapid change in development of technology that has not even entered the market, such as quantum technology. It is believed that future quantum technology may force government and entities to review their current day privacy and data protection laws to ensure key legal concepts are relevant.

5.10 Right to Be Forgotten

The Australian privacy laws do not provide a direct right to be forgotten. However, according to the APP 11,[121] a business must take steps to destroy or de-identify personal information. APP 13 also applies where an APP entity must take reasonable steps to confirm and correct any personal information if it is satisfied that the infor-

[116] Ibid.

[117] Ibid.

[118] Privacy Act 1988, sections 26WK, WL, WH.

[119] Smyth, S *Does Australia Really Need Mandatory Data Breach Notification Laws – And If So, What Kind?"* Journal of Law, Information and Science 159 (2013).

[120] Ibid.

[121] Australian Privacy Principles 11.

mation is inaccurate, out-of-date, incomplete, irrelevant, misleading, or an individual requests the entity correct the information.[122] The problem is that Australia does not even come close – when compared against the GDPR. There is little guidance as to how and what steps are to be taken to destroy personal information. APP 11.30 states that 'reasonable steps' only need to be taken by organization should take to destroy or de-identify personal information will depend.[123] Nevertheless, this is subject to a number of limitations and rules. In other words, an organization needs to consider whether the possible adverse consequences for an individual if their personal information is not destroyed or de-identified—a more rigorous steps may be required as the risk of adversity increases. However, practically an organization can consider whether the time and cost associated with destroying or de-identification is to great or the costs are too high, the organization may not necessarily have to undertake this function. Moreover, an organization is not excused from destroying or de-identifying personal information by reason only that it would be inconvenient, time-consuming or impose some cost to do so. Whether these factors make it unreasonable to take a particular step will depend on whether the burden is excessive in all the circumstances.[124]

The APP Guidelines go onto say that, where it is not possible for an organization to irretrievably destroy personal information held in electronic format, reasonable steps to destroy it would include putting the personal information beyond use.[125] However, it must be noted that undertaking such a step only merely parks the personal information within the systems data base, server or some other place, so as it is not readily accessible.[126] Thus, the information is not permanently deleted or removed.

Nonetheless, the Australian Law Reform Commission (ALRC) had proposed a "right to be deleted," which would be analogous to the EU's right to be forgotten.[127] Support for this proposed law varied.[128] Some believe that Australia's current data privacy and defamation laws are sufficient to address Internet privacy concerns.[129] Had the proposal been realised and established, the right to be deleted, would today enable a person to have their personal information deleted from the Internet. Thus, a form of the right to be forgotten would have existed in Australia. The ALRC summarized the complex balance between the need for privacy and commercial and public interest, as complex. The ALRC stated that, calling something a right is of little value if the right is too readily able to be balanced against competing rights or

---

[122] Ibid, 13.

[123] Australian Privacy Principle Guidelines, Chapter 11: Australian Privacy Principle 11—Security of personal information Version 1.0, February 2014.

[124] Ibid.

[125] Ibid.

[126] Ibid.

[127] Kerr, J *What is a Search Engine? The Simple Question the Court of Justice of the European Union Forgot to Ask and What It Means for the Future of the Right to be Forgotten* Chicago Journal of International Law: Vol. 17: No. 1, (2016).

[128] Ibid.

[129] Ibid.

values. It is inevitable that rights and values will sometimes clash, so there would seem to be no alternative to qualifying the rights. Once it is accepted that privacy and freedom of speech are both important rights and will sometimes clash, then it seems inevitable that each right must sometimes be qualified.[130] The balancing test involves evaluating competing and often incommensurable rights, interests and values. In particular, breaching someone's privacy might be justified because doing so is in the public interest, and therefore justified. In the state of South Australia, the court ruled that Google is effectively a publisher and has responsibility for the content in which its systems and search engines provide to the public.[131]

To date the right has had little consideration by the courts of the Commonwealth or any of the States or Territories (Victoria, New South Wales, Queensland, Western Australia, Tasmania, Northern Territory and Australian Capital Territory). An exception to date has been in South Australia. That is, in 2015, the state of South Australia provided the first insight and consideration of the principle. In *Duffy v Google Inc*[132] the court while not referring directly to the right to be forgotten, did argue that because Google Inc. published information and personal data of individuals, that they had responsibility for its content. Two articles concerning Dr. Duffy were published on the Ripoff Report website in December 2007. Another two more articles were published in August 2008, with a further article in December 2008 and one in January 2009. Dr. Duffy claims that the articles and comments (*the Ripoff Report material*) contained defamatory imputations. The pleaded imputations include that she stalks psychics; obsessively and persistently harasses psychics; fraudulently and/or maliciously accesses other people's electronic emails and materials; spreads lies; threatens and manipulates other people; is an embarrassment to her profession; misused her work email address for private purposes and engaged in criminal conduct.[133] Furthermore, other websites, namely Complaints Board, 123 People, Is This Your Name and Wiki Name, published material concerning Dr. Duffy ostensibly derived from the Ripoff Report material.[134] In July 2009, Dr. Duffy became aware that searches for her name on Google's websites resulted in the display of extracts from and hyperlinks to the Ripoff Report material. Two months later, in September 2009, Dr. Duffy notified Google of the Ripoff Report material that she claimed was defamatory of her and being republished by Google, and of extracts from the Ripoff Report material and some of the secondary material that she claimed was defamatory of her and being published by Google. Even though the material on the Internet contained little personal information defined by the Privacy Act 1988, Dr. Duffy requested material be removed. Google declined the request.[135] A number of issues arose that include, but not limited to:

---

[130] Australian Law Reform Commission, *Serious Invasions of Privacy in the Digital Era,* (2014) 150.

[131] *Duffy v Google Inc* [2015] SASC 170.

[132] Ibid.

[133] Ibid.

[134] Ibid.

[135] Ibid.

(1) Was Google the "publisher" of the snippets produced by its search engine, the "auto-complete" suggestions and the articles to which the snippets linked?
(2) Had Google justified the imputations made?
(3) Did Google have a defence of qualified privilege?[136]

The court, by focusing on the three points above looked to the first of these issues, Kourakis CJ[137] holding that a plaintiff in a defamation case had to prove:

> "the defendant participates in the publication to a third party of a body of work containing the defamatory material and the defendant does so knowing that the work contains the defamatory material. That knowledge is presumed conclusively in the case of a primary participant but may be rebutted by a second participant who does not know and could not reasonably have known of the presence of the material. The onus was on the defendant to establish that it did not know or could not reasonably have known that the publication contained the defamatory statement[138] In the case of dissemination of personal information through the Internet, the issue is whether Google's role as facilitator through its search engine is sufficiently proximate to the display of the search results to constitute participation in the publication of their contents.[139]

Kourakis CJ said that the:

> concept of "passive medium" was apt to mislead because the nature of electronic media is that it is pre-programmed to fulfil a purpose.[140] Google submitted that an intention to publish had to be proved and that it could not have intended to publish any snippet when there are over 60 trillion constantly changing webpage and over 100 billion searches a month.[141]

However, Kourakis CJ further held that:

> "Google participated in the publication of the paragraphs about Dr Duffy produced by its search engine because it intended its search engine to do what it programmed it to do".[142] It was not necessary to prove that Google had knowledge of or adopted the contents of its search results.[143] Having considered a number of authorities Kourakis CJ concluded:"Google's search results are published when a person making a search sees them on the screen … It is Google which designs the programme which authors the words of the snippet paragraph. Google's conduct is the substantial cause of the display of the search result on the screen".[144]

Google was a participant in the publication of the snippets. Google did not have any practical ability to review their contents before they are displayed. It did not have advance knowledge of the contents of search results. Even so, this was a sec-

---

[136] Masnik M, *Right To Be Forgotten Now Lives In Australia: Court Says Google Is The 'Publisher' Of Material It Links To*, https://www.techdirt.com/articles/20151028/09424232657/right-to-be-forgotten-now-lives-australia-court-says-google-is-publisher-material-it-links-to.shtml, accessed 28 October 2017.

[137] *Duffy v Google Inc* [2015] SASC 170, 102–187.

[138] Ibid.

[139] Ibid.

[140] Ibid.

[141] Ibid.

[142] Ibid.

[143] Ibid.

[144] Ibid.

ondary publisher of search results and knowledge of their defamatory contents should not be attributed to it until notice is given.[145] Kourakis CJ went on to say:

> Google was liable for the re-publication of the Ripoff Report pages to which it provided hyperlinks. This was because Google's facilitation of the reading of these pages was both "substantial and proximate.[146]

The material was published by Google to persons who had a legitimate interest in having the information and that Google's conduct was reasonable. It could be argued that the elements of the right to be forgotten are beginning to form in Australia. Although, it remains to be seen whether higher courts in Australia adopt the same position in similar and other cases related to peoples' personal information generally, and the personal information defined by the Privacy Act 1988. From the above case law and the Australian Privacy Principles some believe the right to be forgotten now exists in Australia. However, the acceptance is debatable in Australia as to whether the right fully exists. This is unlikely to occur until the issue is considered by the High Court of Australia.

5.11 Data Portability

Data portability is fast becoming an accepted right that will allow a data subject to request that their personal data and information be moved from one data controller or organization to another. The Privacy Act does not include an equivalent right to data portability as the EU does (see Chap. 3). Nonetheless, and while no similar right in Australia, to date, APP 12.1 provides that if an APP entity holds personal information about a data subject, the entity must, on request by the individual, give the individual access to the information. However, APP 12.2 and APP 12.3 provide exceptions to APP 12.1.[147] Moreover, APP 12.5 provides that the entity must take

---

[145] Ibid.

[146] Ibid.

[147] Australian Privacy Principle – 12.1, 12.2, 12.3. If the APP entity is an organization then, despite subclause 12.1, the entity is not required to give the individual access to the personal information to the extent that: the entity reasonably believes that giving access would pose a serious threat to the life, health or safety of any individual, or to public health or public safety; or (b) giving access would have an unreasonable impact on the privacy of other individuals; or (c) the request for access is frivolous or vexatious; or the information relates to existing or anticipated legal proceedings between the entity and the individual, and would not be accessible by the process of discovery in those proceedings; or (e) giving access would reveal the intent ions of the entity in relation to negotiations with the individual in such a way as to prejudice those negotiations; or (f) giving access would be unlawful; or (g) denying access is required or authorized by or under an Australian law or a court/ tribunal order; or (h) both of the following apply: (i) the entity has reason to suspect that unlawful activity, or misconduct of a serious nature, that relates to the entity's functions or activities has been, is being or may be engaged in; (ii) giving access would be likely to prejudice the taking of appropriate ac on in relation to the matter; or (i) giving access would be likely to prejudice one or more enforcement related activities conducted by, or on behalf of, an enforcement body; or (j) giving access would reveal evaluative information on generates within the entity in in connection with a commercially sensitive decision – making process.

reasonable steps to give access in a way that meets the needs of the entity and the data subject. In other words, should an APP entity refuse to give access to the personal information because of subclause 12.2 or 12.3; or to give access in the manner requested by the data subject, the entity must take such steps (if any) as are reasonable in the circumstances to give access in a way that meets the needs of the entity and the individual.[148]

Data portability is becoming increasingly more important as it provides greater controls to data subjects. The right can be found at the intersection between data protection and other fields of law (competition law and intellectual property law).[149] It constitutes, a valuable case of development and diffusion of effective user-centric privacy enhancing technologies and a first tool to allow individuals to enjoy the immaterial wealth of their personal data in the data economy.

5.12 Loss or Damage and Enforcement

The Commissioner in Australia, similar to its counterparts in Singapore and the EU has an expanded set of powers. The Commissioner[150] can make a determination following an investigation of a complaint. The Commissioner can issue a declaration to an organization requesting that any interference with the privacy of an individual and must not repeat or continue such conduct cease. The declaration is a good tool because it allows the commissioner to also require the organization to ensure the conduct is not repeated and specify an amount for compensation for the loss or damage. Nonetheless, an individual does have the option of taking the complaint to either the Australian Appeals Tribunal, Federal Court or Federal Circuit Court.[151] In addition, the Commissioner is responsible for the enforcement of the Act. The Commissioner undertakes an investigation of complaints and ensures the results are publicly available.[152] A determination under section 52 applies to an organization or small business and they must not continue to repeat the conduct. If the determination relates to a commonwealth government agency, the agency must ensure there is no repeat of the conduct and establish steps to ensure it does not occur again. Failure to do so, the individual can apply to the Federal Court or the Federal Circuit Court for an order directing the agency to comply.[153]

---

[148] Ibid, APP 12.5.

[149] De Hert, P., Papakonstantinou, V., Malgieri, G., Beslay, L., Sanchez, I *The right to data portability in the GDPR: Towards user-centric interoperability of digital services*, Computer Law & Security Review, Volume 34, Issue 2, (2018), pp. 193–203.

[150] Privacy Act 1988, section 52.

[151] Privacy Act 1988, section 55A.

[152] Ibid.

[153] Privacy Act 1988, section 62.

The Commissioner can apply to the Federal Court or Federal Circuit Court for an Order where it has been determined that an organization has contravened the Act.[154] However, sections 25 and 25A of the Act do permit an individual to recover compensation or other remedies where a civil penalty order is made against an organization for a contravention of Part IIIA, Credit reporting.[155] Nevertheless, before an Order is obtained the Commissioner's office will first investigate the matter to ascertain any interference with privacy and consider what enforcement action to be taken.[156] The Commissioner will assess the extent of the evidence provided to determine whether to proceed, the availability and credibility of witnesses and what, if any evidence might be excluded by the court. Apart from attempting to keep these matters out of the courts, the process encourages a conciliatory approach.

5.13 Impact Assessment

Impact assessments are an important process and step in strengthening the regulation of personal information and data. An impact assessment is a systematic assessment of a project that identifies the impact (risk) that the project might have on the privacy of individuals, and sets out recommendations for managing, minimizing or eliminating that impact.[157] Furthermore, they are an important tool, within the current legal and policy framework that strengthen the protection of privacy and personal information, and should be part of the overall risk management and planning processes of APP entities. An impact assessment can assist entities to:

- describe how personal information flows in a project;
- analyze the possible impacts on individuals' privacy;
- identify and recommend options for avoiding, minimizing or mitigating negative privacy impacts;
- build privacy considerations into the design of a project; and
- achieve the project's goals while minimizing the negative and enhancing the positive privacy impacts.[158]

[154] Privacy Act 1988, section 80 W.

[155] Ibid, section 25.

[156] Review is undertaken against either the *Privacy regulatory action policy* or *PCEHR (Information Commissioner Enforcement Powers) Guidelines 2013*.

[157] Office of the Australian Information Commissioner, Australian Government https://www.oaic.gov.au/agencies-and-organisations/guides/guide-to-undertaking-privacy-impact-assessments, accessed 30 August 2018.

[158] Ibid. An assessment also assist an entity to demonstrate its compliance with its privacy obligations and its approach to managing privacy risk in the case of a future complaint, privacy assessment or investigation relating to the privacy aspects of a project. APP 1.2 requires APP entities to take reasonable steps to implement practices, procedures or systems that will ensure that the entity complies with the APPs. A PIA can assist in identifying the practices, procedures or systems that will be reasonable to ensure that new projects are compliant with the APPs.

At an organizational level, these assessments are designed to assist in identifying the risk of collecting, using and management personal; data and personal information, as defined under the law. These assessments also force organizations to better understand the risks of privacy breaches, from the practices they implement. While not full proof in managing every risk, they are arguably effective. The assessments have reinforced the risk management approach to data protection and privacy.

5.14 Additional Legislation and Standards

Australia, upon recommendation from the Australian Productivity Commission is intending to introduce new legislation sometime in 2018 to better manage Consumer Data Rights. While the Consumer Data Right initiative focuses on the Banking sector, consumer data includes personal information or data that has been created from consumers' online transactions, internet activity or data purchased about a consumer.[159] Currently, this data and information cannot be used to identify a person, and does not count as personal information or data. If implemented this will broaden out the scope of the right to personal data and information, used for commercial purpose.

Australia's do Not Call Register Act 2006,[160] began operating in 2007, regulating telemarketers so as to minimize the intrusive nature of unsolicited telephone calls to the general community. It applies to home and mobile phones including fax machines. Interestingly, fax machines are rarely used in the modern era, as the email has all but made the fax machine outdated technology. The Australian Communications and Media Authority has responsibility for the implementation of the Do Not Call Registry. The Registry[161] provides Australians' the option to opt-out of receiving calls from telemarketers. The process of opting out is limited to calls only, and does not include text (sms) messages. There is an 8-year period for which the registration of numbers remains valid. The registration extends to individuals,

[159] Australian Government, The Consumer Data Right will give consumers the right to safely access certain data about them held by businesses. They will also be able to direct that this information be transferred to accredited, trusted third parties of their choice. The right will allow the consumer to access data about themselves in a readily usable form and a convenient and timely manner. It will also allow consumers better access to information on the products available to them. Both individual and business customers will be entitled to the Consumer Data Right, https://static.treasury.gov.au/uploads/sites/1/2018/05/t286983_consumer-data-right-booklet.pdf, accessed 4 August 2018.

[160] Supporting laws include, Do Not Call Regulations 2017, specifies the types of calls that are not telemarketing calls and the people deemed to be nominees of a relevant account-holder. Telecommunications (Telemarketing and Research Calls) Industry Standards 2017, sets out the minimum requirements for those making telemarketing and research calls to Australian numbers, including when and how they can make certain calls. Fax Marketing Industry Standards 2011.

[161] Chesterman, S (2018) *Data Protection Law in Singapore Privacy and Sovereignty in an Interconnected World*, Second Edition Academic Publishing, pp. 281–283.

sole traders and sole traders. Singapore has a similar Registry, however, it is out of scope of this book to compare how the respective Registries operate.[162]

In addition, the Australia Government have also established reviews into credit reporting, which could see the transformation in the access to data of individuals and entities for potential lenders.[163] One of the possible changes could see that a bank of credit data will constitute part of an individual's consumer data, which they have a right to access. The proposal will encourage competition for small businesses and retail customers with positive credit histories.[164] Nonetheless, some of the major legislation that supports the Privacy Act 1988, includes and not limited to:

- Privacy Regulations 2013;
- Cyber Crime Act 2001;
- Criminal Code Act 1995;
- Spam Act 2003; and
- Telecommunication (Interception and Access) Act 1979.

The laws pertaining to cybercrime and the criminal code become important when discussing the use of personal data in criminal activity.

Australia has been considering new legislation that will require organizations such as Google, Apple, Facebook and telecommunication providers like Telstra, that will be compelled to hand over sensitive data or grant systems access to Australian authorities—or face fines of up to $10 million under proposed new laws.[165] In December 2018, the Australia government passed law allowing law enforcement agencies to decrypt information on applications such as sms and Whatsaap. The legislation, will assist law enforcement agencies to crack down on criminals and terrorists using encrypted services to conduct activities outside the reach of Australian spies and law enforcement.[166] However, some in the community have raised concerns that the proposed legislation, by compelling organization to hand over personal and other information, change or install systems and software that the current protections of encryption and safety mechanisms in place regarding data protection and privacy will be compromised.[167] Moreover, it is not clear as to the level of independent oversight or controls that will be placed on law enforce-

[162] Ibid.

[163] Australian Government, Mandatory Comprehensive Credit Reporting, http://sjm.ministers.treasury.gov.au/media-release/110-2017, accessed 16 September 2017.

[164] Ibid.

[165] Nick Whigham, *Australia's proposed new cyber legislation to give police greater access to content on smartphones; Proposed new powers will see the government crackdown on Australians using encrypted services outside the reach of spies and law enforcement* (2018), https://www.news.com.au/technology/online/security/australias-proposed-new-cyber-legislation-to-give-police-greater-access-to-content-on-smartphones/news-story/36683241a8799aaadf9b2dcbf3f938fa, accessed 4 October 2018.

[166] Ibid.

[167] Paul Karp, *Tech giants warn Coalition bill opens customers up to cyber attack*, The Guardian, (2018), https://www.theguardian.com/australia-news/2018/aug/20/tech-giants-warn-coalition-bill-opens-customers-up-to-cyber-attack, accessed 4 October 2018.

ment agencies in accessing these applications. One such example is the rise in identity theft, whereby individuals, personal data, defined by the law, is stolen and used elsewhere – by someone else, nor the data subject themselves. This is another area best described as a watching brief – to ascertain whether privacy and personal data will be compromised from this legislation.

5.15 Conclusion

Australia's privacy laws reflect the common law approach to legislative drafting. Australia's regulatory framework for privacy is multilayered. Apart from the legislation and common law on privacy, the adoption of the APP underpins the law by strengthening how entities manage personal information. The APP aim is to promote and ensure entities manage personal information in an open and transparent way as possible. Arguably, this approach goes some way to reducing the level of privacy breaches.

Australia, unlike other jurisdictions refer to privacy and not data protection. The Privacy Act is 30 years old and has set the course for protecting individual's privacy and personal data in Australia. The Act has a unique feature by including credit reporting, which cannot be seen in other jurisdictions laws that are discussed in this book. The right to be forgotten is not absolute in Australia. More importantly, the concept of privacy over the Internet at common law and the legislation, to date, arguably is limited in what areas of privacy are covered. The current limited approach does not fully account for intrusions on the personal privacy or the behavior of the general community and major industry organizations such as those that have a permanent presence on the Internet and media. This is what sets Australia apart from the EU and Singapore, sitting between the two jurisdictions. Australia's framework aims to strike more of a balance between being business friendly and protecting the rights of data subjects.

Moreover, the courts have recognized that fluid nature of privacy law, and acknowledged that it is an area which is not settled, and therefore, it has been difficult to develop a comprehensive common law approach to privacy breaches and remedies. To date, rather than fully develop a tort, the courts have adopted the principle of 'confidence' as part of determining an intrusion of privacy.

Consent has become an important concept and element of Australia's laws. Consent provides the data subject with a level of control and ownership over their personal information. The issue is whether consent varies greatly between the jurisdictions discussed throughout this book. In addition, the question arises whether consent in its current form is adequate and flexible enough for data subjects to control their personal information defined by the law, as technology changes and evolves.

Notwithstanding the importance of consent as a key concept of privacy and data protection law, the definition of personal information is also emerging as another important legal concept. The inclusion of sensitive personal information into this definition has on the one hand, highlighted the importance for data subjects have a greater level of control and ownership over this information. However, as discussed in Chap. 3, the EU has adopted an all-encompassing approach to defining personal information (data). In other words, they treat all personal data that is defined by the law with the same level of importance. Australia has adopted a broad definition of personal information, and there are significant differences in the way jurisdictions have defined this important identifying information and data.

The important role of the Regulator in managing privacy issues and breaches cannot be underestimated. Australia, by installing an Information and Privacy Commissioner has gone some way to strengthen the oversight and governance arrangements for privacy. The role the commissioner plays interacting and providing key information to both government and industry is likely to increase over the coming years as people become more aware of privacy, and breaches of privacy become increasingly common. The Commissioner is provided extensive power to oversee the implementation and enforce the Act along with the APPs. Nonetheless, the Commissioner is subject to funding from the federal government, and funding can from time to time fluctuate significantly, depending on government priorities. The Commissioner has a major role in the development of codes of practices and guidelines that provide the community with valuable information regarding privacy.

Finally, the APPs have gone someway to addressing the core principles of consent, collection, retention amongst others, and serve as an important component to the risk-based regulatory framework. However, they will require constant review and updating in light of the fluid technology space society is currently in.

References

Butler, D (2014) *The Dawn of the Age of the Drones: An Australian Privacy Law Perspective*, University of NSW Law Journal 434

Chesterman, S (2018) *Data Protection Law in Singapore Privacy and Sovereignty in an Interconnected World*, Second Edition Academic Publishing, pp 281–282

De Hert, P., Papakonstantinou, V., Malgieri, G., Beslay, L,. Sanchez, I (2018) *The right to data portability in the GDPR: Towards user-centric interoperability of digital services*, Computer Law & Security Review, Volume 34, Issue 2, pp. 193–203

Greenleaf G (2015) *Going against the flow: Australia enacts data retention law* Privacy Laws & Business International Report, 134, pp. 26–28

Hofman, D., Duranti, L., How, E (2017) *Trust in the Balance: Data Protection Laws as Tools for Privacy and Security in the Cloud Algorithms* MDPI

Kerr, J (2016) *What is a Search Engine? The Simple Question the Court of Justice of the European Union Forgot to Ask and What It Means for the Future of the Right to be Forgotten* Chicago Journal of International Law: Vol. 17: No. 1

Masnik M, (2015) *Right To Be Forgotten Now Lives In Australia: Court Says Google Is The 'Publisher' Of Material It Links To*, https://www.techdirt.com/articles/20151028/09424232657/right-to-be-forgotten-now-lives-australia-court-says-google-is-publisher-material-it-links-to.shtml, accessed 28 October 2017

Rivette, M (2010) *Litigating privacy cases in the wake of Giller v Procopets*, Media and Arts Law Review

Smyth, S (2013) *Does Australia Really Need Mandatory Data Breach Notification Laws – And If So, What Kind?* Journal of Law, Information and Science 159

Chapter 6
India

Abstract India has neither prepared or implemented specific data protection or privacy laws. Since 2011, the Indian Parliament has presided over a Privacy Bill, and today there continues to be little progress on implementing dedicated privacy laws. In 2017, India released a White Paper in relation to a data protection framework for the country, which sought community comment in relation to future privacy and data protection. The White Paper highlights many of the principles that other nations and the EU have currently adopted in their respective data protection laws. India has based the community feedback by looking to the EU, United States of America, Australia, Canada and Singapore, to assist and guide the development of specific privacy laws in their country. However, the current approach is neither close to the EU, Singapore or Australia's model. The current approach taken by India sits well outside what is being considered the global standard, but has similarities to Indonesia. Arguably, one of the dilemmas for India in the continued delay in establishing specific data protection laws, may come at a cost to their Internet economy. This is because, India has one of the largest Internet economies in the world, that has developed from their online outsourcing industry. Arguably, India, as it has sought, and is seeking to continue to position itself as an attractive destination for business and data processing (Kessler, D, Ross, S, Hickok, E *A Comparative Analysis of Indian Privacy Law and the Asia-Pacific Economic, Cooperation Cross-Border Privacy Rules*, National Law School of India Review, Vol. 26, No. 1 (2014), pp. 31–61). The courts in India have interpreted data protection in accordance with the right to privacy in accordance with Article 19 and 21 of the Constitution of India (*Justice K S Puttaswamy, and ANR v Union of India and Ors*, No. 494 of 2012). Chapter 6 will demonstrate how India's current approach is far from specific or coherent and rather based on industry sectors. Due to the limited scope of India's privacy and data protection laws, this Chapter only discusses the key concepts and principles such as Personal Information, Right to be Forgotten, Data Controller, Public and Private, Consent and Collection. The Chapter will also discuss the principles and concepts similar to other chapters and include Cross-Border Transfer, Retention, Enforcement, Commissioner, Controller Functions, Codes of Practice and Standards, along with a brief outline of the Proposed New Privacy and Protection Law and Supporting Laws. The Chapter concludes by summarizing the key principles of the proposed data protection and privacy laws.

© Springer Nature Singapore Pte Ltd. 2019

R. Walters et al., *Data Protection Law*,
https://doi.org/10.1007/978-981-13-8110-2_6

6.1 Introduction

To date, India has neither prepared nor implemented specific data protection or privacy laws. The Information Technology Act 2000[1] (IT Act) is the principal legislation regulating data and privacy across India. The IT Act aims to provide legal recognition for transactions carried out by means of electronic data interchange and other means of electronic communication, commonly referred to as electronic commerce, which involve the use of alternatives to paper-based methods of communication and storage of information and to facilitate electronic filing of documents with the government agencies.[2] Upon implementation Davan Duggal believed that the IT Act was established to assist India as the country faced many challenges in the cyberspace sphere and its regulation in a very bold, prompt and decisive so as it could become an IT superpower in the years to come.[3] Appropriate for its time, 18 years on, and India, like many other countries have, and are beginning to rethink their policy strategy and legal framework to protect privacy and personal data over the Internet.

The Information Technology (Reasonable Security Practices and Procedures and Sensitive Personal Data or Information) Rules, 2011 (the Rules) underpin the implementation to the IT Act. They have been conferred under the exercise of the powers by clause (ob) of sub-section (2) of section 87 and read with section 43A of the IT. The IT Act and the Rules operate hand in hand. The IT Act predominantly deals with electronic signatures and electronic commerce. The Rules provide an organization that outsource functions relating to collection, storage or handling of sensitive and personal information under contractual obligations. The organization if located outside of India is not subject to the requirements of collection, consent or disclosure provided they do not have direct access to the data subject. Nonetheless, the right to privacy across India has not gone without its detractors, or lack of recognition or understanding, most notably dating back to the 1960s.

In the case of *Kharak Singh vs The State of U.P.* [4] the question for consideration before the court was whether 'surveillance' under Chapter XX of the U.P. Police Regulations constituted an infringement of any of the fundamental rights guaranteed by Part III of the Constitution. Regulation 236(b). This provision permitted surveillance by 'domiciliary visits at night' was held to be violate of Article 21. The word 'life' and the expression 'personal liberty' in Article 21 were elaborately considered by this court in Kharak Singh's case.[5] Even though the majority found that

[1] Information Technology Act 2000, http://www.dot.gov.in/sites/default/files/itbill2000_0.pdf, accessed 16 December 2017.

[2] Pavan Duggal, India's information Technology Act, http://unpan1.un.org/intradoc/groups/public/documents/apcity/unpan002090.pdf, accessed 30 November 2018.

[3] Ibid.

[4] *Kharak Singh vs The State of U.P.* (1964) 1 SCR 332.

[5] Ibid.

the Constitution contained no explicit guarantee of a 'right to privacy', the majority read the right to personal liberty expansively to include a right to dignity. The court held that:

> an unauthorised intrusion into a person's home and the disturbance caused to him thereby, is as it were the violation of a common law right of a man—an ultimate essential of ordered liberty, if not of the very concept of civilization.[6]

The right to dignity is considered a fundamental right in many other nation states, including the EU. Some 8 years later, in 1972, the Indian Supreme Court was required to decide on whether there was a right to privacy as a result of wiretapping, by the state. In *R. M. Malkani vs State of Maharashtra* the petitioner's voice had been recorded in the course of a telephonic conversation where he was attempting blackmail. It was argued that the defense to the right to privacy under Article 21 had been violated. The Indian Supreme Court declined the plea holding that the telephonic conversation of an innocent citizen will be protected by courts against wrongful or high handed interference by tapping the conversation.[7] The protection is not for the guilty citizen against the efforts of the police to vindicate the law and prevent corruption of public servants.

Later in *Govind vs. State of Madhya Pradesh*, the court evaluated whether the constitutional validity of Regulations 855 and 856 of the Madhya Pradesh Police Regulation which provided for police surveillance of habitual offenders including domiciliary visits and picketing.[8] The Supreme Court desisted from striking down these invasive provisions holding that—it cannot be said that surveillance by domiciliary visit, would always be an unreasonable restriction upon the right of privacy.[9] It is only persons who are suspected to be habitual criminals and those who are determined to lead criminal lives that are subjected to surveillance.[10] The court went on to argue that under the Constitution:

> too broad a definition of privacy will raise serious questions about the propriety of judicial reliance on a right that is not explicit in the Constitution. The right to privacy will, therefore, necessarily, have to go through a process of case by case development. Hence, assuming that the right to personal liberty, the right to move freely throughout India and the freedom of speech create an independent fundamental right of privacy as an emanation from them it could not he absolute. It must be subject to restriction on the basis of compelling public interest. But the law infringing it must satisfy the compelling state interest test. It could not be that under these freedoms that the Constitution-makers intended to protect or protected mere personal sensitiveness.[11]

The above approach was also reinforced in the case of *R. Rajagopal v. State of Tamil Nadu*,[12] where the Supreme court stated that the right to privacy is implicit in

---

[6] Ibid.

[7] *R. M. Malkani vs State of Maharashtra 1975.*

[8] *Govind vs. State of Madhya Pradesh 1975* (1975) 2 SCC 148.

[9] Ibid.

[10] Ibid.

[11] Ibid.

[12] *R. Rajagopal v. State of Tamil Nadu* (1994) 6 SCC 632.

the right of life and liberty guaranteed to the citizens of this country by Article 21. The court argued that the right of privacy:

> is implicit in the right to life and liberty guaranteed to the citizens of this country by Article 21. It is a "right to be let alone". A citizen has a right to safeguard the privacy of his own, his family, marriage, procreation, motherhood, childbearing and education among other matters. None can publish anything concerning the above matters without his consent—whether truthful or otherwise and whether laudatory or critical. If he does so, he would be violating the right to privacy of the person concerned and would be liable in an action for damages. Position may, however, be different, if a person voluntarily thrusts himself into controversy or voluntarily invites or raises a controversy.[13]

Furthermore, the court went onto say that the:

> rule aforesaid is subject to the exception, that any publication concerning the aforesaid aspects becomes unobjectionable if such publication is based upon public records including court records. This is for the reason that once a matter becomes a matter of public record, the right to privacy no longer subsists and it becomes a legitimate subject for comment by press and media among others. We are, however, of the opinion that in the interests of decency [Article 19(2)] an exception must be carved out to this rule, viz., a female who is the victim of a sexual assault, kidnap, abduction or a like offence should not further be subjected to the indignity of her name and the incident being publicized in press/media.[14]

Arguably, the court had taken a cautionary approach by holding that the right to privacy...will necessarily have to go through a process of case-by-case development. Nonetheless, in 1997, the Indian court was called upon to consider whether wiretapping was an unconstitutional infringement of a citizen's right to privacy. In *PUCL v. Union of India*[15] the court held:

> The right to privacy—by itself—has not been identified under the Constitution. As a concept it may be too broad and moralistic to define it judicially. Whether right to privacy can be claimed or has been infringed in a given case would depend on the facts of the said case. But the right to hold a telephone conversation in the privacy of one's home or office without interference can certainly be claimed as a 'right to privacy'. Conversations on the telephone are often of an intimate and confidential character. Telephone conversation is a part of modern man's life. It is considered so important that more and more people are carrying mobile telephone instruments in their pockets. Telephone conversation is an important facet of a man's private life. Right to privacy would certainly include telephone-conversation in the privacy of one's home or office. Telephone-tapping would, thus, infract Article 21 of the Constitution of India unless it is permitted under the procedure established by law.[16]

Up until this period, India had not conclusively determined what the right to privacy constituted. India was far from determining whether the right to privacy had anything to do with people's personal data or personal information—over the Internet.

It was not until a decade after the turn of the century that an Indian court determined there was no conclusive definition of privacy. In *Shri S. K. Chaurasiya vs*

[13] Ibid.

[14] Ibid.

[15] *PUCL v. Union of India* AIR [1997] SC 568.

[16] Ibid.

Central Vigilance Commission[17] the Commission stated that there was no clear definition in the Right to Information Act 2005, of what constitutes an invasion of privacy. India was for the first time considering how the right to privacy, from the invasion of privacy to ones' personal information would be impacted by the law. Due to the lack of specific data protection laws in the country, India would turn to its former colonialists for guidance. The court went onto to say that therefore, one must look elsewhere and in this and other cases be guided by the United Kingdom's Data *Protection Act 1998.*[18] The appellant Shri Chaurasiya had asked for information on names, designations and places of posting of those against whom cases had been registered and those who have been cleared. This information qualified as personal information. The court in referring to section 2 of the United Kingdom's *Data Protection Act 1998* of which, titled 'Sensitive Personal Data', reads as follows, in this Act "sensitive personal data" means personal data consisting of information as to;

(a) The racial or ethnic origin of the data subject.
(b) His political opinions.
(c) His religious beliefs or other beliefs of a similar nature.
(d) Whether he is a member of a Trade Union.
(e) His physical or mental health or condition.
(f) His sexual life.
(g) The commission or alleged commission by him of any offence.[19]

The court went onto to say that:

> If we were to construe privacy to mean protection of personal data, this would be a suitable starting point to help define the concept. As will be clear from the above, although indeed the clearance or refusal to clear names is a public activity and cannot, therefore, claim disclosure under that portion of Sec. 8(1)(j), it nevertheless provided information which will qualify for invasion of privacy as defined in section 2(h).[20]

Even in light of the above case, to date, India, has adopted a sectorial approach that span across healthcare, telecommunication, banking and securities. Despite the lack of specific privacy legislation in India, Kessler, Ross and Hickok note there have been work undertaken towards the realization of a comprehensive privacy law.[21] For example, the Personal Data Protection Bill was introduced in Parliament some time ago. However, the Bill prompted by concerns over the misuse of the selling and use of personal data for direct marketing purposes unfortunately lapsed in Parliament. Four years later in 2010, the Department of Personnel

---

[17] *Shri S. K. Chaurasiya vs Central Vigilance* Commission [2010].

[18] Ibid.

[19] Ibid.

[20] Ibid.

[21] Kessler,. D, Ross., S, Hickok, E *A Comparative Analysis of Indian Privacy Law and the Asia-Pacific Economic, Cooperation Cross-Border Privacy Rules*, National Law School of India Review, Vol. 26, No. 1 (2014), pp. 31–61.

and Training (DoPT) published an "Approach Paper for a Legislation on Privacy," which a group of officers for the purpose of developing a "conceptual framework that could serve the country's balance of interests and concern on privacy protection, and security.[22] The Paper reviewed privacy laws in thirteen jurisdictions and set forth recommendations for a privacy regime that included on data protection. Notably, the paper defined privacy as "the expectation that confidential personal information any individual to Government or non-Government entity should not to third parties without consent of the person and sufficient safeguard be adopted while processing and storing of the information."[23] Moreover, they note that at the time there was little to no recognition of privacy across India. One of the issues identified was that fact that the Government often discloses personal information of citizens as part of its transparency efforts.[24] Another factor cited as a driving force for privacy legislation in India was the trend towards centralization of government. In this regard, the paper highlighted the privacy concerns posed by the Unique Identification Project. The paper pointed out that there has been an increase in the use of personal data by private sector organizations.[25] The paper recommended a statute defining broad principles for the processing of collected personal industry bodies defining detailed and sector specific guideline by member organizations who collect and use personal data.[26]

Kessler, Ross and Hickok go onto to say that in 2011, there were three significant privacy events in India. Firstly, a Press Information Bureau release from the Ministry of Personal and Public Grievances stating that "The Government proposes to provide protection to individuals privacy.[27] Secondly, news reports at the time indicated by the Niira Radia Tapes scandal, the Government drafted a "Rig 2011" which sought to create a statutory right to privacy.[28] The third came at the end of December 2011, whereby the Planning Commission of the Government of India constituted a Group of Experts on Privacy to study privacy regimes from different jurisdictions, to analyse current programmes and projects being implemented in India from a privacy perspective. The aim was to formulate recommendations for the Department of Personnel and Training for incorporation in the proposed draft Bill on Privacy. A year later in 2012, the Committee published the Report of the Group of Experts on Privacy (known as the "Report"). Though not officially accepted by the Government of India, news items indicate that the Department of

---

[22] Government of India. Ministry of Personnel, PG & Pensions, Department of Personnel Training, Approach Paper for a Legislation on Privacy, (2010) http://ccis.nic.in/WriteReadData/CircularPortal/D2/D02rti/aproach, accessed 20 November 2018.

[23] Kessler,. D, Ross., S, Hickok, E *A Comparative Analysis of Indian Privacy Law and the Asia-Pacific Economic, Cooperation Cross-Border Privacy Rules*, National Law School of India Review, Vol. 26, No. 1 (2014), pp. 31–61.

[24] Ibid.

[25] Ibid.

[26] Ibid.

[27] Ibid.

[28] Ibid.

Personnel and Training has incorporated in the upcoming draft of the Privacy Bill recommendations found in the 2012 Report.[29]

Yet, it was not until 2017, when India began the task of seeking community input into the development of specific data protection/privacy laws. The White Paper of the Committee of Experts on a Data Protection Framework for India noted that the digital revolution has permeated India along with many other countries.[30] Recognizing its significance, and that it promises to bring large disruptions in almost all sectors of society, the Government of India has envisaged and implemented the Digital India initiative.[31] This initiative involves the incorporation of digitization in governance; healthcare and educational services; cashless economy and digital transactions; transparency in bureaucracy; fair and quick distribution of welfare schemes to empower citizens.

Notwithstanding the above,[32] across India, the state uses personal data for purposes such as the targeted delivery of social welfare benefits, effective planning and implementation of government schemes, counter-terrorism operations, amongst others. Furthermore, India is considered the largest outsourcing country alongside the Philippines in Central and South East Asia. However, this is no different to many other countries including the EU who collect, analyse and use personal data for similar purposes. Moreover, in most countries and the EU, which are discussed in this book, the collection and use of data is usually backed by law, though in the context of counter-terrorism and intelligence gathering, it appears not to be the case.[33] The rise of the right to privacy throughout India has largely been influenced by the actions of the state towards its citizens. It has only been recently where the courts have begun to decide on privacy in the context of personal data. Until the process of community input into the White Paper has concluded, and specific data protection or privacy laws are implemented in India, or a draft bill is circulated, will the local and international community have a better understanding of the direction India will adopt.

---

[29] Ibid. The "Approach Paper for a Legislation on Privacy" and "The Report of the Group of Experts on Privacy" contain analysis of the privacy protections found under the Information Technology Act and the Information Technology (Reasonable security practices and procedures and sensitive personal data or information) Rules 2011. The Approach Paper notes that even though the Information Technology Act protects personal data to some extent, the provisions are not comprehensive enough as they speak only to digital data. The Report of the Group of Experts on Privacy notes that the Rules fall short of meeting the standards defined by the National Privacy Principles in the Report as the Rules do not address or require anonymization of data when appropriate, do not require Body Corporate to provide notice of changes in purpose of collection or use, do not address the destruction of data, require Body Corporate to provide notice of breach of information to affected individuals, require Body Corporate to provide notice to changes in its privacy policy, and require Body to conduct an external audit on all policies and practices to ensure accountability.

[30] Ministry of Electronics and Information Technology, White Paper of the Committee of Experts on a Data Protection (2017).

[31] Ibid.

[32] Ministry of Electronics and Information Technology, White Paper of the Committee of Experts on a Data Protection (2017).

[33] Ibid.

6.2 Personal Information

The IT Act does not specifically define personal information in the same way as other countries, namely Australia. The 43A Rules define personal information to mean any information that relates to a natural person. This includes either directly or indirectly, in combination with other information available or likely to be available with a body corporate, and is capable of identifying such a person.[34] Rather than the principal legislation define sensitive data, it has been left to the Rules.[35] Sensitive data or information includes similar elements to that of the other countries and the EU. Sensitive data constitutes a person's:

- password;
- financial information such as bank account or credit card or debit card or other payment instrument details;
- sexual orientation;
- medical records and history;
- Biometric information;
- any detail relating to the above clauses as provided to body corporate for providing service; and
- any of the information received under above clauses by body corporate for processing, stored or processed under lawful contract or otherwise provided that, any information that is freely available or accessible in public domain or furnished under the Right to Information Act 2005 or any other law for the time being in force shall not be regarded as sensitive personal data or information for the purposes of these rules.[36]

Biometrics include technologies that measure and analyze human body characteristics, such as fingerprints, eye retinas and irises, voice patterns, facial patterns, hand measurements and DNA. Furthermore, the Aadhaar (Targeted Delivery of Financial and other Subsidies, Benefits and Services) Act 2016[37] enables the Government to collect identity information from citizens including their biometrics. Thus, reinforcing the earlier point that India's laws are far from being coherent.

---

[34] The Information Technology (Reasonable Security Practices and Procedures and Sensitive Personal Data or Information) Rules, 2011, section 2.

[35] Ibid.

[36] Ibid, section 3.

[37] The Aadhaar Act also provides for Aadhaar based authentication services wherein a requesting entity (government/public and private entities/agencies) can request the Unique Identification Authority of India (UIDAI) to verify/validate the correctness of the identity information submitted by individuals to be able to extend services to them.

6.3 Right to Be Forgotten

The right to be forgotten is not formally codified in India. However, it appears to be covered under "right to privacy" in accordance with Article 21. The Information Technology Act 2000 and, Information Technology Rules 2011, provides for the collection of information by corporates and individuals under Rule 5. Rule 5 (7) allows "the provider of information [to a body corporate] shall, at any time while availing the services or otherwise, also have an option to withdraw its consent given earlier to the body corporate."[38] On 31 May 2014, Medianama.com became the first Indian website to be asked by an individual to remove a link. The website owner declined a request.[39]

Since the *Google*[40] case in the European Union, India has begun considering what the right to be forgotten would mean to its people and the nation state. In 2017, there were two cases that considered the right to be forgotten. The first was where the Karnataka High Court in *Sri Vasunathan v The Registrar General*[41] which recognized the right to be forgotten and safeguarded the same in sensitive cases involving women in general and highly sensitive cases involving rape or affecting the modesty and reputation of the person concerned, in particular.[42]

At the Supreme Court in *Justice K.S. Puttaswamy (Retd.)& Anr v Union of India & Ors*[43] the court stated that:

> the impact of the digital age results in information on the internet being permanent. In the digital world preservation is the norm and forgetting a struggle. People are not static; they are entitled to re-invent themselves and correct their past actions. It is privacy which nurtures this ability and removes the shackles of unadvisable things which may have been done in the past. Privacy removes the shackles of unadvisable things which may have been done in the past, and correct the past. It could be as simple as an academic writing about the law and sociology, having a particular ideological view at a particular time, and over time that view has changed. That academic may wish to erase the past position that is widely available on a blog, on the internet.[44]

Arguably, India is far from considering the right to be forgotten in the same context, for example, as the EU. To begin with, privacy has not fully matured across India, particularly over the Internet. It is our view that until this has been settled, it is unlikely the right to be forgotten will be fully accepted.

---

[38] The Information Technology Act 2000 and Information Technology Rules 2011.

[39] The Right to be Forgotten poses a legal dilemma in India, https://www.livemint.com/Industry/5jmbcpuHqO7UwX3IBsiGCM/Right-to-be-forgotten-poses-a-legal-dilemma-in-India.html, accessed 2 October 2018.

[40] Case C-131/12 *Google Inc. v. Agencia Espanola de Proteccion de Datos, Mario Consteja González*, 95–96.

[41] *Sri Vasunathan v The Registrar General*, 2017 SCC Online Kar 424.

[42] Ibid.

[43] *Justice K.S. Puttaswamy (Retd.)& Anr v Union of India & Ors.*, (2017) 10 SCALE 1.

[44] Ibid.

6.4 Grievance Officers

Data controllers, as they are referred to in other jurisdictions are not appointed in India to provide a point of contact within an organization that handles personal data and privacy. Rather they use Grievance Officers (GO), which have a limited role and deal with complaints in relation to sensitive information. The GO also has a role to respond to data subject's request to and correct their personal information.[45] The officer is required to redress the grievances within one month from the date of receipt of grievance. Even though such officers are appointed they are not accountable to the level that is required, for example, in the EU.[46] The White Paper discusses the importance of controllers at lengthen, and argues:

> control over data, in such systems, refers to the competence to take decisions about the contents and use of data. The entity that has control over data is responsible for compliance with data protection norms and is termed a data controller. In addition to the data controller, other entities which take part in the processing of data are often identified and defined. For instance, a data processor is an entity which is closely involved with processing, which however, acts under the authority of the data controller. Identification of all entities participating in the entire cycle of data processing is not the only method of allocating responsibility. There are various models which have evolved in this regard in other jurisdictions. Each operates at a different level of specificity in identifying the entities involved in processing.[47]

The model, as highlighted in Chap. 3, identifies how the EU have adopted a very prescriptive and multilayered approach to ensure that the personal data collected and used by an organization, is done so undertaken where there is a clear line of responsibility.

6.5 Public and Private

The current legal framework is limited in its reach and section 43 of the IT Act only applies to a body corporate or an individual who is in possession of, dealing or handling any sensitive personal data or information in a computer resource which it owns, controls or operates.[48] The IT Act does not apply to the government sector. The proposed *Data (Privacy and Protection) Bill 2017,* which has not been approved, and is some way from being implemented, is set to expand the reach of the data protection and privacy laws to the government sector.

---

[45] The, Information Technology Rules 2011, section 5(9).

[46] Ibid.

[47] Ministry of Electronics and Information Technology, White Paper of the Committee of Experts on a Data Protection (2017).

[48] The Information Technology Act 2000.

6.6 Consent and Collection

Generally, India does not require consent for the processing of general personal data. However, the rules surrounding sensitive personal data are somewhat different and require consent. Consent can be obtained in a number of forms, and includes letter, fax or email. Electronic consent via tick box such as an 'I Agree' tab is also permitted. Even so, the collection of sensitive data has a stricter measure applied to data subjects of the purpose for which data that it is being collected. The collection of sensitive data can only be undertaken according to the function of the organization concerned and must be for the specific purpose.[49] Rule 5(7) allows a data subject to withdraw their consent. In the case of a provider of information not providing or later withdrawing their consent, the body corporate has the option not to provide goods or services for which the said information was sought.[50] Moreover, there is no clear principle of accuracy in either the IT Act or the Rules. However, the IT Act requires that an electronic record must represent accurately the information originally generated when it was sent or received, for the purposes of its retention.[51]

Nonetheless, India have, like many other nation states identified that consent is increasingly becoming one of the most important concepts related to data protection and privacy. That is, allowing an individual to have autonomy over their personal information allows them to enjoy a level of informational privacy.[52] Firstly, informational privacy may be broadly understood as the individuals ability to exercise control over the manner in which her information may be collected and used. Secondly, consent provides a morally transformative value as it justifies conduct, which might otherwise be considered wrongful.[53] The Indian court in *Justice K.S. Puttaswamy (Retd.) & Anr. v. Union of India & Ors*[54] reinforced the importance of consent and held that:

> the right to privacy would encompass the right to informational privacy, which recognizes that an individual should have control over the use and dissemination of information that is personal to him or her.[55]

Thus, as discussed throughout this book, consent is an important concept of data protection and privacy law, because the concept provides data subjects with not only control over their personal data or personal information defined by the law, but also a greater level of ownership.

---

[49] The The Information Technology Act 2000 and, Information Technology Rules 2011, section 5(2) & (3).

[50] Ibid.

[51] Information Technology Act 2000, section 7.

[52] Ministry of Electronics and Information Technology, White Paper of the Committee of Experts on a Data Protection (2017), p. 78.

[53] Ibid.

[54] *Justice K.S.Puttaswamy(Retd) vs Union Of India* 26 September, 2018.

[55] Ibid.

6.7 Cross-Border Transfer

The cross border transfer of data in India, is no different to other states that are sending data outside the country. Rule 7 states clearly that the transfer of information by a body corporate or any person on its behalf may transfer sensitive personal data or information including any information, to any other body corporate or a person in India, or located in any other country, that ensures the same level of data protection that is adhered to by the body corporate as provided for under these Rules.[56] The transfer may be allowed only if it is necessary for the performance of the lawful contract between the body corporate or any person on its behalf and provider of information or where such a person has consented to data transfer. This is somewhat limited to sensitive personal information. The processing and transfer of personal data under Rule 7 must be read in accordance with Rule 4 which requires a company who is handling personal data to prepare a privacy policy.[57]

Moreover, where personal or sensitive data may require it being transferred to a foreign jurisdiction, specific conditions must be met. The transfer can only be undertaken provided that the transferee entity has standards in place that are not lower than those set by IS/ISO/IEC 27001, and meets the performance of a contract between the Indian organization and data subject. In other words, a contract between an organization and data subject that does not specify a transfer of data to other jurisdiction cannot be undertaken. However, this does not apply if the data subject has provided consent, before the transfer occurs.

A company is restricted from disclosing sensitive data to a third party. However, they can disclose this information provided they have obtained consent from the data subject.[58] Where there is a contract between the data subject and the company, consent is not necessarily required. In addition, if the company privacy policy adequately reflects that data subjects consent, and the privacy policy is understood by the data subject, the disclosure could occur.

India views the principle of territoriality ordinarily connoting the jurisdiction of a state over an act committed both within its territory, and can be exercised over the same acts which take place outside the state but have consequences within the State.[59] This view espoused by India is consistent with other jurisdictions discussed in this book.

Nevertheless, India are concerned that any future legislation developed specifically for data protection and privacy will need to consider that any strict notion of territoriality could fail to adequately protect Indian residents and citizens as a large number of actions which the state may have a legitimate interest in regulating could fall outside the scope of the law, if not drafted accurately.[60]

---

[56] The Information Technology Rules 2011, 7.

[57] Ibid, 4 and 7.

[58] The Information Technology Rules 2011, 6.

[59] Ministry of Electronics and Information Technology, White Paper of the Committee of Experts on a Data Protection (2017), p. 24.

[60] Ibid, 25.

6.7.1 *Data Localization*

India, through their sectorial approach has a limited approach to data localization. Based upon the Information Technology Act of 2000, Rule 7 of the Information Technology (Reasonable Security Practices and Procedures and Sensitive Data or Information) Rules of 2011 has been interpreted by the Ministry of Communications and Information Technology as requiring companies located in India to obtain the consent of Indian citizens before transferring their sensitive personal data or information abroad. Moreover, this requires is to be met provided the receiving country has the same level of data protection and only if it is necessary for the performance of lawful contracts.[61] The Public Records Act 1993 also prohibits the transferring of public records outside of India, except for public purposes without the prior approval of the Central government.[62]

6.8 Retention

A body corporate or person holding sensitive personal data cannot retain that information for longer than is required.[63] Section 7 of the IT Act clarifies ways of retaining electronic documents, records and information. The electronic record is to be retained in the format it was originally generated. The regulatory requirements setting out the retention for data is sporadic and there is no consistent approach due to the sectorial laws. The retention requirements for ISPs and UASL licenses must comply with the Telegraph Act 1885. However, the current law and position adopted throughout India is arguably contrary to the current review on data protection. The view with regard to retention is that the government will be burdened with the task of prescribing different retention guidelines for different categories of data, and may not end up performing this task satisfactorily.[64] While this concern is warranted, it is argued that without adequate guidelines detailing how the retention of personal data and information can be retained, India risk leaving a black hole in their regulatory framework.

---

[61] Selby, J *Data localization laws: trade barriers or legitimate responses to cybersecurity risks, or both?* International Journal of Law and Information Technology, (2017) pp. 213–232.

[62] Ibid.

[63] The Rules, section 5(4).

[64] Ministry of Electronics and Information Technology, White Paper of the Committee of Experts on a Data Protection (2017), p. 118.

6.9 Enforcement

The enforcement provisions within the IT Act and the Rules is limited. Section 43 provides for penalties for the damage of computers and their systems.[65] That is, any person that without permission of the owner of the computer or its systems changes that computer or the systems supporting it, or a network has committed an offence. Furthermore, individuals are also required to have permission form the owner of the computer and its systems to access or secure the access to the computer, its system or network. Permission is also required for the downloading, obtaining copies or extracts of any data held or stored within the system. Arguably the reference to data would include all types of personal and commercial data. Moreover, an individual found to have damaged a computer, its systems, networks and data (personal and commercial) would be liable to pay damages by way of compensation.[66] Importantly, India's privacy requirements as found under section 43A of the ITA Act and sequent Rules is closely aligned with the APEC Cross-Border Privacy.[67] However, Kessler, Ross and Hickok note that while they are not perfectly aligned. India would generally need to expand a few aspects of its privacy specifically.[68] The limited alignment of APEC principles, goes some way for India to continue to strengthen and look outside the state for those principles and concepts that will provide certainty within their future privacy legal framework.

[65] Information Technology Act 2000.

[66] Ibid, for the purposes of this section,—(i) "computer contaminant" means any set of computer instructions that are designed—(ii) "computer data base" means a representation of information, knowledge, facts, concepts or instructions in text, image, audio, video that are being prepared or have been prepared in a formalized manner or have been produced by a computer, computer system or computer network and are intended for use in a computer, computer system or computer network; (iii) "computer virus" means any computer instruction, information, data or programme that destroys, damages, degrades or adversely affects the performance of a computer resource or attaches itself to another computer resource and operates when a programme, daia or instruction is executed or some other event takes place in that computer resource; (iv) "damage" means to destroy, alter, delete, add, modify or rearrange any computer resource by any means.

[67] Kessler,. D, Ross., S, Hickok, E *A Comparative Analysis of Indian Privacy Law and the Asia-Pacific Economic, Cooperation Cross-Border Privacy Rules*, National Law School of India Review, Vol. 26, No. 1 (2014), pp. 31–61.

[68] The specific issues identified include, to the Collection requirement, the protection to include jail personal information, rather than be limited personal information. The Purpose requirement could be expanded from limit use of collected information, to include both data transfers and disclosures. The APEC requirement and Correction Principles would likely require the large modifications. This expanded section could contain descriptions of processes for individuals to confirm that a Body Corporate or control of an individual's personal information, and under what circumstances the individual can obtain information from the Body Corporate, as how and under stances an individual can request changes to that information. The Opt Out right, where individual to provide information and can withdraw consent, it W to provide examples of when such a right applies and d such as APEC's example of a company that is centralizing resources data and does not need to provide an opt-out. The Redress Mechanism that India requires would need to include the third party accountability in the redress of discrepancies. Finally, Disclosure of Information Collection requirement, the protections would need personal information, rather than be limited to sensitive information.

The Adjudicating Officer is responsible for overseeing any breaches of the laws. The officer is also responsible for determining the level of the injury or damages as part of a claim, provided the claim does not exceed R50 million. The Secretary of the Ministry of information Technology in each of the state governments are appointed as the Adjudicating Officer. Interestingly, the office has the power (s) to summon a person for examination, demand the production of documents and electronic records. The banking, telecommunications and medical sectors have their own regulator who is responsible for overseeing enforcement.

The compromising of critical systems or information, targeted scanning or probing of critical networks and systems, identity thefts, spoofing and phishing attacks must be reported to the Computer Emergency Response Team. In addition, the unauthorized access of IT systems or data, defacement or intrusion of a website or malicious code attacks are to be also reported.[69] The information obtained through notification can be used under a court order to identify individuals. More importantly section 72 provides the basis for the penalties to be imposed where there has been breach the confidentiality and privacy by an individual or entity.[70] In other words, any person who, in pursuance of any of the powers conferred under the IT Act, rules or regulations, who has secured access to any electronic record, book, register, correspondence, information, document or other material without the consent of the person concerned discloses can be punished by imprisonment for a term which may extend to 2 years, or with fine which may extend to one lakh rupees, or both.[71] Section 73 imposes penalties also apply for publishing or make available digital signatures certificate without the approval from the relevant Certifying Authority.[72]

There is no formal complaints mechanism established. As highlighted above, India has established adjudicating officers who are responsible for hearing and deciding on cases where there have been breaches of the IT Act. The officer can adjudicate on issues related to access to computer, computer system or computer network; downloads, copies or extracts any data; computer virus; causes disruption and denial of access.[73] The officer must not be below the level of Director or an equivalent officer within the state government who has the required experience in information and technology. The power of the office is that equivalent to that of a civil court. The fragmented approach taken by Indian, makes it extremely complex to understand where the responsibility lies.[74]

[69] Information Technology Act 2000, 70B.

[70] Ibid, 72.

[71] Ibid.

[72] Ibid, 73, any person who contravenes the provisions of sub-section (1) shall be punished with imprisonment for a term which may extend to 2 years, or with fine which may extend to one lakh rupees, or with both.

[73] Ibid, 43, 44 and 45.

[74] Ibid, 46(5).

6.10 Commissioner

Unlike the other jurisdictions discussed throughout this book, India does not have a dedicated Commissioner overseeing the implementation of data protection or privacy laws. The Ministry of Electronics and Information Technology are responsible for administration of the IT Act and the Rules. It is a shared responsibility with the cyber Tribunal, High court and Supreme courts for enforcing the IT Act.[75] The Tribunal has the same powers and is considered to be a civil court, to assist in the enforcement of the IT Act.

India has not appointed a Commissioner and therefore, to date, there are no specific regulatory powers to investigate and audit public or private organizations. Nevertheless, India see that a co-regulatory approach is the major way forward. With that in mind, India is of the view that a Commissioner or a similar role will be required.[76] Back in 2012, a Group of Experts on Privacy was constituted by the erstwhile Planning Commission under the Chairmanship of Justice AP Shah (Justice AP Shah Committee).[77] The report of the committee recommended a detailed framework that serves as the conceptual foundation for a privacy law, with enforcement undertaken by Privacy Commissioners set up by statute. Nevertheless, the IT Act, unlike the laws of the EU, Australia and other countries, does not specify the core principles of data protection and privacy. However, the IT does provide provision for consent, disclosure, retention, controller, breach and enforcement.

6.11 Controller Functions

The controllers are required to supervise the activities pertaining to data management of the certifying authority, by establishing the standards to be maintained.[78] In other words, India consider that the role of the controller can be limited to certain sectors. For example, in banking regulation where systemically important financial institutions seem to require additional forms of regulation.[79] This approach is adopted, would follow the exemption by Australia in relation to small business that have an annual turnover of $3 million or less (see Chap. 5). The current review underway in India, questions the role of the controller in data protection. India argue that due to the breadth of a data protection law, its effectiveness can come to depend on the ability of a regulatory body to have adequate awareness and monitoring capacity of actual data protection practices so that it can identify and effectively

[75] Ibid, 48(1).

[76] Ministry of Electronics and Information Technology, White Paper of the Committee of Experts on a Data Protection (2017), p. 118.

[77] Ibid.

[78] Information and Technology Act 2000, 17 and 18.

[79] Ibid.

address data protection risks.[80] However, it was noted that not all processing activities pose risks of similar gravity and the nature or volume of the data being processed or the form of the processing operations themselves may require greater scrutiny and oversight. However, as highlighted in Chap. 3, the EU approach is comprehensive in making the controller a central feature and role having responsibility for the collection and use of personal data.

6.12 Codes of Practice and Standards

A company which handles general or sensitive personal data must develop a privacy policy. The policy must outline how the organization is to collect, retain and use the data. In addition, the policy must outline the specific purpose for why that organization requires the data, and highlight the security measures that will or have been established to protect the data.[81] Moreover, the management of personal data, particularly in transferring it outside of India, cannot be a lower standard set out in IS/ISO/IEC 27001.[82]

The IT does not require codes of practice to be implemented by industry sectors, associations or individual organizations. Although the IT does require the application of procedures and practices to ensure electronic signatures are secure. The procedures and practices must prescribe how the organization will secure electronic signatures. This extends to licenses being issued to certify the specification and form of electronic signatures. This process is very different to other jurisdictions, who require the central regulatory authority, not the controller to approve an operational matter. Under the Rules, a body corporate handling and processing sensitive personal data is required to have its security practices and procedures certified and audited by an independent auditor. The auditor must be approved by the central government at least once every year, or when there is a significant upgrade in its computer resource.[83]

Rule 8 of the SPDI Rules requires that an organization needs to establish certain security practices that a body for the purpose of protecting 'sensitive' personal data.[84] These security practices and standards should be supplemented by a comprehensive documented information security program and information security policies. There is also guidance on what standards are appropriate. For instance, ISO 27001 and the use of code of practice has been determined as being necessary for organizations to have in place. Other jurisdictions discussed in this book do not specify the use of ISO 2700 or any other standard, rather they leave it to industry to determine what is suitable for them.

[80] Ministry of Electronics and Information Technology, White Paper of the Committee of Experts on a Data Protection (2017), pp. 167–172.

[81] The Information Technology Rules 2011, 4.

[82] Ibid, 7.

[83] Ibid, 8(4).

[84] Ibid, 8.

6.13 Proposed New Privacy and Protection Law & Supporting Laws

The *Data (Privacy and Protection) Bill 2017*[85] has yet to be approved. The Bill represents many similarities to other countries and sets the framework for the future collection, collation, retention and enforcement of privacy laws across India. However, it will have its own distinct features that, as discussed below, will separate their proposed laws to those of the EU, Singapore and Australia. The draft proposal aims to expand the current obligations of data and privacy protection beyond the current requirements of body corporates to include the government sector.[86] The draft proposal expands on the current definition of personal data, and will define general and sensitive personal data. The proposal will also deal with the collection, storage, processioning, security, disclosure, accuracy, sensitive personal data and intelligence provisions.[87] It is proposed to establish a Commission and Commissioner. If approved the proposal will differ from the other countries and the EU legal framework.[88] It will also ensure security and create a duty of confidentiality and secrecy, around the interception of communications. Chapter five of the proposed Bill has been prepared to deal with surveillance by the state on its citizens. This is another area of digression from the EU and other countries discussed in this book.

Nonetheless, the proposal is seeking to establish a data fiduciary which will purportedly operate to include any entity that alone or together with others determines a way in which personal data will be processed.[89] Such a proposal has not been included into other jurisdictions data protection laws. The proposal appears to be based on the notion of trust that the data principal places with entities that are sharing the personal data. Moreover, data subjects in India, expect that their personal data will be used fairly and in a manner that their personal interest will also be considered for the foreseeable future.[90] In other words, the proposed provision will exist where a fiduciary (person or business) has an obligation to acting in a trust worthy manner and for the purpose stated, and in the interest of the data subject—the contract.

Moreover, there has been some criticism[91] towards the proposal. Amba Kak is of the view that the proposal in its current form has many loopholes. Kak argues that the requirement to store a copy of all personal data within India, creating broad

---

[85] Draft Data Privacy Bill, http://164.100.47.4/BillsTexts/LSBillTexts/Asintroduced/889LS%20AS.pdf, accessed 16 December 2017.

[86] Ibid.

[87] Ibid.

[88] Ibid.

[89] Lakshmikumaran and Sridaran, *Data Principal and Data Fiduciary in the Personal Data Protection Bill 2018*, https://www.lexology.com/library/detail.aspx?g=f0522766-30c6-4c07-ab5a-fb924a74f5cc&utm_source=lexology+daily+newsfeed&utm_medium=html+email+-+body+-+general+section&utm_campaign=australian+ihl+subscriber+daily+feed&utm_content=lexology+daily+newsfeed+2018-11-27&utm_term, accessed 22 November 2018.

[90] Ibid.

[91] Bhattacharya, A *India's first data protection bill is riddled with problems* (2018) https://qz.com/india/1343154/justice-srikrishnas-data-protection-bill-for-india-is-full-of-holes, accessed 25 November 2018.

permissions for government use of data, and the independence of the regulator's adjudicatory authority.[92]

The proposal for every data fiduciary (any entity processing personal data) shall ensure the storage, on a server or data center located in India, of at least one serving copy of personal data to which this Act applies. Experts believe this may become a big hurdle for existing companies to operate in India, and new ones to set shop. It will particularly impact foreign firms such as Facebook and Twitter, which already have millions of users in India but store their data at remote locations.[93]

Furthermore, they are of the view that mandating localisation of all personal data as proposed in the Bill is likely to become a trade barrier in the key markets.[94] However, this issues will not be unique to India as other countries have either implemented localization controls, or likely too. Even though the proposal aims to improve transparency and accountability, this authority—comprising a chairperson and six other members appointed by the central government—would hardly operate autonomously. Finally, the proposes Bill provides excessive powers (to) the central government, especially under Section 98 which not only states that the central government can issue directions to the authority, but also that the authority shall be bound by directions on questions of policy in which the decision of the central government is final.[95]

Justice B.N. Srikrishna makes the point that the proposed laws that the data fiduciaries must only be allowed to share and use personal data to fulfil the expectations of the data principal in a manner that furthers the common public good of a free and fair digital economy.[96] Srikrishna further argues that a regime based on the principles mentioned above and implemented through the relations described above will ensure individual autonomy and make available the benefits of data flows to the economy.[97] More importantly, the proposal aims to achieve twin objectives of protecting personal data while unlocking the data economy have often been seen as conflicting with each other.[98] Specifically, it is proposed to have set up a false choice between societal interests and individual interests, a trade-off between economic growth and data protection. They argue that both are designed to achieve the constitutional objectives of individual autonomy, dignity and self- determination. Therefore, India is aiming that the protection of personal data assists in facilitating the growth of the digital economy. However, they note that each of them is motivated by distinct intermediate rationales—the former ensuring the protection of individual autonomy and consequent harm prevention and the latter seeking to create real choices for citizens.[99]

---

[92] Ibid. Amba Kak Policy Advisor software company Mozilla in India.

[93] Ibid.

[94] Ibid.

[95] Ibid. Shweta Mohandas, Programme Officer at the Centre for Internet and Society.

[96] Chairmanship of Justice B.N. Srikrishna *A Free and Fair Digital Economy Protecting Privacy, Empowering Indians* http://meity.gov.in/writereaddata/files/Data_Protection_Committee_Report.pdf, accessed 25 November 2018.

[97] Ibid.

[98] Ibid.

[99] Ibid.

India, through the development of their proposed data protection laws have arguably identified the need to balance the right to protect personal data and subsequently privacy, but, balance this with their broader ecommerce economy. This, like other nation states is a fine line, and could lead to India adopting more elements of Singapore's model. Although, this will be contentious, and will ultimately come down to whether India have a stronger view about protecting privacy over the Internet, or see it more important to protect over current and future economic drivers—in the digital economy.

A possible solution to these criticisms and concerns, is for India to work closely with the EU, Australia and Singapore to ensure that they not only meet their needs, but also ensure they continue down the path of adopting similar data protection laws to those that have been discussed throughout this book.

The current sectorial laws across India are vast and varied. Although they have been divided by Financial, Health and Telecommunications sectors. Below outlines some of the sectorial laws that provide, in part, some level of protection of personal data. It is out of scope of this book to analyse and explore these laws.

Financial sector	Health sector	Information technology and telecommunications sector
Banking Regulation Act, 1949	The Indian Medical Council (Professional Conduct, Etiquette and Ethics) Regulations, 2002	The Indian Telegraph Act, 1885
Credit Information Companies (Regulation) Act, 2005	Pre-conception and pre-Natal	The Telecom Regulatory Authority of India Act, 1997 Information Technology Act, 2000
Credit Information Companies Regulation, 2006	Diagnostic Techniques (Prohibition of Sex Selection) Act, 1994	The Information Technology (Reasonable security practices and procedures and sensitive personal data or information) Rules, 2011
The Insolvency and Bankruptcy Code, 2016 and the regulations framed thereunder such as the Insolvency and Bankruptcy Board of India (Information Utilities) Regulations, 2017	The Mental Health Act, 1987	The Information Technology (Intermediaries Guidelines) Rules, 2011
		The Information Technology (Guidelines for Cyber Cafe) Rules, 2011
Payment and Settlement Systems Act, 2007		The Information Technology (Electronic Service Delivery) Rules, 2011
Reserve Bank of India Act, 1934		

6.14 Conclusion

India like every other country has also been touched by the internet and digital bug. India does not currently have a coherent single piece of legislation that deals with data protection and privacy. The current legal framework, and particularly the IT Act has a dual role. On the one hand the laws protect personal data, however this is limited. On the other hand, it provides the basis for protecting elements of commercial and personal data from cybercrime.[100] They have, to date, taken a sectorial approach only because of the need for them to protect one of their main economic activities (outsourcing from other countries). In recognizing its significance, India promises to bring large disruptions in almost all sectors of society, the Government of India has envisaged and implemented the Digital India initiative.[101] This initiative involves the incorporation of digitization in governance; healthcare and educational services; cashless economy and digital transactions; transparency in bureaucracy; fair and quick distribution of welfare schemes to empower citizens. More importantly, this shift will see large amounts of Indian citizen's personal data and personal information being collected, used and traded both within and outside the state. That is, it is well understood both the public and the private sector are collecting and using personal data at an unprecedented scale and for multifarious purposes.[102] While data can be put to beneficial use, the unregulated and arbitrary use of data, especially personal data, has raised concerns regarding the privacy and autonomy of an individual.[103] Some of the concerns relate to centralization of databases, profiling of individuals, increased surveillance and a consequent erosion of individual autonomy.[104]

Arguably, as the courts continue to recognize privacy as a right, particularly in relation to personal data, this is likely to force the right to be forgotten—to gain even greater recognition. The ongoing lack of a coherent legal framework will only be problematic for India as the EU and other countries continue to harmonize the legal principles and norms of data protection and privacy. This may, if it continues, have a significant economic impact to India, particularly in maintaining their status as an outsourcing powerhouse in the global economy.

However, in understanding India and their recognition of privacy, as discussed earlier in this chapter, Indian citizens have traditionally related privacy with surveillance. Importantly, the proposal will see India adopt the OECD data protection

[100] Bharuka, D, *Indian Information Technology Act 2000, Criminal Prosecution Made Easy for Cyber Pshycos*, Journal of the Indian Law Institute, Vol. 44, No. 3 (2002), pp. 354–379.

[101] Ministry of Electronics and Information Technology, White Paper of the Committee of Experts on a Data Protection (2017).

[102] Ibid.

[103] Ibid.

[104] Ibid.

concepts and principles. For the past 5 years India has been developing news data protecting and privacy laws that are based on the EU, Australia and other regional countries laws. In doing so, India have highlighted the need to strengthen the current concepts of consent, collection and commissioner for example, to include additional concepts enforcement, controllers, amongst others. India has also looked to the United Kingdom for guidance on jurisprudence in relation to privacy. However, there appears along way to go before the actuality of the laws will be approved and implemented across the country. Furthermore, the right to be forgotten is in its infancy within India. There is a lot of work to do, to codify the right in the same way as the EU, if that is the way India choose to formally strengthen the existence of the right. They may well choose leave the decision to the judiciary, to decide on whether the right exists on a case by case basis. This will be an area of the law to watch.

References

Bhattacharya, A *India's first data protection bill is riddled with problems* (2018). https://qz.com/india/1343154/justice-srikrishnas-data-protection-bill-for-india-is-full-of-holes, accessed 25 November 2018.

Devashish Bharuka, D, *Indian Information Technology Act 2000, Criminal Prosecution Made Easy for Cyber Pshycos*, Journal of the Indian Law Institute, Vol. 44, No. 3 (2002), pp. 354–379

Goodwin, T (2015) *The Battle is for Customer Interface*, TechCrunch. https://techcrunch.com/2015/03/03/in-the-age-of-disintermediation-the-battle-is-all-for-the-customer-interface/, accessed October 2018, cited in *Justice K.S. Puttaswamy (Retd.) v. Union of India & Ors.* 2017 (10) SCALE 1, Per S.K. Kaul, J. at paragraph 17.

Kessler, D, Ross, S, Hickok, E. *A Comparative Analysis of Indian Privacy Law and the Asia-Pacific Economic, Cooperation Cross-Border Privacy Rules*, National Law School of India Review, Vol. 26, No. 1 (2014), pp. 31–61

Selby, J *Data localization laws: trade barriers or legitimate responses to cybersecurity risks, or both?* International Journal of Law and Information Technology, (2017) pp. 213–232

Chapter 7
Indonesia

Abstract Indonesia is a relatively new country in regulating data protection and privacy. In 2016, the Indonesian Parliament approved the Electronic Information and Transactions Law No. 11 of 2008 (EIT). In the same year the Minister of Communication and Information (MCI) Regulation No. 20 of 2016 on Personal Data Protection in the Electronic System (PDP) was also established. This is the first law in Indonesia that goes some way to regulating personal data and privacy, although, it is restricted to data in electronic form. Regulation No. 20 of 2016, implements Regulation No. 82 of 2012 on Implementation of Electronic Transactions and Systems (Aditya Rahman A, *Indonesia Enacts Personal Data Regulation, Privacy Laws and Business*, Data Protection and Privacy Information Worldwide, Iss 145, 2017). However the current framework, is sectorial and is similar to India's model. The current approach is far from achieving the same level or data protection for Indonesian citizens, to their counterparts in Singapore, Australia or the EU.

Even though privacy originated in Western thought, it is gaining traction at various levels throughout South East Asia, including Indonesia. For Indonesia, the development of privacy is considered a fundamental right, and the archipelago state is a signatory to relevant international legal instruments such as the 1966 International Covenant of Civil and Political Rights (ICCP). Furthermore, the awareness and understanding of privacy is also being strengthened from the use and application of Internet technology by Indonesians. The ICCPR has been ratified by Indonesian Law Number 15, 2005. Therefore, it is argued that privacy does apply to some level in and across Indonesia.

The development and evolution of democracy across Indonesia has also been an important part of the acceptance of the concept of privacy. Like many other nation states, the development of information communication and technology is a tool of governance within government and the private sector. Information communication is also the focus of a recent national program to enhance the adoption of the Internet, into the economy. Indonesia, is and has been looking to the EU and other neighboring states in the development of specific data protection laws. Indonesia has also followed many other countries by establishing a dedicated cybersecurity agency to oversee the protection of commercial and personal data.

© Springer Nature Singapore Pte Ltd. 2019

R. Walters et al., *Data Protection Law*,
https://doi.org/10.1007/978-981-13-8110-2_7

7.1 Introduction

The Indonesian legal system is based on civil law that is also a mixture of customary and religious law. Indonesia is the largest economy and has the largest population amongst ASEAN member states.[1] There are three kinds of laws enacted to regulate citizens in Indonesia. First, customary law, for example, the citizens of Indonesia still use this in their daily lives. Second, civil law, which is influenced by Dutch colonialization, and third, Islamic law.[2] Interestingly, dating back to Dutch colonization, inhabitants of the Indonesian archipelago have been divided for legal purposes into various "population groups" (*golongan rakyat, bevolkingsgroupen*), based primarily on racial origin.[3]

The history of human rights within the Indonesian Constitution can be traced back to the original 1945 Constitution. That Constitution did not distinguish between human rights and the rights of citizens. However, there were a few arrangements to guarantee the rights of citizens. This condition can be understood in light of that fact that there was no Universal Declaration of Human Rights (UDHR) at the time of Indonesian independence in 1945; therefore, there was no guidance to arrange human rights in the Constitution. Citizens' rights in the original 1945 Constitution consisted of 6 (six) provisions (articles and sections), namely relating to: equality before the law, freedom of speech, the right to religion and the right to education. However, it is argued that this history has little to no influence or impact on the recognition and implementation of privacy as a human right, or on the establishment of data protection law. Indonesia's 1945 Constitution does not explicitly mention privacy. However, Article 28G (1) protects the right to dignity and "to feel secure", concepts that are often related to the right to privacy. It provides that every person shall have the right to protect his / herself, family, honour, dignity and property.[4]

Indonesia established its Human Rights Law Number 39 in 1999.[5] That law does not specifically refer to privacy. However, it adopts the 1948 Universal Declaration of Human Rights. Indonesia also ratified the International Covenant on Civil and Political Rights 1966 (ICCPR). However, not every right guaranteed under ICCPR has been implementing by domestic legislation, including Article 17 of the ICCPR, concerning privacy. Article 7(2) of The Law on Human Rights specifically states

---

[1] Association of South East Asian Nations, https://asean.org/asean/asean-member-states/, accessed 20 December 2018.

[2] Ahmad, A *Law and Development in Changing Indonesia*, IDE Asian Law Series No. 8 Law and Development in Asian Countries, Institute of Developing Economies (IDE), JETRO, (2001).

[3] Ibid.

[4] The 1945 Constitution of the Republic of Indonesia, As amended by the First Amendment of 1999, the Second Amendment of 2000, the Third Amendment of 2001 and the Fourth Amendment of 2002, https://www.ilo.org/wcmsp5/groups/public/%2D%2D-ed_protect/%2D%2D-protrav/%2D%2D-ilo_aids/documents/legaldocument/wcms_174556.pdf, accessed 20 December 2018.

[5] Human Rights Law Number 39 in 1999.

that the 'provisions set forth in international law concerning human rights ratified by the Republic of Indonesia, are recognized under this Act as legally binding in Indonesia.[6] Article 2 states that the Republic of Indonesia acknowledges and holds in high esteem the rights and freedoms of humans as rights which are bestowed by God and which are an integral part of humans, which must be protected, respected, and upheld in the interests of promoting human dignity, prosperity, contentment, intellectual capacity and justice.[7] This affirmation of human rights is commendable in theory. However, it is well understood that in practice, like many other nation states, including Australia and the EU, actual full implementation of human rights, even the ICCPR, may never be achieved.

Sinta Dewi highlights how, following the fall of President Suharto in 1998, Indonesia underwent extensive political and social reform. Indonesian's now enjoy freedom of expression, freedom of information; and the government is trying to lay down the checks and balances between executive and legislative powers.[8] Dewi argues that one of the most fundamental changes in political reform was through a Constitutional Amendment,[9] which resulted in changes to all branches of government, as well as additional human rights provisions. For the first time, the media and public were provided the ability and freedom to criticize the government. This was a fundamental step towards accepting and adopting some of the democratic principles that can be seen in Indonesia today.

For Sinta Dewi, the awareness of privacy in Indonesia was also influenced by two major drivers. The first involved the rise in and development of information communication technology (ICT), which was adopted, like in many other countries as a tool of, and for, governance, both in the public and private sectors. The second was the rise in the use and awareness of the Internet across the country. Throughout the Asian region, Indonesia ranks 5th behind China, Japan, India, and South Korea in the top 20 countries with the highest number of Internet users.[10] The increase in Internet usage has affected privacy, and that usage is increasing every year. Indonesia, has a population of about 250 million people and will become a large

---

[6] Greenleaf G, Dewi Rosadi, S *Indonesia's data protection Regulation 2012: A brief code with data breach notification,* Privacy Laws & Business International Report, Issue 122, (2013), pp. 24–27.

[7] Law No. 39 Year 1999, Article 67 Everyone within the territory of the Republic of Indonesia is required to comply with Indonesian legislation and Indonesian Law, including unwritten law and international law concerning human rights ratified by Indonesia. Article 68 Every citizen is required to participate in measures to defend the state in accordance with prevailing legislation. Article 69 (1) Everyone is required to respect the human rights of others, and social, national, and state morals, ethics and order. (2) Every human right gives rise to the basic obligation and responsibility to uphold the human rights of others, and it is the duty of government to respect, protect uphold and promote these rights and obligations.

[8] Dewi, S *Balancing Privacy Rights and Legal Enforcement: Indonesia Practices*, Presented at The 2011 IAITL Legal Conference Series, Lecturer at Faculty of Law, Department of Law and Technology, University of Padjadjaran Bandung, Indonesia.

[9] Ibid.

[10] Internet World Stats, Usage and Population Statistics, www.internetworldstats.com, accessed 15 November 2018.

e-commerce market.[11] In 2003, the court affirmed the growing importance of privacy. In *006/PUU-I/2003, KPKPN v KPK*, the court stated that:

> privacy rights is derogable rights therefore the State can place restriction however to "prevent the abuse of authority through wiretapping and recording, laws and regulations on wiretapping and recording procedures are needed".[12]

Even though this case concerned the application and use of powers by law enforcement agencies using technology and the impact to privacy, the case demonstrates that privacy was becoming more important as a wider societal issue within the Indonesian community.

More recently in 2006, the Indonesian Constitutional Court in *Judgment No. 012, 016/PUU-IV/2006*,[13] affirmed the importance of privacy as a fundamental right within the state. The court stated that:

> According to Article 28F Constitution, there is a guaranty regarding privacy rights that cannot be breached by whatever means since it has been considered as a human right. The right to privacy is a fundamental human right and recognized under Article 17 International Covenant on Civil and Political Rights (ICCPR) which has been ratified by Indonesia since 2005 (Law no. 12 of 2005). Personal information of the Applicant has been intercepted and recorded according to Article 12 (1) letter a without his permission whereas, according to the Law, all applicant's communication both privately or publicly shall be protected.[14]

In referring to an earlier court decision, above, in *Court Judgment No.012/ PUU-IV/2006*, it was noted that the:

> Right to privacy under Article 28F Constitution is NOT one of non-derogable rights. This right can be restricted according to Article 28J Constitution and Article 73 Law no. 39 of 1999 concerning Human Rights. Furthermore, Article 42(2) letter b Law No. 36 of 1999 (Telecommunication Law) states: "(2) For the purposes of criminal prosecution, the telecommunications services operator may record the information transmitted and/or received by the telecommunication services operator and may provide the information required on the basis of: … the request of an investigator for certain criminal offenses- in accordance with prevailing laws.[15]

Even though there was no specific reference to personal data, the court decisions highlight the growing importance of privacy within Indonesia. Some two years later, in 2008, the government of Indonesia implemented the *Information and Electronic Transaction Act*.[16] Apart from demonstrating that the government was

---

[11] Ibid.

[12] Dewi, S *Balancing Privacy Rights and Legal Enforcement: Indonesia Practices*, Presented at The 2011 IAITL Legal Conference Series, Lecturer at Faculty of Law, Department of Law and Technology, University of Padjadjaran Bandung, Indonesia.

[13] *Judgment No. 012, 016/PUU-IV/2006.*

[14] Ibid, right to privacy under Article 28F Constitution is not one of non-derogable rights. This right can be restricted according to Article 28J Constitution and Article 73 Law no. 39 of 1999 concerning Human Rights. Furthermore, Article 42(2) letter b Law No. 36 of 1999 Telecommunication Law.

[15] Ibid.

[16] Ibid.

committed to support the development of information technology, it also sought to protect the flow of information used by Internet systems and infrastructure. Importantly, this was considered a watershed moment across Indonesia, whereby, the word *privacy* was introduced more formally in the law. Article 26 states that:

(1) As otherwise stipulated by the laws and regulations, the use of any information by means of electronic media relating to someone's personal data shall be carried out with the approval from the person concerned.
(2) Every person whose right is infringed as referred to the article (1), may file a law-suit for the loss incurred based on this law.[17]

Sinta Dewi highlights that the term privacy in the elucidation of Article 26 divided privacy into three categories. These include:

(a) the right to enjoy individual life and is free from any and all kind disturbance;
(b) the right to communicate with any other persons without being spied; and
(c) the right to control the access of person's personal data.[18]

Privacy is also regulated under the *Freedom of Information Act 2008*. However, information relating to the personal information person is exempted from this Act, such as personal information relating to the history and conditions of their family members, health conditions (physical and mental) and financial conditions.[19]

The Constitutional Court decision of *Judgement No. 5/PUU-VII/2010* affirmed the recognition of the right to privacy in Indonesia in accordance with Article 28(G).[20] The Constitutional Court recognised the importance of:

> restricting communications surveillance powers to prevent misuse and ultimately the violation of the right to privacy. Law no. 39 of 1999 on Human Rights" contains in Article 32 are similar in function, whereby Article 12 of the United Declaration on Human Rights 1948 enshrining the right to privacy and contains an acknowledgement in Article 2 that human rights are "...an integral part of humans, which must be protected, respected, and upheld in the interests of promoting human dignity, prosperity, contentment, intellectual capacity and justice."[21]

Unlike most of the other jurisdictions discussed in this book that have a single Act or Regulation that provides the overall framework for data protection, Indonesia's legislative framework can be best described as being disconnected and sectorial. The introduction of the Human Rights Court Law 2000, has created a

[17] Electronic Information and Transaction 11/2008, Article 26. Note that the EIT and its amendement is only stipulated in article 26 that has not been used by people to claim due to very general and not any details articles how the mechanism to make a complaint will operate.

[18] Dewi, S *Balancing Privacy Rights and Legal Enforcement: Indonesia Practices*, Presented at The 2011 IAITL Legal Conference Series, Lecturer at Faculty of Law, Department of Law and Technology, University of Padjadjaran Bandung, Indonesia.

[19] Ibid.

[20] *Judgement No. 5/PUU-VII/2010.*

[21] Sinta Dewi Rosadi, LLB (Unpad), LLM (Washington College of Law, American University), Ph.D (Unpad), Associate Professor in Law at Faculty of Law University of Padjadjaran, Bandung, Indonesia, provided input and verified the information in this section.

Human Rights Court in the country. This new court is likely to become important in the future in order to determining what will and will not constitute privacy over the Internet. It is asserted that for Indonesia, this progressive innovation is a stand-out in the region because its neighbours of Australia and Singapore do not have a similar court. Nevertheless, so far, it has not been relevant to privacy issues.[22] Moreover, Indonesia is preparing further laws (not sectorial) to manage data and privacy issues related to personal data, to ensure they account for the developments in other countries, particularly the EU and the recent implementation of the GDPR.[23]

That said, Indonesia has been slowly developing its data protection and privacy laws, even though this has often not been evident. In 2010, the *Public Information Disclosure Act*[24] was implemented. Apart from providing the right to access information held by government and other organisations, the law aims to protect personal data and information. Article 4 notes that every individual has the right to obtain Public Information pursuant to the provisions of this law.[25] Every individual has the right to:

1. see and to know about Public Information;
2. attend public meetings that are open to the public in order to obtain Public Information;
3. get a copy of the Public Information by applying for it pursuant to this Law; and/or
4. disseminate Public Information pursuant to the regulations of the laws.[26]

Every Public Information Applicant has the right to request Public Information, and has to state the reason for such a request. Every Public Information Applicant also has the right to file a suit in court if he/she is obstructed from obtaining, or fails to obtain Public Information pursuant to the provision of this Law.

Graham Greenleaf highlights that the law is very specific and does not, in the same way as in Australia, establish a 'public interest' test.[27] Article 17, provides exceptions that protects personal data, and that includes but are not limited to:

- the history and condition of a member of the family;
- the history, condition and care, physical medical treatment, and physic of an individual;
- the financial condition, assets, income and bank account of an individual;

[22] Greenleaf G *Asian Data Privacy Laws: Trade & Human Rights Perspectives*, Oxford University Press, (2014) p. 381–382.

[23] Greenleaf G *The Legal and Business Risks of Inconsistencies and Gaps in Coverage in Asian Data Protection Laws* Session II Materials, *Asian Business Law Institute (ABLI) Data Privacy Forum*, Singapore, (2018).

[24] Public Information Disclosure Act no 14 of 2008.

[25] Ibid.

[26] Ibid.

[27] Greenleaf G *Asian Data Privacy Laws: Trade and Human Rights Perspective*, Oxford University Press, (2014) p 383.

- evaluation results of the capability, intellectuality and recommendations on the capability of an individual; and/or
- personal notes of an individual pertaining to his/her formal education and non-formal education activities.[28]

This provision implies the direct applicability as part of Indonesia's national law of every human right treaty it has ratified.[29] Furthermore, the rights and freedoms protected under the law also include, but are not limited to, freedom of speech, conscience, religion, assembly and association. In addition the law protects equality, sex, race, color, religion, ethnic, or social origin.[30] These rights and freedoms are consistent with the laws of other countries discussed in this book, such as, Australia, EU and Singapore. Today most of the data protection laws in those countries include: sex, race, colour, religion, ethnic or social origin as part of the definition of personal information and data

7.2 Definition of Personal Information

Defining *personal information* has been outlined in both Government Regulation 82/2012[31] and MCI Regulation 20/2016.[32] Article 1 of Regulation 20/2016 defines personal data as individual data that is stored, maintained and kept for correctness and protected for confidentiality.[33] Additionally, the 'particular individual data' means any correct and actual information that relates to any individual and is identifiable directly or indirectly to be changed under the laws and regulations. This definition could be either viewed restrictively or very broadly, and could include those other elements of personal data that other countries have defined as sensitive personal data.[34] Firstly, as the process of collecting personal information is not voluntary, data collectors must disclose the information only to the authorized person/institution. Secondly, there must be a mutual understanding between the owners of the personal information and the data collectors to use the information only for a specific purpose. In addition to requiring the confidentiality of personal information, transparency of information is also required to consider the public interest, so that the dissemination of personal information must align with the concept of

---

[28] Ibid.

[29] Ibid.

[30] Harkrisnowo H, Juwana H, Oppusunggu Y *Law and Justice in a Globalized,* World Editors Faculty of Law, Universitas Indonesia, Indonesia (2016).

[31] Implementation of the Electronic System and Transaction 82/2012.

[32] Electronic Information and Transaction 11/2008.

[33] MCI Regulation 20/2016, Article 1.

[34] Harkrisnowo H, Juwana H, Oppusunggu Y *Law and Justice in a Globalized,* World Editors Faculty of Law, Universitas Indonesia, Indonesia (2016).

confidentiality.[35] Despite the definition of personal information, there is unfortunately no law in place pertaining to data protection that explains what 'sensitive information' might constitute.

7.3 Public and Private

The protection of personal data applies to both the public and private sectors.[36] The importance of any data protection and privacy laws applying in both the public and private sectors, demonstrates how the countries studied are moving towards greater legal convergence and harmonization. Doing so, ensures that they not only meet their sovereign needs, but also comply with internationally agreed concepts and principles of transparency and accountability, as prescribed by the OECD. However, Indonesia, like most South East Asian countries, is not a member of the OECD. Arguably, it is important that with the ever increasing movement of large quantities of personal data between public and private sector organizations, countries that do not apply their law to both sectors, could be forced to do so in the future.

7.4 Controller or Officer

A notable difference between Indonesia's laws and those of other states examined in this book, is the absence of any legal requirement for a Data Protection Officer to be appointed – in Indonesia. The Electronic System User is defined as including any person, state administrator, business entity, and the public that uses the benefit of goods, services, facilities, or information that are made available by an Electronic System Provider (ESP).[37] The ESP has also been afforded certain obligations under the PDP Regulation. The ESP must secure certification for their electronic systems, have an internal policy and security procedure.[38] However, not having a controller or designated officer does dilute the legal framework, resulting in a lack of a designated person within an organization who is responsible for managing personal data. It also raises potential problems for the Regulator/Commission in not having a formal channel in which to engage in direct interaction and dialogue in order to resolve ongoing management issues.

---

[35] Ibid.

[36] Aditya Rahman A '*Indonesia Enacts Right to be Forgotten and Comprehensive Personal Data Regulation'*, 145 *Privacy Laws & Business International Report, (*2017) p. 1–4.

[37] Protection of Personal Data in the Electronic System Regulation, Article 1.

[38] Electronic Transactions Regulation, Article 31 and 32.

7.5 Commissioner, Agency[Regulator], Principles and Codes

Indonesia recently established the National Cyber Security Agency. This new agency combines the capabilities of the Indonesia Security Incident Response Team on Internet Infrastructure and the State Cipher Agency to better coordinate the country's efforts in cybersecurity. Article 35 of the PDP provides the relevant Minister with the power to create and appoint a supervisory agency and Sectorial Supervisors.[39] The Minister also has the power to request data and information from the Electronic System Providers (ESP). However, this is limited to the protection of personal data only. The ESP is responsible for securing information and certification of their systems. The ESP must also establish an internal policy for the organization, along with security procedures and facilities for their electronic systems. These processes form part of the overall certification of the electronic systems that are used for actions related to personal data.[40] The certification process requires the authorizing authority to undertake a series of inspections and tests to ensure the system is competent and functioning according to the law.[41] Arguably, this resembles the authorization or approval process that other countries have adopted to approve organizational policies, procedures and guidelines.

The powers afforded to the head of the new national Cyber Security Agency have yet to be fully determined. A key strategy will be to collaborate closely with enforcement agencies to track cybercrime and identify perpetrators. The principles applied by Indonesia reflect those of modern day data protection laws. Article 2, provides that personal data is to be:

- Respected as privacy;
- Confidential;
- Subject to consent;
- Relevant for the purpose of collection, processing, analysis, storage, display, published, transmitted and dissemination;
- For the Electronic System to be viable;
- Act in good faith towards the management and notification to the data subject;
- Make available internal regulations;
- Be responsible for Personal Data held by Users;
- Provide Data Subjects with easy access and correction to Personal Data; and
- Provide Personal Data that is integrated, accurate, and valid, as well as updated.[42]

The principle of making available internal regulations appears to apply, not only to government but also to the private sector (or those sectors that are obliged to

[39] Protection of Personal Data in the Electronic System Regulation, Article 35.

[40] Certification in accordance with the Electronic Transactions Regulation, which is the Electronic System Worthiness Certification.

[41] Electronic Transaction Regulation, Article 1.

[42] Protection of Personal Data in the Electronic System Regulation, Article 2.

consider and manage personal data). The promotion of this requirement also ensures that a co-regulatory approach is taken. In addition to the above, the following are also considered important principles for the implementation and management for personal data across Indonesia;

- Transparency, to allows data owners with access to their data, and to ensure that data subjects are informed of any breach.[43]
- A lawful basis for processing data, by obtaining consent from the data owner.
- Setting limitations, so that data is only processed in accordance with the needs of the Electronic Systems Operator.
- Data minimization, pertaining to the collection, storage, use and disclosure of data that is relevant.
- Retention of data, governed by the retention period set by the Supervisory Agency and or a Sectorial Regulator.[44]

All personal data must be stored in accordance with the electronic system's security procedures.[45] The ESP is required to provide an audited track record of activities related to enforcement, dispute settlement verification, inspection, and any other examinations required. Furthermore, there is a requirement to establish security measures for components of electronic systems; establish and implement mitigation procedures; ensure confidentiality in all operations; and for the systems to be functioning properly. Moreover, employees of the ESP are to secure and protect the facilities, infrastructure of the electronic systems used to collect, store, use and process personal data.[46] To strengthen the principle of transparency, the ESP is required to make contact information available and fulfil all data information requests from the Minister of Communications.[47] Nevertheless, to date, neither the EIT or MCI Regulation provide for the adoption of codes of practice.

7.6 Cross Border Transfer

All Electronic System Providers that are either part of, or service government or regional government agencies and are located within Indonesia or outside Indonesia must coordinate with the relevant Minister or agency.[48] That is, any transmission of personal data must be in the form of a personal data transmission plan that outlines the country of destination, where the personal data will be disseminated. Furthermore, as part of the plan, the organization is required to report on the transmission activity. However, the data and disaster recovery centers that collect, store, analyze and use

[43] Protection of Personal Data in the Electronic System Regulation.

[44] Ibid.

[45] Protection of Personal Data in the Electronic System Regulation, Article 18.

[46] Electronic Transactions Regulation, Article 18, 19 and 20.

[47] Protection of Personal Data in the Electronic System Regulation, 28.

[48] Protection of Personal Data in the Electronic System Regulation, Article 22.

personal data, which includes all of the public sector agencies, must have their servers stationed on the territory of Indonesia.[49] Apart from the national interest and other public interest involved in collecting personal data such as health data, arguably, Indonesia has adopted a limited and localized approach to data protection law. The Electronic Information Law provides for extraterritorial coverage, and is also applicable to any legal actions conducted outside the Indonesian jurisdiction that are potentially detrimental to Indonesia, and includes an Indonesian citizen, foreign citizen, an Indonesian legal entity and a foreign legal entity.

The Electronic Information Law also seeks to protect the national interest, and includes, but is not limited to, losses in connection with the national economy, strategic data protection and the dignity, protection, defence and sovereignty of the nation, citizens and legal entities. The current approach has little relevance to the protection of personal data – of the data subject, but rather seeks to protect state interests. This is very different to the other jurisdictions discussed throughout this book.

7.7 Right to Be Forgotten

Zeller *et al* argue that Indonesia is moving very slowly towards recognising the right to be forgotten.[50] To date, the right to be forgotten has not been considered by any court in Indonesia.[51] The EIT, introduced the concept of the 'right to be forgotten'.[52] Article 26 (3) states:

> A controller of an electronic system must delete an electronic information and/or electronic document under his control which is no longer relevant if that deletion is requested by a related person through a decision of a court.[53]

In October 2016, the Parliament of Indonesia passed the revised EIT[54] law that now enables data subjects to request that their personal data be deleted. Article 26 section 3 requires that each Operator Electronic System to delete irrelevant information and records of an individual's personal data and information, under its control,

[49] Protection of Personal Data in the Electronic System Regulation, Article 7.

[50] Zeller, B., Trakman, L., Walters, R., Dewl Rosadi, S *The Right to be Forgotten – the European Union and Asia Pacific Experience (Australia, Indonesia and Singapore),* European Human Rights Law Review (under review).

[51] Hak untuk dilupakan Direvisi UU ITE Masih Belum Berlaku, https://tekno.kompas.com/read/2016/11/29/09250047/, Accesses 10 July, 2018.

[52] Protection of Personal Data in the Electronic System Regulation, Article 26.

[53] Verified by Sinta Dewl Rosadi, LLB (Unpad), LLM (Washington College of Law, American University), Ph.D (Unpad), Associate Professor in Law at Faculty of Law University of Padjadjaran, Bandung, Indonesia. Indonesian interpretation – "Setiap Penyelenggara Sistem Elektronik wajib menghapus Informasi Elektronik dan/atau Dokumen Elektronik yang tidak relevan yang berada di bawah kendalinya atas permintaan Orang yang bersangkutan berdasarkan penetapan pengadilan.

[54] Electronic Information and Transactions Law No. 19/2016.

but only upon the direction and request that has been issued through a court order.[55] However, it is likely that the application of the principle underlying the right to be forgotten in the revised EIT Act will pose practical problems for its effective implementation. This is because of the gaps in the law, whereby, the EIT does not clearly define personal data. These gaps will arguably cause difficulties in determining what exactly constitutes personal data. Without properly defining what personal data is, there is no baseline or benchmark that clarifies the boundaries establishing when personal data begins and concludes. This is an area of the law that Indonesia will need to consider in light of the practices of Australia, the EU and Singapore who have defined personal data or personal information.[56] This comparison is complicated by the fact that definitions of personal data and information differ.[57] It is outside the scope of this chapter to compare competing definitions of personal data (see Chap. 11).

7.8 Consent

Consent in Indonesia is similar to other jurisdictions discussed throughout this book. It requires prior consent of the person to whom the personal data applies. The process for obtaining consent by an electronic system provider is through a Standard form in Bahasa Indonesia, and agreement sought by the personal data owner.[58] This consists of the type, purpose and details of the personal data owner. Article 9 (2) strengthens the position of data owners, as they, upon providing consent, can also request that their personal data be treated as confidential. Furthermore, where consent has not been formally provided for the disclosure of personal data, any person who collects this type of data, including an Electronic Systems Provider, must maintain confidentiality.

A minor in Indonesia is considered a person under the age of 21 years. There are significant variances between jurisdictions and how they determine who is a minor and who is not. Nonetheless, a minor must have approval from one or both parents.[59] The Consent Standard Form is considered under the Indonesian Civil Code to be an agreement or contract.[60] Under Indonesian law, the Civil Code prevails over all other laws, when it comes to minors.

Unique to Indonesian law is the display and publication of personal data in accordance with Article 21, which requires that consent is obtained before any

[55] Zeller, B., Trakman, L., Walters, R., Dewl Rosadi, S *The Right to be Forgotten – the European Union and Asia Pacific Experience (Australia, Indonesia and Singapore)*, European Human Rights Law Review (under review).

[56] Ibid.

[57] Ibid.

[58] Protection of Personal Data in the Electronic System Regulation, Article 6 and 9.

[59] Ibid, Article 37.

[60] Indonesian Civil Code, Article 330.

personal data is displayed or published. This also includes any personal data that is held within an Electronic System that is either displayed, published, transmitted, disseminated, or, accessed by different Electronic System Providers and Users.

Consent is required from the data owner, for that person's personal data to be manipulated for use when that personal data will be displayed or published.[61] The use and manipulation or the changing of personal data can only be undertaken for the purpose for which that data has been collected, processed and analyzed. What this means is that personal data collected for health purposes cannot be manipulated and used, without the consent of the person to whom the data pertains commercial purposes related to consumer behavior.

Rather than the MCI specify consent for the processing of personal data, the EIT states that prior consent from the data subject must be obtained. The EIT does not distinguish between sensitive or general personal data.[62] The data subject must be informed of the purpose to which the data will be processed, and consent can only apply to the scope that the actual processing will entail. In other words, the processing of personal data may be limited to biometrics, so that only that data can be processed under such a consent. Other data, such as health records could not be used, where the data subject has not provided consent. Furthermore, consent must be in writing, whether electronically or in hard copy.

7.9 Collection

The collection, processing, retention, display, publication, dissemination and destruction of personal data can only be achieved through an 'Electronic System' that has acquired the relevant certification.[63] In addition, the collection of personal data is restricted by the Supervisory Agency and sectorial Supervisor as they deem to be relevant. On the one side, this provides a broad approach and quite some discretion to the agency, as to what is deemed to be relevant personal data. On the other side, it can to some degree contradict the first point which appears to apply a level of relevancy to personal data, before it is certified. The Electronic Systems Provider must obtain consent from the individual to which the personal data applies. This includes, but is not limited to, any data collected which must be verified with the person to whom the data was supplied (the data subject).[64]

The verification processes not only ensure the accuracy of the data; it also validates the data. Furthermore, the system that is used to collect and store that data must have two key components. The first is that the system has to be inter-operable and secondly, it must have compatible performance. Inter-operability constitutes the

[61] Protection of Personal Data in the Electronic System Regulation, Article 24.

[62] Electronic Information and Transactions on the Amendment to Law No. 11 of 2008, Articles 27 & 28.

[63] Protection of Personal Data in the Electronic System Regulation, Article 4.

[64] Protection of Personal Data in the Electronic System Regulation, Article 10.

ability of Electronic Systems that have different performance capabilities to operate together. Compatibility requires one Electronic System to be compatible with another Electronic System.[65]

Article 14 requires that all personal data that is processed and analyzed must be Personal Data that have been verified for accuracy. Accuracy and verification also applies to personal data that is stored in an Electronic System.[66]

7.10 Retention [Storage]

The retention or storage of data must be encrypted and kept f for a period of at least 5 years, but only where there are no laws that require a shorter or longer period of retention.[67] This requirement is something that other jurisdictions do not prescribe. The personal data must also be stored in accordance with the procedures of the Electronic System security.[68]

7.11 Breach

There is no requirement for any authority to be informed of a breach. A dispute can be dealt with through the complaints management process that requires the individual to make a complaint to the Director General of Application and Informatics. However, it will be up to the director General to determine whether a dispute resolution panel needs to be formed to address the issue. Thus, there appears to be an umpire in place to resolve complaints promptly. The umpire can also categorize the complaint, and dismiss low level complaints that may not have any bearing on the overall legal framework or impinge a person's right.

7.12 Enforcement

Electronic System Providers must surrender Personal Data in the Electronic System or Personal Data that is generated by an Electronic System when requested by the relevant law enforcement agency. [69] Sanctions for breaches of data privacy are found under the relevant legislation and are, essentially, fines. Imprisonment may be imposed in severe instances such as in the event of intentional infringement. The

[65] Protection of Personal Data in the Electronic System Regulation, Article 11.

[66] Protection of Personal Data in the Electronic System Regulation, Article 15.

[67] Ibid.

[68] Protection of Personal Data in the Electronic System Regulation, Article 18.

[69] Protection of Personal Data in the Electronic System Regulation, Article 23.

EIT Law provides a range of criminal penalties that assist in the management of cybercrime pertaining to elements of personal and data generally. The EIT penalties range from Rp. 600,000,000 fine to Rp. 800,000,000 and/or 6–8 years' imprisonment for unlawful access.[70]

Furthermore, failure to comply with Reg. 82 is subject to administrative sanctions (which do not eliminate any civil and criminal liability) and can result in (1) a written warning; (2) administrative fines; (3) temporary dismissal; or (4) expulsion from the list of registrations (as required under the regulation). However, to date there is no complaints mechanism established by law of the newly formed Cyber Security Agency.

7.13 Supporting Laws & Proposed New Data Protection Laws

The supporting legal framework for data protection law continues to be fragmented across Indonesia. Currently, there are other Regulations that have a direct or indirect impact on data protection across Indonesia, including;

- MCI Regulation No. 9 of 2017, on content Providing Services operation on Cellular Mobile Network;
- MCI Regulation No. 4 of 2016, on the information Security Management System;
- Financial Services Regulation No. 77/POJK.01/2016 on Information technology – Based Lending Services;
- MCI Regulation No. 36/2014 on the Registration Procedure of the Electronic System Operator;
- Government Regulation No. 46 of 2014 on Health Information System;
- Decree of Head of SKK Migis PTK-008 on the Information and Telecommunications Technology Management over Production Sharing contract Contractors;
- Government Regulation No. 96 of 2012 on the Implementation of Public Services; and
- Bank of Indonesia's Regulation No. 9/15/PBI/2007 on the Implementation of Rick Management in the Utilization of Information Technology by the bank.

The MCI Regulation 20/2016 on Protection of Personal Data in Electronic System came into effect and provided the basis for the Electronic Information and Transaction 11/2008 and the Implementation of the Electronic System and

[70] Electronic Information and Transactions on the Amendment to Law No. 11 of 2008. Additional penalties include – Rp. 800,000,000 fine and/ or 10 years imprisonment for interception/wiretapping of transmission; or Rp. 2,000,000,000 to Rp. 5,000,000,000 and/or 8–10 years' imprisonment for alteration, addition, reduction, transmission, tampering, deletion, moving, hiding Electronic Information and/or Electronic Records.

Transaction Laws 82/2012. The MCI is supported and underpinned by the following laws:

- The Electronic System Provider (Operator) is the person, state administrator, business entity and public entity that provides, manages, operates, and electronic system for their own interest or the interest of others.[71]
- Law Number 39 of 2008 concerning The State Ministries (State Gazette of the Republic of Indonesia Number 166 of 2008, Supplement to the State Gazette of the Republic of Indonesia Number 4916);
- Regulation of the Government Number 82 of 2012 concerning Electronic System and Transactions (State Gazette of the Republic of Indonesia Number 189 of 2012, Supplement to the State Gazette of the Republic of Indonesia Number 5348);
- Regulation of the President Number 7 of 2015 concerning Organization of the State Ministries (State Gazette of the Republic of Indonesia Number 8 of 2015);
- Regulation of the President Number 54 of 2015 concerning The Ministry of Communications and Informatics (State Gazette of the Republic of Indonesia Number 96 of 2015); and
- Regulation of the Minister of Communications and Informatics Number 1 of 2016 concerning Organization and Working System of the Ministry of Communications and Informatics (State Gazette of the Republic of Indonesia Number 103 of 2016).

The above laws only seek to support the legal framework and confirm the current sectorial approach taken by Indonesia in relation to data protection and privacy. It is out of scope to discuss all the above laws in any detail.

7.13.1 Proposed New Data Protection Law

Indonesia is travelling the same road as many other countries and are in the midst of developing specific Data Protection Law.[72] Arguably the proposed laws reflect the laws and introduce key concepts and principles of data protection that have been established by the EU. They are looking to the international community to develop a framework that meets their own sovereign needs, but also, considers extensively the international direction that data protection laws have taken.

Mark Innes points out that the proposed draft laws will provide a broad definition of personal data to likely include: any data about a person that can identify auto-

[71] Electronic Information Transaction 11/2008, Article 1.6a, Protection of Personal Data Regulation Article 1.6.

[72] Innes M, *Indonesia: Government Pushes Draft Data Protection Law* Global Compliance News, (2018) https://globalcompliancenews.com/indonesia-draft-data-protection-law-20180518/, accessed 12 August 2018.

matically the person, and any data about a person, when combined with other information directly or indirectly obtained through electronic and/or non-electronic systems, can identify a person.[73]

7.13.1.1 Defining Personal Data

The proposal aims to define personal data that can identify automatically the person, any data about a person, when combined with other information directly or indirectly obtained through electronic and/or non-electronic systems, that can identify a person. If adopted, the proposal seeks to have a broad definition and could go some way to allowing Indonesians to have all forms of Internet activity constituted as relating to personal data. This proposal would exceed the current definition of personal data in many other jurisdictions. Moreover, Indonesia aims to provide for both general and specific data.[74] Apart from the standard data that can identify an individual, such as name and date of birth, general data at this stage is likely to constitute personal data that can be obtained from the public domain or that has been disclosed under an identity document (identity card number, photo, telephone number, email address). This proposal has many similarities to Singapore's use and application of personal identity cards. While, at this stage, there appears to be no reference to sensitive data in Indonesian law, specific data is likely to include this area of data within the definition of personal data. Furthermore, at this stage in the development cycle, personal data will include specific data that requires special protection, including based on religion/beliefs, health, physical and mental conditions, biometrics, genetics, sex life, political views, criminal records, child data, and personal financial information, as is already noted above.

7.13.1.2 Controller and Processor

Mark Innes believes that the proposed law will differentiate between a party that collects Personal Data (and obtains consent from the Data Owner) and manages the data processing, being a Personal Data Controller, and a party that processes the Personal Data on behalf of a Personal Data Controller, being a Personal Data Processor. Similar to other jurisdictions, particularly the EU, the proposed laws will aim to place a greater focus and responsibility on Personal Data Controllers as the parties who should obtain consent from the Personal Data Owner. Almost half of the relevant provisions under the Draft Law relate to Personal Data Controllers. Processing is likely to include, but not limited to, acquiring and collecting, processing and analyzing, storing and displaying, fixing and renewing, announcing and delivering, distributing and disclosing, and deleting and/or destroying – personal data. Deleting and destroying personal data, depending on the framework that

[73] Ibid.

[74] Ibid.

supports these concepts, may provide all the hallmarks of the right to be forgotten (to erasure). However, when, how and to what extent this will apply, has yet to be made available to the public.

Notwithstanding the above, before the right to be forgotten can be fully adopted in Indonesia, further work will be required. Consideration will need to be given to whether information that is requested to be deleted is personal data in light of the definition of personal data; and personal data that has entered the public domain or has become a matter of public interest that an individual can request be deleted.[75] Further consideration also needs to be given to how the personal data of a public figure, such as government officers or public figures, cannot be requested to be deleted.[76] Preliminary work in Indonesia suggests that it is looking to the EU for guidance in the development of its data protection laws, while taking account of the significant role that religious background plays in the development of rights-based laws. The right to be forgotten across Indonesia has a long way to go for it to be entrenched and fully accepted, not only in law, but also by the government and the broader community.[77] This will require a shift to accepting that the legal concept ought to play a more important role in Indonesian society as its citizens continue to embrace and use modern day technology – and given that their personal data and information is becoming increasingly exposed to privacy issues.[78]

7.13.1.3 Consent

As highlighted throughout the book, there are many concepts that have begun to play an ever increasing role to the overall management and regulation of personal data. None more so than the concept of consent. However, as the proposal currently stands in Indonesia, consent is limited in its scope. The proposal will require consent to be obtained in writing for any personal data defined by the law, although the specifics and detail on how consent is to apply remains unclear. Nonetheless, Innes points out that consent should only be given after a Personal Data Controller provides the following information:

- the legality of the Personal Data management – what this means is not clear at the moment; or
- the purposes for which the Personal Data will be managed; or
- the types of Personal Data that will be managed; or

---

[75] Ibid.

[76] Sinta Dewi Rosadi, Quo Vadis Perlindungan Data Pribadi Data pribadi dalam Revisi UU ITE http://www.hukumonline.com/berita/baca/, accessed 12 July, 2018.

[77] Zeller, B., Trakman, L., Walters, R., Dewl Rosadi, S *The Right to be Forgotten – the European Union and Asia Pacific Experience (Australia, Indonesia and Singapore),* European Human Rights Law Review (under review).

[78] Ibid.

- the retention period of the Personal Data; or
- details on the information that will be collected; or
- the period of the Personal Data management by and the deletion policy of the Personal Data Controller; or
- the right of the Data Owner to revise and/or retract any consent.[79]

However, there are likely to be exceptions, whereby specific personal data may be managed by a controller or processor, without the written consent of the data subject. These exceptions are proposed to only operate when and for:

- the Data Owner's data security protection,
- medical purposes by doctors, other medical staff,
- law enforcement purposes, and
- as required under laws and regulations.[80]

In addition, the proposal is framed in a manner that allows specific personal data to be processed without written consent when that data has come into the public domain due to the actions of the data subject. It remains to be seen how this provision would operate and how it is to be applied, because there are many situations in which a person's personal data and information is available on social media, and that the individual has not consented to having the information available.

7.13.1.4 Data Transfer

Innes highlights that Personal Data Controllers under the proposed laws are likely to be required to obtain consent from Data Owners in order to transfer Personal Data to a third party within Indonesia; otherwise that third party cannot use the Personal Data, except for the intended use that has been approved by the Data Owner.[81] For this to be achieved, the controller will need to ensure that the receiving country's laws are subject to similar principles and standards as Indonesian laws. However, there is likely to be the need for a contract to be established, or for an international bilateral agreement to be put in place.

7.13.1.5 Commission

It is contented that the Proposal will introduce a Commission. Any Commission established will have responsibility for ensuring that Personal Data Controllers comply with the provisions of the law and encourage individuals and entities to establish a risk management framework that will strengthen the protection of personal data and enhance the protection of privacy over the Internet. The proposed

[79] Ibid.

[80] Ibid.

[81] Ibid.

Commission, if established, will bring Indonesia in line with neighboring countries, such as Singapore and Australia. The proposal would also see the Commission have the power and responsibility for monitoring compliance, receiving complaints, facilitating dispute resolution, and providing guidance to Personal Data Owners in the event of any breach of the law.

7.13.1.6 Enforcement & Breach Notification

The enforcement and particular notification of breaches of data protection laws, is being accepted by many other states. It enhances the transparency and accountability of individuals and entities collecting and using personal data. It also helps to promote the idea of a reasonable standard of data security and to encourages the revision of applicable rules of application.[82] Innes notes that the draft law proposes that Personal Data Controllers also have an obligation to notify Personal Data Owners if their Personal Data has been inadvertently disclosed.[83] The Draft Law does not state when the notice should be made (however Regulation 20 requires that a notice be made within 14 days after the data breach is known). Currently this obligation does not extend to Personal Data Processors, even though a breach of Personal Data is usually by the Personal Data Processor.[84] Innes argues that it remains to be seen whether the Personal Data Controller would be held liable for not notifying the Personal Data Owner about a breach.

7.13.1.7 Deletion – Destroying Personal Data

The draft law proposes that personal data will be able to be deleted or destroyed, when applicable. However, the proposal is framed in a manner that distinguishes between Personal Data deletion and Personal Data destruction. Deletion is applicable to Personal Data that is processed electronically, while destruction is applicable to Personal Data that is not processed electronically.[85] In other words, a controller is likely to destroy personal data:

(a) that no longer has usage value,
(b) that has an expired retention period,
(c) if there are indications of a leak in the Personal Data management system caused by that particular Personal Data,

---

[82] Schwartz, P., Janger, E *Notification of Data Security Breaches*, Michigan Law Review, Vol. 105:913 (2017).

[83] Innes M, *Indonesia: Government Pushes Draft Data Protection Law* Global Compliance News, (2018) https://globalcompliancenews.com/indonesia-draft-data-protection-law-20180518/, accessed 12 August 2018.

[84] Ibid.

[85] Ibid.

(d) if there is a written request from the Personal Data Owner to destroy it (no court order is required under the Draft Law but a Personal Data Owner may need to seek a court order to request a Personal Data deletion given requirements under the Electronic Information and Transaction Law and Regulation 20), or
(e) that is not related to any dispute resolution proceeding.[86]

Furthermore, a controller is likely to have to delete personal data when:

(a) that data is no longer needed to achieve the purpose of the Personal Data management,
(b) if the Personal Data Owner has revoked his consent related to the management of the Personal Data, through a written request to the Personal Data Controller, or
(c) if the Personal Data Controller uses the Personal Data for purposes that are not in line with the consent or the Draft Law.[87]

This proposal, if realized, will enhance and strengthen the control over personal data by data subjects. The ability to have one's personal data destroyed or deleted also reinforces the development and acceptance of privacy by Indonesia, particularly the right to be forgotten.

In summary, the proposed new laws have similarities to the laws in Singapore, the EU and Australia. The proposal will provide citizens with a level of protection from direct marketing. It is also proposed to include provisions that will limit the use and installation of visual data processing devices in a public facility that could threaten an individual's privacy. What these devices might constitute has yet to be clarified.

7.14 Conclusion

The development, awareness and understanding of privacy in Indonesia is beginning to take hold following the national elections in the late 1990s. Since then, there has been wider acceptance of media and other rights and freedoms within the country, as it has, in part, embraced democracy. As highlighted by Sinta Dewi, privacy awareness in Indonesia has been driven by several factors, such as the changing political architecture and social change in which Indonesian people are becoming more aware of their rights. Today, privacy, Internet technology and personal data being collected and used by individuals and entities across Indonesia is arguably, better understood.

The first comprehensive data protection and privacy laws in Indonesia were established in 2016. Privacy is only regulated by the EIT and Indonesia is still drafting the comprehensive Personal Data Bill. While Indonesian courts have not made

[86] Ibid.

[87] Ibid.

a decision on the right to be forgotten, it appears at this early stage of the draft PDP, that the right is likely to be afforded by law. Indonesia, like many other countries, has looked to the EU for guidance on the development of the PDP and other relevant data protection laws. However, as Andin Rahman highlights, one of the major issues in applying the PDP, as modified, will be the ability for it to be adequately enforced.[88]

The PDP restricts the ability of an individual to submit formal complaints against an ESP, to only in relation to personal data leaks. This restrictive approach does not appear to hold the ESP to full account for any other non-compliance. Even so, a person can commence a lawsuit against an ESP where the use of personal data has been unlawful. The PDP has reconciled the long standing issue for Indonesia and Indonesians by clarifying the meaning of personal data, since it was initially defined by the Citizens Administration Law 2006. However, compared to other countries, that definition is very broad. The current PDP does limit the regulation of personal data to only that data, which is in electronic form. The newly formed Cyber Security Agency will strengthen Indonesia's approach to managing cyber security and data. Finally, Indonesia has provided a very broad approach to compensation for the loss or humiliation arising from the misuse of personal data, although there is very little specific guidance on how such compensation is determined.[89]

Should the proposed draft laws be fully implemented, Indonesia will have come a long way in bringing their data protection [legal] framework into the twenty-first century. The proposal will resemble the EU legal framework, and strengthen the rights to citizens in protecting their personal data and information. Doing so, will also demonstrate how Indonesia is adopting most, if not all, of the concepts and principles espoused by the OECD, such as accountability and transparency. However, it is argued that the longer Indonesia takes to move towards fully implementing specific data protection laws, it may miss the opportunity to form strategic economic partnerships in the new digital economy. Finally, the current day laws of Indonesia do not reflect the models of Australia, the EU or Singapore. However, this will change should the proposed new laws should they be realized. At the time of writing this book, the indications are that the proposed PDP will not be settled until sometime in late 2019, at the very earliest.

---

[88] Aditya Rahman A, *Indonesia enacts Personal Data Regulation,* Privacy Laws Business, (2017).

[89] Article 26 (2) of the Law Concerning Electronic Information and Transactions 11 2008 states that unless provided otherwise by Laws and Regulations, use of any information through electronic media that involves personal data of a Person must be made with the consent of the Person concerned. Any Person whose rights are infringed as intended by section (1) may lodge a claim for damages incurred under this Law.

References

Aditya Rahman A, *Indonesia Enacts Personal Data Regulation, Privacy Laws and Business*, Data Protection and Privacy Information Worldwide, Iss 145, (2017)

Ahmad, A *Law and Development in Changing Indonesia*, IDE Asian Law Series No. 8 Law and Development in Asian Countries, Institute of Developing Economies (IDE), JETRO, (2001)

Dewi, S (2011) *Balancing Privacy Rights and Legal Enforcement: Indonesia Practices*, Presented at The 2011 IAITL Legal Conference Series, Lecturer at Faculty of Law, Department of Law and Technology, University of Padjadjaran Bandung, Indonesia

Greenleaf G (2014) *Asian Data Privacy Laws: Trade & Human Rights Perspectives*, Oxford University Press, p. 381 – 382

Greenleaf G (2018) *The Legal and Business Risks of Inconsistencies and Gaps in Coverage in Asian Data Protection Laws* Session II Materials, *Asian Business Law Institute (ABLI) Data Privacy Forum*, Singapore

Greenleaf G, Dewi Rosadi, S (2013) *Indonesia's data protection Regulation 2012: A brief code with data breach notification,* Privacy Laws & Business International Report, Issue 122, (2013), pp. 24–27.

Harkrisnowo, H, Juwana H, Oppusunggu Y (2016) *Law and Justice in a Globalized,* World Editors Faculty of Law, Universitas Indonesia, Indonesia

SchwartZ, P., Janger, E (2017) *Notification of Data Security Breaches*, Michigan Law Review, Vol. 105:913

Zeller, B., Trakman, L., Walters, R., Dewi Rosadi, S (2018) *The Right to be Forgotten – the European Union and Asia Pacific Experience (Australia, Indonesia and Singapore),* European Human Rights Law Review (under review).

Chapter 8
Malaysia

Abstract Today's Internet technology provides access to information and allows people to do things, no matter where they are located in the world that, would not have been possible even three decades ago. This is no different for the people of Malaysia. This chapter provides and overview of the Malaysian data protection and privacy legal framework. Malaysia makes for an interesting comparison with the other countries in this book because it has been heavily influenced by Western practices, yet it is a country that is predominantly Islamic. Malaysia's legal system is common law, inherited and influenced by the United Kingdom. It is also heavily influenced by Islamic law. Malaysia have had to find a balance between its religious and cultural distinctiveness and those of other states, in regulating data protection and privacy. This sets Malaysia apart from its neighbour Singapore, and Australia along with the EU. Yet, as highlighted in Chap. 1, Islamic law treats the concept of privacy as a very important part of society. Nonetheless, the road to the current day data protection and privacy laws in Malaysia has not been smooth sailing. Similar to Indonesia and Singapore, Malaysia is grappling with the idea of what the right to be forgotten might look like.

In 2000, the Malaysian government introduced a draft Personal Data Protection Bill based on European standards of data protection (Madieha Azmi I *Personal Data Protection Law: The Malaysian Experience*, 16 Info. & Comm. Tech. L. 125 (2007)). This Bill never made it to Parliament due to heavy opposition from the communication and multimedia industry (Ibid). In a surprise move, the Malaysian government redrafted the Bill but with some relaxation of the data protection provisions (Ibid). Three years later, Malaysia established the Personal Data Protection Act (PDPA) in 2010. The Act makes a distinction between personal data and sensitive personal data, such as medical history, religious beliefs, political opinions and the commission or alleged commission of any offence. It also provides for the processing of personal data, which requires explicit consent. The PDPA regulates the processing of personal data by the user in a commercial transaction, including throughout the Malaysian territory. Consistent with other Chapters, only key definitions, concepts and principles will be examined.

© Springer Nature Singapore Pte Ltd. 2019

R. Walters et al., *Data Protection Law*,
https://doi.org/10.1007/978-981-13-8110-2_8

8.1 Introduction

Malaysia like many other countries in the South East Asian region have had a complex history. Malaysia have been heavily influenced by Western, Central Asian and Asian practices, culture and values. The Malaysian legal system combines the common law with Islamic law. Thus, it can be argued that in relation to data protection and privacy Malaysia is faced with not only cultural and religious tensions, but also tensions between legal families. However, it must be noted that the English law has been widely accepted by Malaysia. This was reaffirmed in *Yong Joo Lin v Fung Poi Fong Terrell Ag*[1] where the court held:

> that principles of English law were for many years accepted in the Federated Malay States where no other provision was made by statute, and that the qualification contained in s 3(1) was in fact the statutory recognition of the judicial practice of resorting to English law to fill the lacunae in the local law.[2]

In adopting the common law of the United Kingdom, Malaysia's development of privacy can be best described as being similar to other common law countries of Singapore and Australia. The evolution of the law of privacy in Malaysia, to date, has largely focused on the principles and concepts of defamation,[3] nuisance,[4] trespass[5] and breach of confidence.[6] These principles are widely applicable to protect aspects of privacy. In the case of *Public Prosecutor v Lee Sin Long*[7] the court noted that a warrant was required before a search could take place of residential premises.[8] Even though this is viewed as a procedural step in allowing for enforcement activities to be undertaken, in relation to privacy, the court highlighted that the privacy of a person in his home must be respected, and cannot be disturbed unless first shown proper authority that reasonable cause for interference is warranted.[9]

It must be noted that the concept of privacy was not fully accepted from this affirmation of privacy and remained very much in its infancy across Malaysia. During this early period, the concept of privacy largely related to a person's property

---

[1] [1941] MLJ Rep 54.

[2] Ibid.

[3] *JB Jeyaretnam v Goh Chok Tong* [1985] 1 MLJ 334. In 2011, the case of *Shaharuddin bin Mohammad v Malayan Banking Bhd* [2011] 7 MLJ 589 The court also discussed the same issue by stating - to the effect that the publication of a notice in that case was pursuant to a court order and therefore any defamatory imputation in the notice was justified and gave rise to no liability in defamation, at [13].

[4] *Ong Koh Hou v Perbadanan Bandar & Anor* [2009] 8 MLJ 616 (HC).

[5] *Sin Heap Lee-Marubeni Sdn Bhd v Yip Shou Shan* [2005] 1 MLJ 515, [13]. The court held that the trespass committed by the appellant was a continuing act of trespass from the time when the respondent was not in the possession of the land to the time he came into possession of the land and in fact continuing after he came into possession of the said land.

[6] *Worldwide Rota Dies Sdn Bhd v Ronald Ong Cheow Joon* [2010] 8 MLJ 297.

[7] [1949] 1 MLJ 51.

[8] Ibid.

[9] Ibid.

and self. However, the protection of self, did not include or require the need for an individual to protect their personal data or information from technology.

Later in 1987, in the case of *Re Kah Wai Video (Ipoh) Sdn Bhd*[10] the court highlighted how search and seizure undertaken by law enforcement agencies, violated Article 13 of the Malaysian Constitution. Article 13 is considered to provide that no person shall be deprived of property in accordance with the law.[11] Thus, it is argued that an implied right to privacy had, by this case received constitutional validity.

Notwithstanding the above, the evolution of privacy in Malaysia did not begin to shift and develop into a more ridged legal concept until the 2000's. In *Ultra Dimension Sdn. Bhd v. Kook We Kuan,*[12] the court argued that the:

> right to privacy is not recognized under Malaysian law. The case involved appellants who took a photograph of a group of pupils at an open area outside their kindergarten and published it in two local newspapers with the caption "Bonus Link Share Your Points". The issue was whether or not the picture amounted to invasion of privacy and breach of confidence.[13]

The court went onto say that:

> the supplying of the photograph for an advertisement had invaded the respondent's privacy. However, the publication of the photograph in the advertisement did not give the respondent a cause of action as the facts of the case did not fall within the boundaries of any recognized and existing tort. Rather, the case focused on whether the issue fell under Malaysian copyright law.[14]

Two years later in 2006, the position in *Ultra* was reinforced by the Malaysian High Court in *Dr Bernadine Malini Martin v MPH Magazines Sdn Bhd & Ors*[15], whereby it was confirmed that an invasion of privacy was not an actionable wrong. In this case, the plaintiff who was a government medical officer who brought an action against the defendants for the tort of defamation.[16] Martin alleged that the publication of her photograph in a bridal gown together with a write-up without her consent was defamatory, and alleged that the advertisement portrayed her as a woman of loose morals and an unsuccessful doctor who resorted to part-time modeling to supplement her income. The court acknowledged that the:

> unauthorized publication of the plaintiff's photograph in a magazine was intended for public circulation and that it was unethical and morally wrong of the defendants to have

[10] [1987] 2 MLJ 459.

[11] Federal Constitution of Malaysia, Article 13, http://www.agc.gov.my/agcportal/uploads/files/Publications/FC/Federal%20Consti%20(BI%20text).pdf, accessed 20 December 2018.

[12] *Ultra Dimension Sdn. Bhd v. Kook We Kuan* [2004] 5 CLJ 285. In this case, a photograph of a group of kindergarten pupils had been published in an advertisement in several local newspapers. In a claim that the supply of the photograph to the newspaper amounted to a breach of privacy, the learned judge had to explore whether invasion of privacy is a recognized tort of action under Malaysian law.

[13] Ibid.

[14] Ibid.

[15] [2006] 2 CLJ 1117.

[16] Ibid.

> published it without her consent for the purpose of their commercial promotion. It was, he found, an unwarranted invasion of the plaintiff's privacy.[17]

Despite this affirmation, the court had not approved damages based on the privacy infringement, since it is not recognized as an actionable wrong under Malaysian law. Nonetheless, the concept of privacy gained further traction in *Maslinda Ishak v. Mohd Tahir Osman & Ors*[18] whereby the court confirmed that the right to privacy exists in Malaysia. The case involved a guest relations officer who was photographed easing herself in a truck by a volunteer reserve corps member.[19] Subsequently, the guest relations officer proceeded to the court and was granted damages for the wrongdoing.[20] There was no consent or approval for the photograph.

The principles of breach of confidence has become such an important principle that in the eyes of the Malaysian courts it appears they have taken the stance that trade secrets include confidential information, which needs to be protected and managed effectively. In *Worldwide Rota Dies Sdn Bhd v Ronald Ong Cheow Joon*[21] the court found that:

(a) the information which the plaintiff sought to protect was of a confidential nature;
(b) that the information in question was communicated in circumstances importing an obligation of confidence; and
(c) that there was an unauthorized use of that information to the detriment of the party communicating it.[22]

Even though the court did not mention personal data or information, it could be argued that confidential information in this case could also have included personal data, as the facts of the case concerned confidential information within a contract. However, an action for breach of confidence in Malaysia has been extended to actions in tort, but not to privacy.[23]

It was not until 2011, whereby the Malaysian High Court in *Lew Cher Phow @ Lew Cha Paw & Ors v Pua Yong Yong & Anor*[24] held that the right to privacy of the plaintiffs should be protected. The plaintiffs and defendants were neighbours whose houses were separated by zinc sheets only.[25] Disagreement arose between them and their relationship deteriorated. One of the plaintiffs was charged and convicted of

---

[17] Ibid.

[18] [2009] 6 CLJ 653.

[19] Ibid.

[20] Bakar Munir, *A Malaysia's Data Protection Law* in Simon Chesterman Data Protection Law in Singapore, Privacy and sovereignty in and interconnected World, Academic Publishing (2018) Chap. 13.

[21] *Worldwide Rota Dies Sdn Bhd v Ronald Ong Cheow Joon* [2010] 8 MLJ 297.

[22] Ibid.

[23] *Coco v A N Clark (Engineers) Ltd* [1969] RPC 41.

[24] 2011 MLJU 1195.

[25] Ibid.

criminal intimidation against the defendants.[26] Sometime later, the defendants installed five CCTV cameras in their house. Camera (number 3) was located at the front porch and pointed directly at the plaintiffs' house, capturing and monitoring images of their front courtyard.[27] The plaintiffs brought an action claiming that their right to privacy had been infringed.

The court reasoned that:

> privacy in the present case was related to a person's right to respect for his private and family life and his home. The fact that there is no specific provision in the Constitution guaranteeing the right to privacy does not preclude a court from holding that such a right exists because privacy is recognized as a fundamental human right internationally, given recognition by international covenants, treaties and regional human rights treaties. The defendants' continuing act of putting the plaintiffs under surveillance represented a failure of respect for the plaintiff's dignity and autonomy. It constituted an intrusive surveillance on the plaintiffs' private and family life and home. The defendants' fear for their safety and security did not justify their actions and did not override the plaintiffs' right to privacy.[28]

However, it must be noted that the judgments made by the Malaysian High Court are subject to review by Higher courts. Therefore, the recognition of privacy now opens the door for the Higher courts to further consider privacy more broadly across Malaysian society, and in the context of personal data. Coincidentally, the position taken by the Malaysian High Court followed the implementation of the *Personal Data Protection Act 2010*. Moreover, and as with the other jurisdictions discussed in this book, the law of privacy in Malaysia as it pertains to personal data defined by the law, has not been settled.

The implementation of the *Personal Data Protection Act 2010 (PDPA)* has arguably strengthened the position of privacy across Malaysia - in personal data, over the Internet. The PDPA is supported by the Personal Data Protection Regulations 2013 and the 2015 Personal data Protection Standards. Moreover, the PDPA has established five general rights for data subjects, these include (1) the right to be informed; (2) the right to access; (3) the right to withdraw consent; (4) the right to prevent the processing of personal data that is likely to cause harm, damage or distress and (5) the right to prevent the processing of personal data for direct marketing purposes.[29] The constitutional protection of privacy[30] along with the PDPA, has emerged as providing a regulatory framework for privacy and personal data protection. The Penal Code also plays a role in understanding and recognizing the importance of privacy in Malaysia. Section 509 of the Penal Code states that whoever intends to insult the modesty of any person, utters any word, makes any sound or

---

[26] Ibid.

[27] Ibid.

[28] Ibid.

[29] Personal Data Protection act 2010, Division 4.

[30] Ibid, The Malaysian Constitution contains no specific provision concerning the right to privacy. One related provision is Article 5, which upholds the individual's right to liberty. Article 5 comes into play when the dispute borders on the right of an accused person to be brought before a magistrate on the grounds of arrest stated in the Article. To date there has not been any specific invocation of Article 5 for the purpose of supporting the right to privacy.

gesture, or exhibits any object, intending that such word or sound shall be heard, or that such gesture or object shall be seen by such person, or intrudes upon the privacy of such person, shall be punished with imprisonment for a team which may extend to 5 years or with fine or with both.[31]

However, privacy has been controversial in Malaysia. Ida Madieha Azmi highlights that, between 2000–2006, there was community concern and backlash against the government for not having specific personal data protection laws in place.[32] Reports of sales of personal data hit the newspaper headlines and opened debate on the need to regulate situations involving violations of data privacy. There were allegations that data pertaining to students was being sold to private institutions, which resulted in public backlash.[33] Thus, the common law, and its historical connections to the United Kingdom (UK), religion, culture and community concern - have all converged to force the establishment of data protection laws in the state.

The development of the PDPA dates back to its initial draft in 1998.[34] It would take more than a decade for the PDPA to be finally approved by the Malaysian Parliament, following considerably criticism and opposition from the communication industry. The PDPA has been heavily influenced by other regional data protection laws, namely the *Hong Kong Personal Data (Privacy) Ordinance 1995*,[35] and the UK.

This Chapter will highlight the core principles and concepts of Malaysia's modern data protection laws. The PDPA aims at regulating the collection, holding, processing and use of personal data in commercial transactions and also to prevent the malicious use of personal information. The PDPA safeguards the personal interests of individuals and makes it unlawful for individuals or entities to sell personal information, or allow such use of data by third parties.[36] This objective is clearly outlined in its preamble, which states that:

> to regulate and protect the process of personal data from being misused through commercial transactions and matters relating thereto.[37]

---

[31] Bakar Munir, *A Malaysia's Data Protection Law* in Simon Chesterman *Data Protection Law in Singapore, Privacy and sovereignty in and interconnected World*, Academic Publishing (2018) Chap. 13.

[32] Madieha Azmi I *Personal Data Protection Law: The Malaysian Experience*, 16 Info. & Comm. Tech. L. 125 (2007) p. 126.

[33] Ibid. The enforcement of MyKAD, a multipurpose identification card, which was undertaken last year, further called into question the security and privacy of an all-embedded card full of personal information if it were to fall into the wrong hands. All these events informed public consensus on the need to regulate the processing and use of personal data-something of which the government continuously assures the public. Another major uptake of all these events is whether industry standards are being developed to ensure good working practices to alleviate consumer concerns on data privacy.

[34] Ibid.

[35] Ibid.

[36] Personal Data Protection Act 2010.

[37] Ibid.

Furthermore, the PDPA is based on, and incorporates the principles set out by the OECD.[38] However, it must be noted that, rather than rely on the OECD and APEC principles, Malaysia has sought to base them on the former EU Data Protection directive 95/46/EC.[39] Therefore, it can be seen that the EU has also significantly influenced the development of data protection law, cutting across the Western and Eastern divide. In other words, the EU data protection laws are having such a profound influence that they are transcending traditional ethnic, cultural and religious boundaries.

Nevertheless, in 2016, the appointment of a Commissioner was done by virtue of the subsidiary law under the PDPA. This new 2016 Order amended the previous Order, i.e. Personal Data Protection (Class of Data Users) Order 2013. The appointment of a Commissioner was introduced to oversee the implementation and enforcement of the PDPA. The Class of Data Users (Amendment) Order 2016 also accompanied the amendments to the PDPA, which expanded the classes of users to Pawn Broker and Money Lender sectors.

8.2 Definitions of Personal Data

The definition of personal data provides a starting point for all Malaysians to have a level of control and ownership over their personal data. Personal data in Malaysia constitutes any information in respect of commercial transactions, which:

(a) is being processed wholly or partly by means of equipment operating automatically in response to instructions given for that purpose;
(b) is recorded with the intention that it should wholly or partly be processed by means of such equipment; or
(c) is recorded as part of a relevant filing system or with the intention that it should form part of a relevant filing system that relates directly or indirectly to a data subject, who is identified or identifiable from that information or from that and other information in the possession of a data user, including any sensitive personal data and expression of opinion about the data subject; but does not include any information that is processed for the purpose of a credit reporting business carried on by a credit reporting agency under the Credit Reporting Agencies Act 2010.[40]

---

[38] Halili Hassan K *Personal Data Protection in Employment: New Legal Challenges for Malaysia* Faculty of Law, Universiti Kebangsaan Malaysia (2012).

[39] Bakar Munir, *A Malaysia's Data Protection Law* in Simon Chesterman Data Protection Law in Singapore, Privacy and sovereignty in and interconnected World, Academic Publishing (2018) Chap. 13.

[40] Amended Act on the Protection of Personal Information 2016, section 4.

In addition to the above, sensitive personal data means any personal data consisting of information as to the physical, mental health or condition of a data subject. Sensitive data also means the political, religious or other belief, the commission or alleged commission of any offence or any other personal data as the Minister may determine by order published in the Gazette.[41] The ability for the relevant Minister to determine what additional things might constitute sensitive data by publishing in a gazette, provides a flexible approach. It is outside the scope of this book to examine any Gazettes the Malaysian Government has established. There is however no new gazette so far in relation to the scope of sensitive personal data.

The processing of sensitive data has a higher burden to meet than general data as defined above. Section 40 only allows sensitive data to be processed when there is 'explicit' consent. Sensitive data can be processed when imposed by the law on the data user in connection with employment, or to protect the interests of the data subject or another person.[42] The rule protects the data user and individual from a person or organization, such as a healthcare professional who passes on data that is confidential. Sensitive data can be used in legal proceedings, obtaining legal advice, exercising or defending legal rights. or in the administration of justice.[43]

According to Zuryati Mohamed Yusoff, the data and confidential information of online consumers fall under the meaning of "commercial transactions" intended by the Act.[44] In other words, if a customer provides their name, address, contact number and some other information to complete a transaction, that data or personal information is protected under the Act. On the other side, the company receiving the information is under an obligation to keep the data and is allowed to use or disseminate the data only with the consent of the data subject. The resulting effect is that data merits similar protection for the reason that it is easily abused and misused through online transactions.[45]

8.3 Consent & Principles

Consent has also emerged both in the common law and statute of Malaysia. Firstly, there is considerable responsibility placed on the data user to ensure consent has been obtained before the processing of an individual's data.[46] Section 6(3) requires that personal data not be processed, unless the personal data is processed for a lawful purpose directly related to an activity of the data user.[47] In addition, the

[41] Ibid, section 4.

[42] Ibid, section 40.

[43] Ibid.

[44] Mohamed Yusoff, Z *The Malaysian Personal Data Protection Act 2010: Legislation Note,* New Zealand Journal of Public and International Law (2011).

[45] Ibid.

[46] Personal Data Protection Act 2010, Section 6(1).

[47] Personal Data Protection Act 2010, section 6(3).

processing of the personal data is necessary for or directly related to that purpose; and the processing of that personal data is adequate but not excessive in relation to that purpose. The data user is able to process the data without any consent where the performance of a contract to which an individual is a party, or when complying with an existing contract. Consent is also required to ensure the best interests of the person are protected such as life, death or security.[48] The general principle of consent also prohibits the processing of any data, unless it is for a lawful purpose directly related to the activity of the data user. For sensitive personal data in relation to physical, mental health or condition, political opinions or religious beliefs, explicit consent[49] has to be obtained.[50]

Secondly, notice and choice[51] has been included to require that a data user can inform an individual that his/her data is being processed.[52] The data user is required to inform the person in writing and to provide an outline of the data that relates to the individual concerned. The person has the right to request access to their data to ensure that it is accurate, and in situations where the data is incorrect, the person can request that it be corrected. Moreover, the data user must inform the person to whom the data relates, to a third party, to whom that data will be disclosed.[53] The person providing the data must be informed whether that data is being provided voluntarily, or subject to specified obligation provided for by law. Thirdly, disclosure[54] is an important part of managing data. Therefore, disclosure of a person's data cannot be undertaken without his/her prior consent. There is an exception to this where a person has withdrawn consent to process personal data.[55]

The collection, collation, use, processing and disclosure of any data has to be secure. This is a requirement under the security principle[56] within Malaysian law. The data user is required to take the necessary steps to protect personal data from any loss, misuse, modification, unauthorized or accidental access or disclosure, alteration or destruction. The principle also requires the data user to ensure any transfer of data is undertaken in a secure manner, and the individual receiving the data is competent. The data user must also provide sufficient guarantees of organizational security governing the processing of data and ensure compliance. In other words, the data user must minimize unauthorized access or use. The data user therefore has a high burden to ensure the safety and security of personal data.

[48] Ibid.

[49] Personal Data Protection Act 2010, section 40(1)(a).

[50] Personal Data Protection Act 2010, section 6(3).

[51] Personal Data Protection Act 2010, Section 7.

[52] Personal Data Protection Act 2010, section 39.

[53] Ibid.

[54] Personal Data Protection Act 2010, Section 8.

[55] Ibid.

[56] Personal Data Protection Act 2010, section 9.

The fifth principle[57] is retention of data.[58] Retaining or storing data has to be secure. Organizations should not retain personal data for longer than is necessary. This is consistent with the other jurisdictions discussed throughout the book. Integrity is another important principle that has been determined by the Commissioner to ensure the legal, policy and operational framework across the public and private divide is robust. Furthermore, the data user is responsible for determining that the data is accurate, complete, not misleading and kept up-to-date.[59] Moreover, a person to whom the data applies can access[60] their personal data in order to ensure it is correct, and up to date except when compliance with a request is refused under the Act.

Apart from requiring consent to process sensitive data, Malaysia has extended the consent requirement to include general processing,[61] the data user refusing to comply with data access[62] and withdrawing consent.[63] In addition, the data user can disclose[64] the personal data of an individual who has provided consent.[65] Nevertheless, the courts have not considered the concept in accordance with the provisions of the PDPA.

In the case *Lee Ewe Poh v Dr Lim Teik Man*[66] the issue was whether consent was provided for the taking and using photographs by a surgeon and displaying these outside of the normal protocol (using the picture for medical evidence only). The court noted that the:

> first defendant contended that the taking of photographs during the course of the procedure without the consent of the patient.[67]

The court when referring to the Emergency Medical Journal argued:

> on the issue of consent stated that an image taken for clinical purposes forms part of patient's health record. Consent to x rays and ultrasound investigations are given implicitly by the patient undergoing those procedures. Similarly, by presenting for treatment and investigation, the patient enters into a tacit agreement to documentation, which includes images as well as written information. An image taken for the purpose of treating a patient must not be used for any other purpose without express consent.[68]

---

[57] Personal Data Protection Act 2010, section 10.

[58] Ibid, section 10.

[59] Personal Data Protection Act 2010, section 11.

[60] Personal Data Protection Act 2010, section 12.

[61] Personal Data Protection Act 2010, section 6.

[62] Personal Data Protection Act 2010, section 32.

[63] Personal Data Protection Act 2010, section 38.

[64] Personal Data Protection Act 2010, section 8.

[65] Personal Data Protection Act 2010, section 39.

[66] *Lee Ewe Poh v Dr Lim Teik Man & Anor* [2011] 1 MLJ 835.

[67] Ibid, 845.

[68] Ibid.

The case highlights how consent will play an important role in protecting a data subject's personal data that has been defined by the PDPA, and in particular when related to sensitive personal data. Abu Bakar argues that consent in the PDPA can mean any freely given specific and informed indication of a data subject's wishes and his/her agreement to his personal data being processed, whether oral or implied.[69]

Accuracy is another principle that is finding its way into the data protection legal framework of Malaysia. Arguably, it is also becoming a central feature of the law to ensure any personal data collected by an organization is accurate.[70] Under the Data Integrity Principle a data user must ensure that personal data is accurate, complete, not misleading and kept up to date[71], and relates to the purpose for which it was collected. It is the duty of the data user to guarantee the accuracy, completeness and correctness of the data collected.[72]

Malaysia requires a privacy impact assessment.[73] A Privacy Impact Assessment (PIA) has been under development to provide a tool to assess the potential effect, risks or impact on privacy of a project, initiative, system, or even scheme which involves the handling of an individual's personal data.[74] This risk assessment tool is used to mitigate or avoid the identified risks through a series of activities. However, the PIA is not being widely implemented and in fact, it was initiated in Australia, Canada, Hong Kong and Ireland. Besides the UK, the other countries have started to reach their PIA maturity and because of that, their established PIA guidelines are being studied by experts and researchers to understand the PIA processes being implemented by those countries. It is argued that the establishment of a PIA guideline will be helpful in assessing the potential risks that might have compromised the privacy of those personal data. In order to design the proposed PIA guideline, it is crucial to conduct a thorough study on this field by analyzing the existing PIA guidelines, research in this area of law. The biggest challenge in this project lies in selecting the best activities and number of PIA steps to be included in the proposed guideline, due to the absence of an international PIA standard. A further challenge is addressing and meeting the different needs and requirements of organizations. In conjunction with that, a comparison and mapping activities will be conducted which, in the end, will result in the selection of the appropriate activities and number of steps needed to comply with the proposed guideline. The proposed guideline

[69] Munir, AB *Personal Data Protection Act: Doing Well By Doing Good* 1 MLJ 1 (2012).

[70] Personal Data Protection Act 2010, section 11.

[71] Ibid.

[72] Mohamed Yusoff, Z *The Malaysian Personal Data Protection Act 2010: Legislation Note,* New Zealand Journal of Public and International Law (2011).

[73] Abdul Razak, F (2013) *Privacy Impact Assessment (PIA) Guideline for Securing Personal Data,* Universiti Teknologi Malaysia.

[74] Ibid.

will then need to be validated by experts in this field. Nevertheless, once implemented the proposal will provide another layer of regulatory oversight to strengthen the management of personal data across Malaysia.

8.4 Commissioner – Agency [Regulator]

Malaysia has dedicated an entire government department to have oversight of the PDPA. The Department of Personal Data Protection has responsibility to enforce and cultivate a data protection culture across Malaysia.[75] The department also has extended responsibility to ensure that there is confidence across the community and business sectors[76] particularly in relation to commercial transactions that are undertaken online.

Malaysia has also appointed a Personal Data Protection Commissioner (the Commissioner).[77] The Commissioner is responsible, not only to advise the relevant Minister on national policy for data protection, but also to implement and enforce the data protection laws.[78] The Commissioner has oversight for the development of operational policies and procedures and to promote associations or bodies representing data users to prepare codes of practice.[79] The Commissioner is appointed for 3 years and can be dismissed by the relevant Minister.[80] The Commissioner, sitting within the government Ministry, can be considered as having diminished separation from the government, unlike some other jurisdictions.[81]

The Commissioner's remuneration and allowances are also determined by the Minister.[82] Furthermore, the Commissioner shall be responsible to the Minister' and 'the Minister may give the Commissioner directions of a general character consistent with the provisions of the Act'.[83] The Commissioner's annual report is only submitted to the Minister, and no further disclosure is required.[84] Graham Greenleaf believes the current process established, significantly reduces the independence of the Commissioner, unlike in other jurisdictions such as Australia.[85] In Australia and

---

[75] Department of Personal Data Protection, Malaysia, http://www.pdp.gov.my/index.php/en/, accessed 10 December 2018.

[76] Ibid.

[77] Personal Data Protection Act 2010, section 47.

[78] Ibid.

[79] Personal Data Protection Act 2010, section 48.

[80] Personal Data Protection Act 2010, section 53–54

[81] For instance, the EU and Australia have established similar, however they are separate from government, but do provide a report to government on the functioning of the respective Commissioner.

[82] Personal Data Protection Act 2010, section 57.

[83] Personal Data Protection Act 2010, section 59.

[84] Personal Data Protection Act 2010, section 60.

[85] Greenleaf G, "*Limitations of Malaysia's data protection Bill*" [2010] ALRS 5; (2010) 104 Privacy Laws & Business International Newsletter 1.

Japan, for example, the legislation has provisions that underpin the independence of the Commissioner. This is an important point because it could be argued not having full independence, may dilute the effectiveness of implementing the laws.

Notwithstanding the above, the Commissioner is also responsible for advising the Minister on national policy for the protection of personal data, and to enforce the legal protection of personal data, including the formulation of policies and operational procedures. Importantly, the Commissioner is required to promote the development and adoption of Codes of Practice, which strengthen the co-regulatory approach to data protection. The Commissioner, like that of its counterparts in other countries, has a coordination and collaboration role to work across government agencies.

Moreover, the Commissioner has a collaborative function and responsibility for data protection outside of Malaysia.[86] Thus, the Commissioner would be expected to liaise with his or her equivalent counterpart in other countries such as Japan, Australia and Singapore. This requirement is no different to the requirements in those countries. More importantly, the Commissioner has responsibility to determine whether any place outside of Malaysia has established a system for data protection that is similar to the process and laws of Malaysia.[87] This role ensures the Commissioner can attend meetings and other forums with their counterparts in other countries.

The powers of the Commissioner are largely administrative, directed at ensuring that they can perform their functions.[88] The Commissioner has the power to collect fees, appoint agents, experts and consultants, enter into contracts, and acquire, purchase, take, hold and enjoy any movable or immovable property.[89] When compared to the Commissioners or their equivalents in other jurisdictions such are Australia and the EU, the Commissioner in Malaysia has little enforcement oversight.

The Advisory Committee (AC)[90] has also been established to provide specialist advice to the Commissioner on all matters related to the personal data protection, across Malaysia. This committee sits under the Commissioner. The AC also provides advice on the administration and enforcement of the PDPA. However, there is no obligation on the Commissioner to act on any advice provided by the AC.

The User Forum Data (the Forum) is responsible for planning, developing and providing Code to protect the rights of users. The Forum also collects, prepares and distributes statistics on personal data protection, while providing the community and industry with a mechanism to lodge complaints, disputes or a grievance. The Forum also provides the procedure for any compensation where there have been breaches of the code.

---

[86] Personal Data Protection Act 2010, section 129.

[87] Ibid.

[88] Personal Data Protection Act 2010, section 49.

[89] Ibid.

[90] Personal Data Protection Act 2010, section 70.

8.5 Public and Private

The PDPA applies to any person who has control over, or authorizes the processing of, any personal data in respect of commercial transactions.[91] Importantly, the PDPA applies to a person that is not physically located in Malaysia, but uses equipment in Malaysia for processing data.[92] However, section 3 provides an exemption to all Federal and State Governments.[93] The PDPA does not apply to any personal data that is processed outside Malaysia, unless that data is to be further processed in Malaysia. In other words, the data could be processed in the first instance in Australia, then sold to an organization in Malaysia, that would then further process that data.

8.6 Extra-territorial Reach

A data user may not transfer personal data to jurisdictions outside of Malaysia unless that jurisdiction has been specified by the Minister.[94] Section 129 of the PDPA provides that:

> A data user shall not transfer any personal data of a data subject to a place outside of Malaysia unless to such a place as specified by the Minister, upon the recommendation of the Commissioner, by notification published in the Gazette.[95]

Furthermore, a data user is not allowed to transfer any personal data outside of Malaysia unless the Minister has specified, on the recommendation of the Commissioner that the personal data is safe to be transferred to that country.[96] Nonetheless, there are exceptions to the above that allow companies to transfer data to another country.[97] However, there is no criteria described that are provided to enable the Commissioner to make a decision on such data transfer. It is for the Minister to decide. In fact, the criteria can be extracted from section 129(2).

A Consultation Paper together with a draft *Personal Data Protection (Transfer Of Personal Data To Places Outside Malaysia) Order 2017* ('Draft Order'), will help determine compatibility of foreign countries laws.[98] The PDPC refers to three criteria which, it states, are considered in drawing up its list. This includes (i) 'places that have comprehensive data protection law (can be from a single comprehensive

---

[91] Personal Data Protection Act 2010, section 2.

[92] Ibid.

[93] Personal Data Protection Act 2010, section 3.

[94] Personal Data Protection Act 2010, section 129.

[95] Ibid.

[96] Personal Data Protection Act 2010, section 129(1).

[97] Personal Data Protection Act 2010, section 129(3).

[98] *Personal Data Protection (Transfer Of Personal Data To Places Outside Malaysia) Order 2017, Public Consultation Paper 1/2017.*

personal data protection legislation or otherwise a combination of several laws and regulations in that place)'; (ii) 'places that have no comprehensive data protection law but are subjected to binding commitments (multilateral-bilateral agreements and others)'; and (iii) 'places that have no data protection law but have a code of practice or national co-regulatory mechanisms'.[99] The current jurisdictions that meet the criteria include all EU member states and those countries that form part of the EEA. The countries within the Asia Pacific Region[100] include Australia, Japan and Singapore.

The White List provides that data users are permitted to transfer personal data to the jurisdictions that have been identified, and will no longer be required to fulfill the prescribed conditions under section 129(3) prior to the transfer of personal data to the said jurisdictions.[101] It must be noted that the 'white list' appears to be a long way off from being endorsed and implemented across Malaysia. For example, the requirement to obtain consent of the Data Subjects prior to transfer of personal data outside Malaysia; the requirement to undertake reasonable precautions and exercise due diligence to ensure that the recipient place will not process personal data in any manner which would contravene the PDPA. However, notable absentees from the current list include Indonesia and Thailand. It is understandable that Indonesia and Thailand would not make the list, as they have not implemented any specific data protection laws. In summary, the PDPA applies when the personal data is first processed in Malaysia before transferring it to a foreign entity. Although, the PDPA will not apply to personal data that is processed outside of Malaysia. That being the case, this element of the law that is different to the EU GDPR, which now applies to EU citizens located in other countries.

8.7 Certificates of Registration

An important feature of the Malaysian laws are the Certificates of Registration that are provided to organizations.[102] The certificate of registration is similar to any other licensing system. The process effectively licenses an organization to collect, store and use personal data. This is unique to Malaysia, where industry sectors are effectively licensed to manage personal data. The Malaysian government has identified a

[99] Personal Data Protection, http://www.pdp.gov.my/images/pdf_folder/PUBLIC_CONSULTATION_PAPER_1-2017_.pdf, accessed 4 January 2018.

[100] *Personal Data Protection (Transfer Of Personal Data To Places Outside Malaysia) Order 2017, Public Consultation Paper 1/2017*. The places jurisdictions on the list include (a) European Economic Area (EEA) member countries United Kingdom, The United States of America, Canada, Switzerland, New Zealand, Argentina, Uruguay, Andorra, Faeroe Islands, Guernsey, Israel, Isle of Man, Jersey, Australia, Japan, Korea, China and Hong Kong.

[101] Personal Data Protection Act 2010, section 129(3).

[102] Personal Data Protection Act 2010, section 14.

number of sectors[103] that must be registered,[104] that include, but are not limited to Utilities,[105] Pawnbrokers[106] and Moneylenders.[107] The sectors covered by the registration process are extensive and enable the government to better control and manage personal data. Data users, whether they belong to one or more categories, are required to take out a registration for each separate category under which they operate. For instance, an organization collecting and using personal data in the education and transportation sector will need to be registered under both sectors. Failure to obtain registration, in collecting and using personal data, may lead to a fine being imposed of up to RM500,000 or to imprisonment for a term not exceeding 3 years, or to both. The certificate or registration must be renewed. If an organization fails to renew their certificate of registration, they are in breach of the PDPA. Note, an organization that has obtained a certificate of registration must display it at the place of business.

In 2017, three separate companies in Malaysia were found to have breached the certificate of registration requirements. They derived from the hotel, education, and employment sectors. Each of the companies had been processing personal data without certification. They were each fined between RM 10,000 to 20,000.[108]

The Commissioner may designate a body as a data user forum in respect of a class of data users. The data user forums can prepare codes of practice to govern compliance with the PDPA which can be registered with the Commissioner.[109] Once

[103] It is for the Minister to decide. In fact, the criteria can be extracted from the Personal Data Protection Act 2010 section 129(2).

[104] Personal Data Protection (Registration of Data User) Regulation 2013, Communications, licensed under the Communications and Multimedia Act 1998 or Postal Services Act 2012 Banking and financial institution, licensed by the Financial Services Act 2013.Islamic Financial Services Act 2013, or Development Financial Institution Act 2002. Insurance, licensed under the Financial Services Act 2013, Islamic Financial Services Act 2013 or Islamic Financial Services Act 2013. Health, licensed by the Private Healthcare Facilities and Services Act 1998, Private Healthcare Facilities and Services Act 1998 or Registration of Pharmacists Act 1951. Tourism and hospitalities services organizations that are licensed by the Tourism Industry Act 1992 or Tourism Industry Act 1992. Transportation, which describes a number of Malaysian airlines. Education providers determined by the Private Higher Educational Institutions Act 1996 and Education Act 1996. Direct selling, a licensee under the Direct Sales and Anti-Pyramid Scheme Act 1993. Services, a company registered under the Companies Act 1965. Real estate, licensed housing developer under the Housing Development (Control and Licensing) Act 1966.

[105] Tenaga Nasional Berhad, Sabah Electricity Sdn. Bhd, Sarawak Electricity Supply Corporation, SAJ Holding Sdn. Bhd Air Kelantan Sdn. Bhd, LAKU Management Sdn. Bhd, Perbadanan Bekalan Air Pulau Pinang Sdn. Bhd, Syarikat Bekalan Air Selangor Sdn. Bhd, Syarikat Air Terengganu Sdn. Bhd, Syarikat Air Melaka Sdn. Bhd, Syarikat Air Negeri Sembilan Sdn. Bhd, Syarikat Air Darul Aman Sdn. Bhd, Pengurusan Air Pahang Berhad, Lembaga Air Perak, Lembaga Air Kuching, Lembaga Air Sibu, Pengurusan Air Selangor Sdn. Bhd.

[106] Pawnbrokers Act 1972.

[107] Moneylenders Act 1951.

[108] Ying Chew K, *Malaysia: Enforcement of the Personal Data Protection Act 2010*, The Personal Data Protection Department is now actively enforcing the PDPA, https://globalcompliancenews.com/malaysia-enforcement-personal-data-protection-20171101/, 20 May 2018.

[109] Greenleaf G, *Asian Data Protection Laws: Trade and Human Rights Perspective*, Oxford University Press, 2014, p. 343.

registered, all data users must comply with the provisions of the code, and non-compliance is an offence under the PDPA.

8.8 Data Officer

Currently, there is no requirement for data users to appoint a data protection officer in Malaysia, often referred to as a data control officer (controller). Section 14 of the PDPA allows the Minister to register a class of data users. In 2016, the Malaysian government has extended the Class of Data Users Order,[110] which expanded the original list of registered data users to include the Malaysia Airlines Berhad, Pengurusan Air Selangor Sendirian Berhad, Pawnbrokers and Money lenders. The 2013 list covers sectors such as insurance, banking, health, transport, real estate, finance, tourism, education, services and utilities.

The data user is responsible for a number of security matters in accordance with section 9 and 2.5 of the Personal Data Protection Code of Practice.[111] Data users are responsible for safeguarding the confidentiality, integrity and availability of personal data. They are required to implement security measures to protect personal data from loss, misuse, modification, unauthorized access, disclosure and alteration or destruction.[112] Their responsibilities also extend to ensuring that the storage and retention of data is not under threat from unauthorized access, particularly when the data is being transferred.

Data processors are appointed by the data user under instruction. Even though the data user is responsible for the overall management of personal data, the processor is obliged to provide sufficient guarantees for the security of the data.[113] To facilitate this security of data with the processor, the data user is required to establish an agreement to ensure that neither itself nor its employees disclose the personal data to any third party. The agreement should include all security measures for that particular industry sector.

8.9 Code of Practice

In accordance with section 23 of the PDPA, a Code of Practice (Code), applies on a sector by sector basis. It is outside the scope of this book to discuss or highlight all the various Codes.[114] One of the most recent Codes issued by the Personal Data

[110] Personal Data Protection (Class of Data Users) (Amendment) Order 2016.

[111] Personal Data Protection Code of Practice – Utilities Sector, section 2.5.

[112] Ibid.

[113] Ibid.

[114] Personal Data Protection Act 2010, section 23.

Protection Commissioner (the Commissioner) has been in the Utilities (Electricity) sector, in 2016. Amongst other things, the Code governs the relationship between data user, subject, processor and employees who process the data. The Code applies to all personal data held by organizations within the utilities sector.[115] The Code expands on the principles of consent, disclosure, notice of choice, security, retention, integrity, and access set out in the PDPA.

The Commissioner has the power to register the Code, however it must be consistent with the Act. According to Bakar Munir, for this step in the process to occur, apart from the Code being consistent with the PDPA, it must also take into consideration the full time period in which the Code is to be in place. [116] Additionally, the code is to specify a specific class of data user forums within the specified time period. Another consideration is that there is no data user forum to develop the relevant code of practice for the class of data users.[117] On the other hand, the Commissioner may refuse to register the code for whatever reason. This will generally be because the Code does not meet the requirements of the Act.

A sectorial approach has been undertaken over the past 2 years in the development of Codes of Practice. In 2016 and 2017 a further two Codes were introduced for the Insurance, Banking and Finance sectors.[118] In addition to the above, the Codes require these sectors to examine their data protection policies and procedures. Nonetheless, due to the sectorial Codes of Practice that have been established, such as in the Utilities and Banking sectors, there may be differences in Code requirements. It is outside the scope of this book to discuss these differences.

Even though the legal framework, similar to other jurisdictions that have established similar data protection laws, provide a self-regulatory model for the development of Codes of Practice, they are enforceable. In other words, the statutory requirement for Codes to be used, adds another dimension to the overall enforceability of data protection laws. Section 29 clearly states that where a data user does not comply with a Code or any provision of a Code can be fined up to RM100,000, or be imprisoned for a year or both.[119] Abu Bakar Munir highlights that to date Malaysia has established and approved three Codes for the Electricity, Insurance, Banking sectors, including the first three plus Legal Practice Sector, Aviation Sector as well as Telecommunications sector.

To underpin and reinforce the co-regulatory model applied by Malaysia, the development of 'standards' have been issued by the Commissioner.[120] The standards aim to assist the data user and apply to security, retention, and integrity amongst

[115] Ibid.

[116] Bakar Munir, *A Malaysia's Data Protection Law* in Simon Chesterman Data Protection Law in Singapore, Privacy and sovereignty in and interconnected World, Academic Publishing (2018) Chap. 13.

[117] Ibid.

[118] Personal Data Protection Code of Practice for the Insurance/Takaful Industry, 2016, and the Personal Data Protection Code of Practice for the Banking and Financial Sector, 2017.

[119] Personal Data Protection Act 2010, section 29.

[120] Personal Data Protection Regulations 2013.

others. Firstly, the security standard guides the management of personal data electronically to restrict access, password protection, protection from malware, viruses and implementation of a backup or recovery system to prevent data loss and theft. The standard (s) encourage the development of internal organizational policy on data protection, and use of standard forms, for the correction of data. This promotes consistency across different industry sectors. Another standard developed has been to manage data retention. This provides the basis for the deletions and removal of personal data, and supports the right to be forgotten. The standard generally applies a 14 day time limit for the personal data to be deleted. The obligation placed on organization for not complying with the standards can result in a fine up to RM 250,000 or imprisonment for up to 2 years.[121]

8.10 Breach and Notification

Malaysia does not require notification of a breach of the data protection laws. In 2018, the Public Consultation Paper (No. 1/2018) was released for public comment by the Personal Data Protection Commissioner (the Commissioner) and entitled *The Implementation of Data Breach Notification* (the DBN Public Consultation Paper), to determine whether a notification mechanism is required.[122] In the DBN Public Consultation Paper, the Commissioner expresses an intention to implement a data breach notification mechanism (DBN) in Malaysia.[123] The DBN is described as a mechanism which will require data users to notify and inform the relevant authorities and affected parties when a data breach has occurred within an organization.[124] By Malaysia adopting this legal process, would see the PDPA harmonize the approach already taken the requirements imposed under the EU GDPR, Singapore, Australia and other jurisdictions. If implemented, it was expected to be implemented in Malaysia by the end of 2018. However, at the time of finalising this book, it had not been fully adopted.

8.11 Enforcement

The Commissioner is responsible for enforcement of the Act.[125] The enforcement responsibility also extends to whether a serious breach of personal data protection principles has occurred through a complaint by any person regarding an act or

[121] Personal Data Protection Standards 2015.

[122] Personal Data Protection Updates – Public Consultation Paper No. 1/2018 – The Implementation of Data Breach Notification, https://www.christopherleeong.com/media/3097/2018-08-clo_pdpa.pdf accessed 16 October 2016.

[123] Ibid.

[124] Ibid.

[125] Personal Data Protection Act 2010, section 48.

practice that contravenes the provision of the Act.[126] Once the Commissioner has received a complaint, an investigation must be undertaken.[127] Although the Commissioner does have the discretion to refuse to undertake an investigation. Nevertheless, once an investigation is completed and the Commissioner has determined that the PDPA has been breached, an enforcement notice is to be issued.[128] In accordance with section 99, a person who is subject to enforcement under the PDPA can appeal to the Appeal Tribunal.[129] However, the decision of the Appeal Tribunal is final. Furthermore, section 100 provides that a decision given by the Appeal Tribunal may, by leave proceed to the Sessions Court. A decision by the Sessions Court is to be enforced in the same manner as a judgment or order to the same effect. The Department of Personal Data Protection has a formal process for individuals to make a complaint in relation to their personal data, or against an organization regarding the laws.

According to Abu Bakar Munir the PDPA has recently created a number of criminal offences namely sections 134 and 135, whereby prosecution can only be instituted with the written consent of the public prosecutor and the Sessions Court has jurisdiction to try any offence under the Act.[130] In addition, section 5(2) provides that subject to sections 45 and 46, a data user who contravenes the personal data protection principles commits and offence. This step arguably raises the stakes between the management of personal data in Malaysia. On the one side the data protection laws are providing a level of privacy protection over the internet. On the other side however, the laws are expanding the field to consider cybercrime and cyber security issues.

8.12 Right to be Forgotten

The right to be forgotten has yet to emerge in Malaysia. Although the right to be forgotten, can only be achieved, when the obligation of the data user (controller) to the data subject that their personal data will be deleted or no longer used. Duryana Binti Mohamed argues that the right to privacy implied by Article 5 of the Federal Constitution. Article 5 specifically pertains to the right to personal liberty.[131] However, the common law has, to date, not decided upon the right. Malaysia's

---

[126] Personal Data Protection Act 2010, section 49.

[127] Personal Data Protection Act 2010, section 105.

[128] Personal Data Protection Act 2010, section 108.

[129] Personal Data Protection Act 2010, section 99.

[130] Bakar Munir, *A Malaysia's Data Protection Law* in Simon Chesterman Data Protection Law in Singapore, Privacy and sovereignty in and interconnected World, Academic Publishing (2018) Chap. 13.

[131] Binti Mohamed D, *The Privacy Right and Right to be Forgotten: the Malaysian Perspectives,* Indian Journal of Science and Technology, Vol 9(S1), (2016) pp. 2–4.

concerns with adopting the right to be forgotten stems from the potential impact the right will have to a third party that is protected by intellectual property law. [132]

Moreover, Malaysia has concerns that such a right will also have an impact on other various professions, such as the freedom of expression that journalists have relied on for decade, so as they can accurately report on individuals. The Constitution of Malaysia states that the fundamental rights include the right to life, personal liberty, free speech, assembly, association, religion, education and property.[133] In Malaysia the breach of privacy is not an actionable wrong.[134] There is no provision within the PDPA that enables a person to request that their personal data be deleted. There is however an obligation of the data user to ensure that all personal data is destroyed or permanently deleted if it is no longer required for the purpose for which it was to be processed by virtue of section 10(2). Therefore, further work is needed by Malaysia should it continue to hold itself out as adopting EU data protection framework (see Chap. 11).

8.13 Retention

Section 10 of the PDPA requires that personal data shall not be kept for longer than is necessary.[135] However, the PDPA does not stipulate the time frame allowed for storage of the personal data but leaves it to the discretion of the data user. Once the data is no longer required for the purpose for which it was processed, the same must be destroyed or permanently deleted.[136] Nevertheless, the data user is required to take all reasonable steps to ensure that all personal data is destroyed or permanently deleted if it is no longer required for the purpose for which it was to be processed.

Furthermore, the *Personal Data Protection Regulations 2013* and *Personal Data Protection Standards 2015,* require personal data collection forms to be destroyed within a period of 14 days, unless such forms contain as "legal value" in connection with the commercial transaction.[137] Data users are also required to dispose personal data where it has been inactive for a period of 24 months. Interestingly, the general limitations for the destruction of legal documents in Malaysia is 6 years and 7 years under revenue laws.[138]

---

[132] Ibid.

[133] Federal Constitution of Malaysia, 1 November 2010, Articles, 5, 10, 11, 12 and 13.

[134] *Ultra Dimension Sdn Bhd v Kook Wei Kuan* [2004] 5 CLJ 285 and *Dr Bernadine Malini Martin v MPH Magazine Sdn Bhd & Ors and another Appeal* [2010] 7 CLJ 525 (CA). in Duryana Binti Mohamed, *The Privacy Right and Right to be Forgotten: the Malaysian Perspectives,* Indian Journal of Science and Technology, Vol 9(S1), (2016) pp. 2–4.

[135] Personal Data Protection Act 2010, section 10.

[136] Ibid.

[137] Personal Data Protection Regulations 2013 and Personal Data Protection Standards 2015.

[138] Ibid.

8.14 Supporting Cyber Security Laws

Data protection law is not stand alone and must be supported by other national laws of the country. In Malaysia, these include but not limited to:

- Computer Crimes Act 1997
- Communications and Multimedia Act 1998
- Digital Signature Act 1997
- Copyright Act 1987 (as amended in 2012)
- Electronic Transactions Act 2006

In 2017, the Malaysian Government were to introduce new cybercrime laws that would strengthen the countries defense against crime related activities that involve the internet.[139] The proposed laws would also strengthen the enforcement of terrorist related activities that are undertaken online such as money laundering and fraud.[140] However, these laws have not yet been approved by the Malaysian parliament.

8.15 Conclusion

The introduction of the PDPA was the first comprehensive personal data protection law in South East Asia. While there are many similarities between the PDPA and Australia, Singapore and the EU, there are however significant differences. For a country with social, economic, cultural and religious differences from its neighbours, Malaysia has emerged adopting many Western data protection concepts and principles. Importantly, the PDPA is based on principles set out by the OECD, and has looked to the EU in the development of their laws.

The PDPA while not having a smooth road to implementation has in many respects provided the Malaysian government with a platform to demonstrate to its citizens that data protection will be taken seriously. However, the current framework can be best described as being narrow and limited in its application. By limiting the laws to commercial activities alone, arguably can create confusion as to its full application.

While limited in its scope and application, the laws do provide the basis for protecting and providing a level of control over one's personal data on the Internet. However, to date, the courts have had little to say in relation to the PDPA. Arguably, though, regulating and protecting personal data, whatever level those protections might be, the resulting effect is that a person's privacy, to some extent, will be protected. The PDPA, as highlighted above, does not extend to federal and states governments. This could be considered a major inconsistency with other jurisdictions.

---

[139] Wong and Partners, https://www.wongpartners.com/-/media/minisites/wongpartners/files/al_wp_newcybersecuritylaw_jun17.pdf, accessed 20 December 2108.

[140] Ibid.

The PDPA has been considered as providing confidence across the community in electronic commerce and business transactions, as a result of the increase in credit card fraud and identity theft. By recently expanding the field to consider criminal offences within the law, Malaysia, along with other states are beginning to recognize that personal data can be used in criminal activities. It is our view the data protection laws of Malaysia closely align with its neighbor Singapore, while considering law from the UK and Australia. Finally, it is our further view that Malaysia, unlike many other countries in Asia have been very effective in establishing data protection laws, as they were one of the first countries to do so. Arguably, this places Malaysia in a good place for further law reform, and ensure that the PDPA keeps pace with other similar laws from across the region and the world.

References

Abdul Razak, F (2013) *Privacy Impact Assessment (PIA) Guideline for Securing Personal Data*, Universiti Teknologi Malaysia.

Bakar Munir, *A Malaysia's Data Protection Law* in Simon Chesterman Data Protection Law in Singapore, Privacy and sovereignty in and interconnected World, Academic Publishing (2018) chapter 13

Binti Mohamed D, *The Privacy Right and Right to be Forgotten: the Malaysian Perspectives,* Indian Journal of Science and Technology, Vol 9(S1), (2016) pp. 2–4

Greenleaf G, *"Limitations of Malaysia's data protection Bill"* [2010] ALRS 5; (2010) 104 Privacy Laws & Business International Newsletter 1

Madieha Azmi I *Personal Data Protection Law: The Malaysian Experience*, 16 Info. & Comm. Tech. L. 125 (2007)

Mohamed Yusoff, Z *The Malaysian Personal Data Protection Act 2010: Legislation Note,* New Zealand Journal of Public and International Law (2011)

Munir, AB *Personal Data Protection Act: Doing Well By Doing Good* (2012) 1 MLJ 1

Ying Chew K, *Malaysia: Enforcement of the Personal Data Protection Act 2010*, The Personal Data Protection Department is now actively enforcing the PDPA, https://globalcompliancenews.com/malaysia-enforcement-personal-data-protection-20171101/, 20 May 2018

Chapter 9
Thailand

Abstract As of 2018 and early 2019, Thailand has not implemented any specific legislation with regard to personal data and privacy protection. Thailand is a constitutional monarchy. The country has had 20 constitutions in total (both interim and permanent ones) since 1932, with the most recent in 2017. Since 2014, Thailand has been ruled by a military government, with elections anticipated sometime in 2019. The National Assembly, Council of Ministers House of Representatives and Senate are made up of people from the local political parties. At the time of writing this book, the most recent draft of Personal Data Protection Bill was approved by the Cabinet in December 2018, (Cybersecurity, data protection bills await NLA approval http://www.nationmultimedia.com/detail/national/30360686, accessed 30 December 2018) and is awaiting further consideration and approval from the National Legislative Assembly.

Currently across Thailand there is a low awareness of privacy over the Internet. The Thai people have generally embraced the use of Internet technology, like many other citizens of other countries. However, their understanding and awareness of any infringement to their individual privacy is – to a large extent – low. The EU's GDPR and the introduction of the extra-territorial concept, will pose challenges for Thailand, the longer they continue to delay the implementation of specific data protection laws. This Chapter will briefly highlight the laws in Thailand pertaining to data used within and by technology. This Chapter outlines the development of the draft bill on Personal Data Protection Act (December 2018 version), which received an approval from the Cabinet. However, due to ongoing political change in Thailand the proposed laws have not progressed. Nonetheless, there have been several iterations of the bill over the past decade. In addition, there will be a brief discussion regarding some of the concerns relating to the proposed draft Bill. This Chapter will conclude by outlining the principles and concept the Thai government have been considering for their new data protection laws. At the time of writing this book the proposed Personal Data Protection Act, had been approved by the National Legislative assembly in February 2018. However, the new laws is awaiting Royal Endorsement and Publication in the Government Gazette.

© Springer Nature Singapore Pte Ltd. 2019

R. Walters et al., *Data Protection Law*,
https://doi.org/10.1007/978-981-13-8110-2_9

9.1 Introduction

There are no current laws in Thailand that specifically manage personal data and protect the privacy of individuals over the Internet in the same way as Australia, India, Indonesia, Japan, Malaysia, Singapore or the EU. There are sectorial laws in areas of the constitution,[1] banking and finance, telecommunications and child protection that do provide some level of protection to personal data. However, it is out of scope to examine these laws. Rather this Chapter will focus on Electronic Transactions Act B.E. 2544, 2001.

As highlighted in Chap. 1, the concept of privacy pervades individualism, liberalism, public-private divide, multiculturalism, religion and customs, along with the rule of law. These underpin liberal democracies widely espoused in the West. Ramasoota, Pirongrong and Panichpapiboon, Sopark argue that these notions of privacy are simply not present in the context of Thailand.[2] Thailand is a Southeast Asian nation with an extensive history of state surveillance. The authors highlight that, from the ancient to the modern period, extensive collection of people's personal information has been a long-standing practice.[3] Ancient Siamese states collected personal information of their commoners' population through registration rolls and a coded wrist-tattooing system.[4] Pirongrong and Sopark believe that in the modern era, the state keeps its population under bureaucratic surveillance through citizen identification cards[5], household registration, passbooks and social welfare cards.

The approach taken towards surveillance is also reflected in the 1991 *Civil Registration Act.* This law was enacted in response to the introduction of computerized record-keeping and data-processing.[6] The law plays an important role in regulating the collection and use of personal information by the Department of Local Administration which houses the Civil Registration Bureau. Furthermore, the law notably allows other government departments to share in the use of civil registration information through requests for copies of information as well as through computer linkages.[7] However, the provisions mainly deal with privacy in conjunction with other rights and legal protections. There are no direct stipulations about violations of privacy per se, since abuses typically have been framed in terms of trespass, defamation, or, breach of trust or confidence instead.[8] Pirongrong and Sopark argue that:

---

[1] Section 32 of the Thai Constitution 2017.

[2] Pirongrong., R, Panichpapiboon, S *Online privacy in Thailand: Public and strategic awareness*, Journal of Law, Information and Science, Vol. 23, No. 1, (2014) pp. 97–136.

[3] Ibid.

[4] Ibid.

[5] Ibid.

[6] Ibid.

[7] Ibid.

[8] Ibid.

> efforts have been underway in the past decade and a half to draft data protection law, along with five other information and communication technology (ICT)-related laws. However, only two laws that have been passed the first *Electronic Transactions Act 2001* and the *Computer Crime Act 2007*. The drafting of the data protection law was influenced by two landmark documents; the 1980 Organization for Economic Co-operation and Development's (OECD) *Guidelines on Protection of Privacy and Transborder Flows of Personal Data* and the 1995 European Union *Directive on Protection of Personal Data and Transborder Flow of Such Data*.[9]

However it must be noted that, despite their efforts to develop specific data protection laws, and implement the above, they currently provide no equivalent alternative to strong and enforceable specific data protection laws. Arguably, as Thailand was initially considering the OECD guidelines, the Thai Government has been looking to the international community for guidance in this area of law, even though its history in this area varies greatly from its other South East Asian counterparts. By Thailand also looking to the EU, and the former Directive 95/46/EC, further confirms that the EU's influence in this area of the law is far reaching. Pirongrong and Sopark go onto to highlight that:

> several data surveillance schemes have been administered in Thailand at least since 1981 without adequate legal safeguards. For instance, the computerized and online civil registration system, the microchip national ID card system, the computerized criminal records database system and surveillance cameras in public areas.[10]

Nevertheless, in 2014, it was estimated that Thailand's capital Bangkok had the highest number of Facebook users in the world, with more than 8 million people.[11] It was also estimated that Facebook's penetration in Thailand was more than 22% compared to the country's populations, with the percentage of overall Internet users estimated to be more than 83%.[12] However, there has been little awareness raised about the privacy implications of these popular applications. The authors in referring to an earlier study in 2007 highlight that the age of users and online privacy awareness are statistically significant, while the number of hours spent using the Internet per week had no correlation with online privacy awareness.[13]

The broader issue for Thailand, is likely to be that, as other countries in the region move closer to the EU, they will be forced to adopt a similar models – whether they agree with the EU model or not. It must be noted that at the time of writing this book Thailand was under military rule. In 2019, they have had their first elections, with a government recently being formed. It remains to be seen as to whether and how the current Thai government view data protection and privacy. Moreover, Thailand initiated polices with regard to information technology back in 1996. This was followed by the Cabinet's approval of the project to develop information technology-related laws in 1998. The six different areas of information

[9] Ibid.

[10] Ibid.

[11] Ibid.

[12] Ibid.

[13] Ibid.

technology included electronic transactions, electronic signatures, electronic fund transfer, computer crime, national information infrastructure and personal data protection.[14] Regarding personal data protection, its legislation process has faced a number of obstructions since the very beginning. For example, there have been many versions of the draft bill by different drafting bodies. However, the bill(s) have continued to be delayed due to many different international governmental conflicts. For instance, one of the most prominent conflicts was between the version proposed by the then Ministry of Information and Communication Technology (MICT) – which was drafted by the National Electronics and Computer Technology Center (NECTEC) – and the version proposed by the Office of the Official Information Commission (OIC). [15] The delay occurred during the process consideration by the Council of the State. Importantly, due to the political uncertainty between 2013 and 2014, the consideration of promising versions of the bill at the parliamentary level were put on hold.[16]

However, only recently there have been calls for Thailand to enact data privacy legislation has returned as the country moves ahead with the digital economy, according to legal and other experts.[17] In May 2018, The Ministry of Digital Economy and Society (MDES) was reported to be revising a 10-year-old draft of the proposed law to catch up with the latest challenges as evidenced by the recent personal data leak that, created fear in the community, which involved telecom operator TrueMoveH.[18] One of the concerns raised in developing the proposed laws is that the security of ID-cards and other personal data of more than 10,000 Thai customers might have been compromised. However, there appeared to be little to no noise from the general community following the leak. Bhume Bhumiratana, a researcher and expert on cybersecurity, said:

> MDES was expected to finalise the new draft for Cabinet approval and enactment by the National Legislative Assembly by the end of this year, even though the European Union's (EU) General Data Protection Regulation law is due to be effective from May 25 this year. The GDPR is said to be the world's new legal standard on data privacy and related regulations with its enforcement affecting other countries, including Thailand, as the personal data of all EU citizens will be protected under the new EU law with binding conditions for companies with EU customers.[19]

More importantly, in 2018, Paiboon Amonpinyokeat, a cyber law expert, said that the Thai people generally still have a low awareness of data privacy issues, which are becoming crucial in their daily life due to the fast-growing development

---

[14] Kamolthamwong, K. et al., Final Report: The Project to Study and Develop the Approach to Personal Data Protection within Asean, https://oer.learn.in.th/search_detail/ZipDownload/58525, (in Thai), pp.59–93, accessed 27 December 2018.

[15] Ibid.

[16] Ibid.

[17] Pornwasin, A *Govt in race against time to update data privacy law 2018* http://www.nationmultimedia.com/detail/national/30344739, accessed 4 November 2018.

[18] Ibid.

[19] Ibid.

of online and mobile banking, e-commerce and other digital services.[20] This, arguably, depends on whether one views that this position is good for the private sector and government in Thailand. However, Ramasoota Pirongrong and Panichpapiboon Sopark argue that the awareness of online privacy in Thailand requires an understanding of four aspects of different cultural attitudes.[21] This includes researching groups under four dichotomies including: collectivism versus individualism, conservatism versus liberalism, localism versus globalization, and freedom of expression versus national security. However, the concept would not be unique to Thailand and arguably, a similar approach can be, and has been applied by other nation states. The authors found that those people who have an individualistic attitude, have a higher online privacy awareness and are more aware of intrusions to their privacy than those with a collectivist attitude.[22] They go on to say that those with a liberal attitude show a higher score in online privacy awareness than those with a conservative attitude.[23] Those who are more inclined toward globalization are also more perceptive and aware of their online privacy and threats to its violation than are those who tend to favour localism.[24] Moreover, those who place more emphasis on freedom of expression than on national security also exhibit higher levels of privacy awareness and threats to privacy than those that think otherwise.

Nevertheless, Pirongrong and Sopark note that since the rise of the Internet, privacy has been almost impossible to sustain. Across Thailand, their study demonstrated that people believe that there is no such thing as 'privacy' in new media.[25] Even when the best laws apply, no one can guarantee privacy in the online world.[26] They argue that:

> Technical defaults in online applications compromise privacy, in many cases without the knowledge of users. Mobile devices such as Blackberry phones are equipped with built-in Global Positioning Systems (GPSs). Once users post anything on Facebook or Twitter, location data will be automatically attached to the end of the post. Yet, very few users seem to care about setting their privacy defaults to prevent such applications from reporting on their location in order to protect their privacy. Apart from built-in surveillance mechanisms in communication devices, the main privacy issue about which civil society is concerned is the marketing surveillance practices (such as consumer profiling, direct marketing through short messaging services, and spam emails), all of which take advantage of new ICTs.[27]

Arguably the study provides a basis for Thailand to better understand the concerns and awareness of privacy over the Internet. However, it must be noted that the study was small in size only interviewing about 800 people, when considering that

---

[20] Ibid.

[21] Ramasoota, Pirongrong and Panichpapiboon, Sopark. *Online privacy in Thailand: Public and strategic awareness*, Journal of Law, Information and Science, Vol. 23, No. 1, (2014) pp. 97–136.

[22] Ibid.

[23] Ibid.

[24] Ibid.

[25] Ibid.

[26] Ibid.

[27] Ibid.

Thailand's overall population is more than 60 million.[28] Therefore, further work is required within Thailand to better understand the Thai people's understanding and awareness of privacy issues over the Internet.

Another important factor that has shaped the way privacy has been received and perceived in Thailand is their understanding of socialism versus democracy. Pirongrong and Sopark found that, despite the fact that democratic revolution took place 80 years ago, the Thai public's understanding of democracy is limited mainly to casting votes at a general election.[29] The realization of citizenship rights and the exercise thereof still needs to be improved, not to mention the lack of understanding about less inherent rights, such as the right to privacy, in the Thai context.[30]

For the Thai government, it may provide the basis for continuing their historical surveillance state activities un-opposed. On the other hand, could this be a sign that other populations of other states around the world will follow the same path – and not be concerned with Internet privacy? That said, in this case, there appears to be a lack of awareness and knowledge in Thailand, and once people become more aware, they may have the same view as those citizens of Europe.

Nonetheless, the protection of the right to privacy is contained in section 32 of the Constitution and section 420 of the Civil and Commercial Code, which protects individuals from wrongful acts by a person who willfully, negligently or unlawfully injures the life, body, health, liberty, property or any right of another person.[31] Section 32 of the 2017 Constitution states that:

> A person shall enjoy the rights of privacy, dignity, reputation and family. Any act violating or affecting the right of a person under paragraph one, or exploitation of personal information in any manner whatsoever shall not be permitted, except by virtue of a provision of law enacted only to the extent of necessity of public interest.[32]

Section 41 also provides that a person and community shall have the right to be informed and have access to public data or information in possession of a state agency as provided by law.[33] Furthermore, section 41 goes onto say that at present a petition to a State agency by an individual is to be informed of the result of its consideration in due time. It also leave open the possibility for an individual to take legal action against a State agency as a result of an act or omission of a government official or employee of the State agency.[34] Furthermore, section 59 of the Constitution provides that the state shall disclose any public data or information in the possession of a state agency, which is not related to the security of the state or government confidentiality as provided by law. Additionally, the state agency shall ensure that

---

[28] Ibid.

[29] Ibid.

[30] Ibid.

[31] Pornwasin, A *Govt in race against time to update data privacy law 2018* http://www.nationmultimedia.com/detail/national/30344739, accessed 4 November 2018.

[32] Constitution of Thailand 2017, https://www.constituteproject.org/constitution/Thailand_2017.pdf?lang=en, accessed 4 November 2018.

[33] Ibid, section 41

[34] Ibid.

the public can conveniently access the data or information.[35] As highlighted there is little recognition of privacy or personal data in Thai law. In this context, the disclosure or transfer of data is considered a wrongful act if it causes harm to the data owner. Not having specific data protection laws in place, as most countries within the region have, could result in Thailand being left out of region's economic activity.

Notwithstanding the above, Thailand is signatory to the International Covenant on Civil and Political Rights 1966[36], the International Covenant on Economic, Social and Cultural Rights 1976[37] and the ASEAN Human Rights Declaration. They all provide a level of protection in relation to personal privacy. The Declaration goes some way to setting the basis for broader consideration and acceptance that personal data and privacy are interrelated and need a level of protection.

Today, the Electronic Transactions Act B.E. 2544, 2001 (EIT) manages the electronic transaction(s) of civil and commercial data within the state of Thailand. Therefore, the discussion regarding Thailand will be limited. The Act ensures there is a standard process for the use and recognition of electronic signatures.[38] The Act does not set out any core principles that have been highlighted in past chapters relating to personal data however, the law goes some way to ensuring data messages are transacted according to the principles of integrity and accuracy. The current framework in Thailand,[39] while limited, requires state agencies to establish an information security policy and a practice statement to ensure notifications are obtained, and approved by the Commission.

9.2 Definitions

The Act[40] defines neither personal information nor sensitive personal information/ data, because the laws only deal commercial and civil electronic transactions. This limited approach is likely to be addressed by the proposed draft Personal Data Protection Bill. However, it remains to be seen whether Thailand adopt a broad or very restrictive approach to the definition of personal data. It should be noted that

---

[35] Ibid, section 59.

[36] The International Covenant on Civil and Political Rights (ICCPR) is a multilateral treaty adopted by the United Nations General Assembly through GA. Resolution 2200A (XXI) on 16 December 1966, and in force from 23 March 1976, ratified 1996.

[37] Thailand signed the Covenant in September 1999, International Covenant on Economic, Social and Cultural Rights, Adopted and opened for signature, ratification and accession by General Assembly resolution 2200A (XXI) of 16 December 1966 entry into force 3 January 1976.

[38] Electronic Transactions Act B.E. 2544 2001, sections 26 to 31.

[39] According to S=section 5 and 7 of Royal Decree Prescribing Rules and Procedures for Electronic Transactions in Public Sector B.E. 2549, Electronic Transactions Act B.E. 2544, 2001, section 35. Translated by Associate Professor Dr. Pinai Nanakorn under contract for the Office of the Council of State of Thailand's Law for ASEAN project.

[40] Electronic Transactions Act B.E. 2544 2001.

the draft bill as of December 2018 adopts a broad definition of personal data. It is our view that, whatever definition is settled, this will provide some level of clarity as to what Thailand aim to achieve from the proposed laws. In other words, should the definition be limited, Thailand could be steering its proposed laws to take on an economic approach similar to Singapore. However, and conversely, if it settles on a broad definition, similar to the EU, the message from Thailand could be that they are placing privacy over the Internet, ahead of the economic benefits, in the trade in personal data.

9.3 Public and Private

Section 35 of the EIT provides that the application of the Act extends to the permission, registration, administrative order, payment, notification or the performance of any act under the law with a State agency.[41] Even though some countries have not fully embraced the idea that their data protection laws should apply to both the public and private sector, Thailand, arguably, needs to seriously consider that any future specific data protection laws are fully transparent, ensuring they apply to both sectors. According to Section 4 of the draft bill (the December 2018 version), certain public sectors will likely be excluded from the enforcement of the personal data protection law. However, it is out of scope to describe what these public sector entities will be, because it has not been settled by the government.

9.4 Retention & Consent

Section 10 of the EIT requires that any information be presented or retained in its original form. Information constitutes an incident or fact regardless of whether expressed in the form of a letter, number, sound, image or any other form.[42] Importantly, this can include personal information of individuals that relate to any incident. Thus, section 10 provides some flexibility as to the format for the retention of data messages provided that the information is capable of being subsequently displayed.[43] However, the current state of affairs in Thailand provide for no consent. The problem has been raised by key individuals within the state. Bhume Bhumiratana believes:

> the lack of a data privacy law has led to consumer abuse, which will become widespread in the coming years due to the advancement of the digital economy and society. For example, banks have used their customer data without specific consent from customers, allowing sales personnel to follow up with customers without authorization after customers open

[41] Electronic Transactions Act B.E. 2544 2001, section 35.

[42] Electronic Transactions Act B.E. 2544 2001, section 10.

[43] Ibid.

> their bank accounts. In the case of the EU's GDPR law, customer consent needs to be specific to prevent data abuse, he said, while Prinya Hom-anek, president and chief executive officer of ACIS Professional Centre, said data abuse is widespread and there is no law to require the removal of such data.[44]

It is necessary to consider the integrity of the information in order to determine its completeness. As highlighted in other Chapters throughout the book, the concept of consent has become very important to the overall framework for managing personal data. It allows an individual to have a level of control over their personal data.

9.5 Commission – Agency [Regulator], Principles, Codes

Thailand does not have a dedicated commissioner that oversees the regulation of data protection or privacy. The Office of the Electronic Transactions Commission has the power to advise Cabinet on policy formulation and promoting the development of electronic transactions. The Commission has a role in supervising business that deal with electronic transactions prescribing security measures, notification, e-registration, and licenses and methods of engaging in electronic transactions in the public sector. The Commission may also appoint sub-commissions to assist in considering operational issues of the Commission.

Section 36 allows the relevant Minister to be the chairperson and Director of the National Electronics and Computer Technology Center (NECTEC), National Science and Technology Development Agency (NSTDA) as a committee member and secretary.[45] Moreover, the Commission has wide powers to develop policy and promote electronic transactions. Additionally, the Commission has a role to oversee, supervise and follow up with the business community on their electronic transactions. The Commission can issue rules or notifications in connection with electronic signatures to perform any other activity in the execution of this Act or other laws. In the performance of an activity under this Act, a member of the Commission is recognized as an official under the Penal Code. The Commission also has a role in advising the Minister on the applications of the laws.[46] The application of the Act extends to both civil and commercial electronic transactions that contain and use data messages.[47]

Thailand's laws do not articulate the core principles of data protection such as accuracy, disclosure, collection, transfer, transparency, security or retention. Thailand does not require codes of practice to be implemented or approved. Even so, sectorial laws may require industry sectors to develop codes, procedures or

---

[44] Pornwasin, A *Govt in race against time to update data privacy law 2018* http://www.nationmultimedia.com/detail/national/30344739, accessed 4 November 2018.

[45] Royal Decree Alternating Provisions to Comply with Transfer of Powers in the Public Sector under the Act for Improvement of Ministry, Office, and Department B.E.2545, Section 102.

[46] Electronic Transactions Act B.E. 2544 2001, sections 36, 37 and 42.

[47] Electronic Transactions Act B.E. 2544 2001, section 37.

guidelines for the management and co-regulation of data. It is out of scope of this book to explore the various sector by sector codes or procedures.

9.6 Enforcement

An individual who operates a service business that involves transacting data messages electronically, without registering or notifying the authorities, could be subject to imprisonment for not exceeding 1 year or a fine of up to THB 100,000.[48] The law requires that an individual must register or notify the relevant government authority that they are conducting a business, which transacts in data. Furthermore, there are penalties where an entity carries on a business without having a license. That is, undertaking the same activity without a license, could result in imprisonment not exceeding 2 years or a fine of up to THB 200,000. Thailand, do not have a formal complaints mechanism, similar to other jurisdictions. It remains to be seen whether the proposed new laws would provide for such a mechanism. However, and while outside the scope of this book, there may be complaints systems established by the banking, telecommunications and other sectors.

In addition, the Computer-Related Crime Act B.E. 2550 (2007) as amended by the Computer-Related Crime Act (No.2) B.E. 2560 (2017), imposes penalties for computer data alterations. For instance, criminal penalties can be established where a third party's computer data has been damaged, impaired, deleted, altered or added either in whole or in part. Offences of this nature can incur a term of imprisonment for up to 5 years or a fine of not more than THB 100,000, or both. Furthermore, an individual who illegally commits any act that causes the working of a third party's computer system to be suspended, delayed, hindered or disrupted to the extent that the computer system fails to operate normally shall be subject to imprisonment for no longer than 5 years or a fine of not more than THB 100,000, or both.[49] Any person sending computer data or electronic mail to another person and covering up the source of the data in a manner that disturbs the other person's normal operation of their computer system shall be subject to a fine of not more than THB 100,000. Additionally, it is also an offence to send computer data or an email(s) to others in the manners which cause a nuisance to the recipients without providing them a convenient option to terminate the reception or to express their intent to refuse the reception. This offence is subject to a fine not exceeding THB 200,000.[50]

[48] Electronic Transactions Act B.E. 2544 2001, section 44 to 46.

[49] The Computer-Related Crime Act B.E. 2550 (2007) as amended by the Computer-Related Crime Act B.E. 2560 (2017), sections 9 and 10.

[50] The Computer-Related Crime Act B.E. 2550 (2007) as amended by the Computer-Related Crime Act B.E. 2560 (2017), section 11.

9.7 Right to Be Forgotten

The right to be forgotten is not entrenched in the law of Thailand. However, there could be an argument that there is currently in existence a limited and implied version of the right. The Law on Obligations[51] provides the right of the creditor to claim for a debtor performance according to the obligation between the parties. Although, a legal relationship, along with the status of the creditor and debtor, must still be clarified. Section 213 states that, if the nature of an obligation does not permit of compulsory performance and the obligation is the performance, the creditor could request for a court' order to be issued ordering the third party to do such act instead of the debtors.[52] In other words, if the legal relationship provided for can be substantiated, an order could be issued requesting the removal of personal data.

Section 420 goes one step further and places an obligation on a person not to injure another person's life, body, health, liberty, property or any right of that person.[53] Furthermore, section 423 states that a person asserts or circulates facts that are injurious to the reputation or the credit of another, or his earnings or prosperity in any other manner, shall compensate that other person for any damage arising therefrom, even if he does not know of its untruth provided that he ought to have known it.[54] A person who makes a communication, the untruth of which is unknown to him, does not thereby render himself liable to make compensation, if he, or the receiver of the communication, has a rightful interest in it. However, this form of tort appears to not have been fully tested in relation to the right to be forgotten. If and when the Personal Information Protection Act comes into effect, whether the laws include the right to be forgotten similar to that of the EU, is uncertain.

Within the December 2018 version of the draft bill of personal data protection law, there is no provision which clearly shows that the draft bill directly adopts the concept of the right to be forgotten. However, sections 33 of the proposed new laws is likely to allow the deletion of personal data if certain conditions are met. In the case that a data controller does not comply with the regulations set out in the draft bill. The data subject has the right to request for the deletion of his or her data.

[51] Civil and Commercial Code 1 January B.E. 2468, sections 194 and 213. https://www.samuiforsale.com/law-texts/thailand-civil-code-part-1.html#193, Daongoen Chinpongsanont. *The Right to be Forgotten, Thailand,* http://ethesisarchive.library.tu.ac.th/thesis/2015/TU_2015_5501040074_4674_3329.pdf, accessed 2 November 2018.

[52] Ibid.

[53] Ibid, section 420.

[54] Ibid, section 423.

9.8 Proposed Data Protection Law

The proposed Persona Information Protection Act (Draft) 2018 of Thailand reflects the common principles of other countries. It is a long awaited area of law that has been under development for several decades. Greenleaf and Suriyawongkul believe that this Bill is likely to be enacted.[55] This is because data privacy has recently become very controversial in Thailand, where it was reported that True Move, exposed an estimated 46,000 customer records (names, addresses, scans of ID cards and passports)[56] to the general public.[57] The authors also note that the implementation of the EU's GDPR and the introduction of the extra-territorial concept, will pose a challenge to all Thai businesses that deal with EU nationals[58] (see Chap. 3). The December 2018 version of the draft bill has been prepared to replicate much of the GDPR. For example, the proposal for the transfer of data to the countries, can only be undertaken when that country has sufficient protection [Section 28]. The proposed laws do set up a framework that provides for:

- Consent;
- Collection;
- Controller;
- Disclosure;
- Personal Data;
- Sensitive Data collection;
- Accuracy;
- Notification; and
- Transfer personal data to a foreign country.[59]

The proposed draft laws aim to protect the personal data and information of individuals. The proposed draft laws also identify the need to establish the Personal Data Protection Commission and the Office of the Personal Data Protection Commission. It would also deal with the transfer of personal data outside of Thailand. The most recent version of the draft proposed laws was approved by the Cabinet in December 2018. However, it remains to be seen whether they will be approved by the legislative body (which currently is the National Legislative Assembly) or not. Haruethai Boonlomjit *et al* provided an overview of the proposed

---

[55] Greenleaf, G., Suriyawongkul, A *Thailand's draft data protection Bill: Many strengths, too many uncertainties* Privacy Laws & Business International Report, (2018) pp. 23–25.

[56] Suchit Leesa-Nguansuk, New data law aimed at ensuring privacy, https://www.bangkokpost.com/business/news/1455534/new-data-law-aimed-at-ensuring-privacy, accessed 5 November 2018.

[57] Ibid.

[58] Ibid.

[59] Ibid.

data protection laws in August 2018.[60] The new law has some key definitions which are similar to data protection laws elsewhere, and include the definition of:

- Personal data is broadly defined as information that is able to directly or indirectly identify a living individual.
- Data controller is a person (whether a natural or legal person) who has authority to make decisions on collection, usage or disclosure of Personal Data.
- Data processor is a person (whether a natural or legal person) who collects, uses or discloses Personal Data in compliance with the orders of data controller.[61]

In addition, the proposal aims to include the concept of extraterritorial, whereby both data controllers and data processors would be regulated, whether or not they are in Thailand. This would be similar, if adopted to the current day data protection laws of the EU.

Greenleaf and Suriyawongkul argue that the transfer of personal data to foreign countries must meet a standard of privacy protection but are otherwise not prescribed by the Act.[62] According to Section 28 of the December 2018 draft bill, the standard of sufficient protection shall be decided by the Personal Data Protection Commission. The proposed law could set 'the same standards as this Act', or higher, or, lower standards. The usual exceptions would be allowed (consent, and where a contract exists). These provisions, in the absence of standards being established, are unlikely to meet the EU adequacy standard.[63] Data exports would also be allowed in 'other cases that could be prescribed in the Ministerial Regulations. Thus, this standard will provide some flexibility in the system and allow the government to determine with ease where data exports can take place. This would allow the possibility of foreign companies certified under APEC Cross Border Privacy Rules (CBPR), being allowed to receive data exports from Thailand, simply because of their APEC-CBPRs compliance.[64] However, Thailand has not indicated any intention to join APEC CBPR. The related concept of certification marks (and recognition of foreign certification marks) as a basis for data transfers has been removed from this version of the PDPB (see Chap. 16).

Specific consent is required from the data subject, in writing or via electronic means, prior to or at the time of collection, use or disclosure of personal data, unless one of the prescribed exceptions applies. A data subject may at any time revoke his/her consent, unless there is a restriction under the law or contract on revoking such

[60] Haruethai Boonklomjit (HK), Natpakal Rerknithi (HK), Anna Gamvros (HK) and Ruby Kwok (HK) on August 6, 2018 Overview of Thailand Draft Personal Data Protection Act, https://www.dataprotectionreport.com/2018/08/overview-of-thailand-draft-personal-data-protection-act/ accessed 12 October 2018.

[61] Ibid.

[62] Greenleaf, G Suriyawongkul, A Thailand's draft data protection Bill: Many strengths, too many uncertainties *Privacy Laws & Business International Report*, (2018) pp. 23–25.

[63] Ibid.

[64] Ibid.

consent. The collection of personal data requires the consent of the data subject before or at the time of collection; that consent must be written, and may be withdrawn at any time.[65] Notice must be given to the data subject about specified matters at the time of collection. The collection must generally be from the data subject. There is a general exception for when another law provides otherwise (the PDPB is inferior to other laws). There are also specific exceptions for collection without consent, including, for example, for a public interest or 'legitimate interests of the controller' (with a test of balancing against the fundamental rights and freedoms of the data subject, as in the GDPR); or as authorized by law; or as prescribed in Ministerial Regulations.[66] Data collection can be 'only carried out to the extent that it is necessary within a lawful purpose of the personal data controller', which implies minimal collection only.[67] The exceptions of the requirement of consent for collecting personal data is proposed to be enumerated in section 24.

Moreover, the proposal is considering how to manage the collection of personal data to bring the laws into line with other countries and ensure any collection is undertaken for the lawful purpose and be directly relevant to, and necessary for, the activities of the data controller. Based on the current draft proposal, the data controller must inform the data subject of the following, prior to or at the time personal data is collected:

- the purpose of the collection;
- the personal data to be collected the period for which such data will be kept;
- to whom the personal data might be disclosed;
- contact information of the data controller; and
- the rights of the data subject.[68]

The above principles and concepts are reflected and widely used in the EU. Furthermore, this information would usually be provided by way of a collection notice. Except under limited circumstances prescribed under the Draft Act, personal data must be collected directly from the data subject (the exceptions are set out in Section 25).[69] Also, the collection of sensitive personal data, such as racial and ethnic origins, religious and philosophical believes, political preference, sexual behaviour, criminal records, medical records, information in relation to labour union membership, biometric data or genetic data, is prohibited, can be undertaken except under limited circumstances prescribed by the Draft Act or ministerial regulation.[70] Furthermore, additional categories of sensitive data shall be determined by the Personal Data Protection Commission. These principles replicate what constitutes

[65] Ibid.

[66] Ibid.

[67] Ibid.

[68] Ibid.

[69] Ibid.

[70] Ibid.

sensitive data in other jurisdictions such as Australia, and more general data in Singapore and the EU. Examples of the permitted circumstances for the collection of sensitive data include where sensitive data is collected to protect or prevent harm to a person's life, body or health, or to comply with any legal requirement on the data controller.[71]

Personal data can only be transferred to another country, where:

- the transfer is made pursuant to any applicable law;
- consent is obtained from the data subject on the condition that the data subject has been informed of the insufficient data protection in the recipient country/international organisation;
- the transfer is necessary to comply with the contract to which the data subject is a party or to carry out what the data subject requests before entering in to a contract
- the transfer is in compliance with the contract entered into between the data subject and the data controller for the interests of the data subject;
- the transfer is to protect, or suppress the danger to the data subject or other person's life, body and health in the case that the data subject is incapable of giving consent; or
- the transfer is necessary to carry out missions for important public interest.[72]

Under the proposal, a data subject would be entitled to access his/her own personal data which is held by the data controller, or to request the data controller to disclose the sources of information where such personal data is collected without his/her consent.[73]

A diluted proposal for the right to be forgotten could arise from these draft laws, when compared with the EU. It is proposed that, in the event that the data controller fails to comply with any provision of the proposed law, a data subject is entitled to request the data controller to delete, destroy, temporarily suspend the use of personal data. The problem with this limited proposal, it may not allow a data subject to request that their personal data be deleted or destroyed at any time.

Greenleaf and Suryawongkul highlight that the administrative structure that is likely to be established will be complex and include:

- The Personal Data Protection Committee;
- The Office of the Personal Data Protection Committee;
- The Secretary-General of the OPDPC;
- The Oversight Committee of the OPDPC; and
- Panels of experts for mediating complaints (Panels).[74]

---

[71] Ibid.

[72] Ibid.

[73] Ibid.

[74] Greenleaf, G Suriyawongkul, A Thailand's draft data protection Bill: Many strengths, too many uncertainties *Privacy Laws & Business International Report*, (2018) pp. 23–25.

However, Greenleaf and Suryawongkul argue that it appears that there will be very little independence from Central Government. Thus, the complex structure could in effect reinforce the current Government's control of the personal data of Thai citizens.

The proposal would also see an increase in fines to THB500,000 and/or imprisonment not exceeding 6 months for data controllers and/or a maximum fine of THB1m and/or imprisonment not exceeding 2 years for data controllers. However any future terms of imprisonment or fines would be subject to whether the offence is committed in order to unlawfully benefit the data controller or another person, or to cause damage to another person. When assessed against current day exchanges rates of the Euro, Australian or Singapore dollar, the fines do not appear to be excessive. Moreover, the proposal may include a provision that requires the data controller to compensate the data owner for any damage caused regardless of intention. The question arises whether such liability would be imposed on the controller personally or on the organization in which the controller is employed.

9.8.1 Potential Issues Concerning the Current Draft Bill – January 2018

The 2018 draft Bill on data protection has generally been seen by the community and industry as a significant step forward for the country, if and when it is fully implemented. However, some have raised concerns over the draft Bill. This section is not an exhaustive discussion regarding industry concerns in relation to the proposed Bill. It is out of scope to review all such concerns. Therefore, this section highlights some of the concerns raised by an organization and key scholars in this area. This section will only discuss concerns raised in relation to consent, processors, cross border transfers, public sector, breach and Commission powers. At the time of writing the book the proposed Personal Data Protection Act had been approved by the National Legislative Assembly, however, the proposed new laws still required Royal Endorcement and publication on the Government Gazette.

9.8.1.1 Consent

The proposal is set to provide for explicit written consent as a legal basis for handling personal data. Thus, the standard for determining the level of consent that is appropriate should be contextual.[75] In circumstances that do not implicate

[75] BSA The Software Alliance Comments on Personal Data Protection Act https://www.bsa.org/~/media/Files/Policy/Data/02062018BSASubmissionThaiPersonalDataProtectionBill.pdf, accessed 5 November 2018.

heightened sensitivity, implied consent may be appropriate.[76] Arguably, Thailand has its own reasons for adopting such an approach. Implied or deemed consent now forms part of other jurisdictions data protection legal framework. One issue is that, with so much of people's lives being conducted over the internet, express consent may not be practicable. This could impose quite a burden on organizations. Conversely, express consent better informs the community of how their personal data is being collected and used.[77] Thus, one of the benefits arising from this approach could be that Thai citizens generally are better informed about how their personal data is being managed. However, express consent through the use of forms is likely to be considered outdated technology, as the internet continues to make advancements, and provides consent online.

9.8.1.2 Processors

The proposed Section 39(1), will enable a processor to handle personal data that has been received from a data controller. One of the concerns raised has been that there is a need to ensure a controller is to provide further guidance and instructions to the data processors, particularly in relation to their legal obligations.[78] For instance, one organization has highlighted that the data processor should be required by contract to notify the data controller of personal data breaches.[79] It is proposed that section 39(2) requires the data processor to notify the data controller of personal data breaches. However, there is no provision concerning about how the data processor is to notify the data controller. The proposal appears not to fully implement the 'notification' principle. However, the notification principle becomes important because it provides organizations with the opportunity to reduce the risk of harm. It also strengthens the OECD principles of accountability throughout the data protection cycle by placing greater responsibility on organizations to manage the risk pertaining to personal data.

9.8.1.3 Cross Border Transfer

It is proposed that the draft Bill will provide the basis to empower the Personal Data Protection Commission with rules to govern the international transfer of personal data. One concern has been raised as to whether the proposal will impose burdensome restrictions on global data transfers. Another concern is to clarify whether the Bill that data allows controllers to be free to transfer data internationally so long as they continue to protect the data or otherwise comply with international practices,

[76] Ibid.

[77] Ibid.

[78] Ibid.

[79] Ibid.

such as a commitment to abide by the APEC CBPR.[80] It is noted that the accountability model, of the OECD *Guidelines Governing the Protection of Privacy and Transborder Flows of Personal Data* and APEC CBPR, endorses an integrated approach to data protection.[81] What this means is that any organization that collects and uses personal data in cross border transactions is required to take steps to ensure that any obligations under the law are complied with. Moreover, the legal profession has highlighted that the Bill is unlikely to provide safe harbour or exemptions that will fit the nature of digital business, such as, big data and cloud computing business.[82] This could be an issue, because Thailand, may seek to obtain an EU adequacy assessment and approval. Based on the current draft Bill, this alone could pose problems, should Thailand make such an application. It should be noted that, as required by Section 28(1), the data controller has to inform the data subject of the insufficient data protection in the country/ international organization to which the data is transferred, and he/she must receive a consent from the data subject.

9.8.1.4 Public Sector

The proposal will include rendering the public sector also responsible for managing personal data. However, exceptions are likely to be provided to legislative organizations and the courts. The exemption could also extend to security agencies and other state agencies that fall within the *Official Information Act 1997*.[83] The approach taken in Thailand to date is in stark contrast to the EU, to a lesser extent Australia. But, it is consistent with most other countries within ASEAN such as its closest neighbours Malaysia and Singapore.

9.8.1.5 Breach

A proposed has been made to establish a personal data breach notification system that could be applicable to all businesses and organizations. The current proposal in Section 36(4) will require the data controller to notify the data subject of the breach of personal data. However, how the notification will be conducted will be prescribed in the rules and procedures imposed by the Commission.[84] Arguably,

---

[80] Ibid.

[81] Ibid.

[82] Suchit Leesa-Nguansuk, New data law aimed at ensuring privacy, https://www.bangkokpost.com/business/news/1455534/new-data-law-aimed-at-ensuring-privacy, accessed 5 November 2018.

[83] Greenleaf, G Suriyawongkul, A Thailand's draft data protection Bill: Many strengths, too many uncertainties *Privacy Laws & Business International Report*, (2018) pp. 23–25.

[84] BSA The Software Alliance Comments on Personal Data Protection Act https://www.bsa.org/~/media/Files/Policy/Data/02062018BSASubmissionThaiPersonalDataProtectionBill.pdf, accessed 5 November 2018.

without clear direction within the proposed draft Bill, this leaves it open for the Commission to decide when, how and in whom data breaches should be notified. Leaving these determinations to a committee would create uncertainty, even though, the EU and other jurisdictions have applied the words "as soon as practicable" to redress such uncertainty. The make-up of the committee could also pose significant issues to competition and intellectual property law that relates to data protection.

9.8.1.6 Commission Powers

The powers afforded to a Commission are very important to ensure that any future laws in this area are effective. BSA Software Alliance notes there are several issues that remain unresolved where broad powers are provided, not only to the Commission, but also to the proposed Expert Committees.[85] They go onto highlight that there is a variety of open-ended powers for the proposed PDPC "to stipulate measures and guidelines for personal data protection", and "to interpret, make enquiries into, and address issues" that will be provided.[86] Of greater concern is the proposed Section 70(2) which purports to grant the Expert Committee an unspecified level of authority to inspect the actions of a data controller and its employees or contractors regarding personal data that adversely affects data subjects. The current proposal appears to provide quite extensive powers to the proposed Expert Committee, once established. On closer analysis of this provision, it appears that the proposal is consistent with other jurisdictions such as Singapore and Australia. What is not clear, is the level of independence the Commission and Expert Panels will be from central government.

9.9 Conclusion

Of the jurisdictions examined in this book, Thailand lags a long way behind in a significant way, with regard to developing any sought of data protection law (s). The current status and stage of Thailand's development of specific data protection laws, are, arguably at a similar stage to that of Indonesia. There have been a number of draft proposed bills. However, none have progressed to the implementation stage. Apart from the political situation, privacy is only a recent phenomenon in Thailand. Thai tradition does not recognize privacy in the same as the West. Nonetheless, the country is slowly adopting, like many other countries and regions across the world, comparable privacy laws to the West.

---

[85] Ibid.

[86] Ibid.

However, a further notable observation is the fact that the proposed draft bill, in its current form, does not meet all the requirements set out in the APEC Privacy Framework. The APEC Privacy Framework establishes a framework that ensures consumers receive notification about the type of data an online product or service will collect. For instance, the proposal does not adequately fulfill the 'notice principle', which enables consumers to make informed decisions about whether they are comfortable with an online service's data collection practices. In practice, the APEC Privacy Framework recognizes that the operator of an online service may use data it has collected from consumers to the extent that such uses are consistent with the terms described in the notification.

Apart from the social policy issues related to privacy protection over the Internet, a broader economic issue for Thailand is the potential to miss the opportunity to build key strategic business partnerships and opportunities because of its lack of specific data protections laws. In other words, ASEAN has an estimated population of more than 600 million, with a combined GDP of US$2.5 trillion reported in 2016. The ASEAN community is currently the sixth largest economic block in the world with total trade amounting to US$3.7 trillion.[87] The ASEAN Economic Community (AEC), which was established in 2015 will allow businesses to capitalize on opportunities in the region as an integrated market reaching over 600 million people, instead of 10 fragmented economies with lesser impact were their markets not to be integrated.[88] Furthermore, should Thailand formulate up to date data protection laws, it could engage with its neighbors, Malaysia and Singapore, that are both developing economic initiatives to advance their data protection regimes within the new digital economy. For instance, Malaysia has set up the world's first Digital Free Trade Zone (see Chap. 7). Singapore, on the other hand, has a Smart City initiative that is seeing the rapid integration of economic activity through technology.

Finally, Thailand can be best described as adopting a watching brief. It will be interesting to see whether data protection and subsequently privacy over the Internet will remain on its national agenda, following the 2019 national elections. Thailand has the opportunity to become one of the leaders in this area of the law, and could help to set the benchmark for data protection throughout the Asian region. A notable drawback is the limited understanding of data protection and privacy, by the general population. However, it may be that the general population has embraced the Internet and do not care about protecting privacy. Therefore, Thailand and its citizens need to begin to better understand whether those who access the Internet have a concern that their personal data is being used in an environment that supports a minimal regulatory approach.

---

[87] Tse Gan, T *Data and privacy protection in ASEAN – what does it mean for businesses in the region?* Deloitte Southeast Asia (2018) p. 3.

[88] Ibid.

References

Greenleaf, G Suriyawongkul, A Thailand's draft data protection Bill: Many strengths, too many uncertainties *Privacy Laws & Business International Report*, (2018) pp. 23–25

Kamolthamwong, K. *et al.*, Final Report: The Project to Study and Develop the Approach to Personal Data Protection within Asean, https://oer.learn.in.th/search_detail/ZipDownload/58525, (in Thai), pp. 59–93, accessed 27 December 2018

Pirongrong., R, Panichpapiboon, S *Online privacy in Thailand: Public and strategic awareness*, Journal of Law, Information and Science, Vol. 23, No. 1, (2014) pp. 97–136

Pornwasin, A *Govt in race against time to update data privacy law 2018* http://www.nationmultimedia.com/detail/national/30344739, accessed 4 November 2018

Chapter 10
Japan

Abstract In Japan, the Act on the Protection of Personal Information 2016 came into effect on 30 May 2017. However, the road to adopting their current day data protection laws was complex and due, in part, to significant pressures from the community, and the need for the government to ensure trade with the European Union was not impacted in any way. Pressure and concerns also came from the Japanese media, and from the Japanese Federation of Bar Association to establish sound concepts and principles to protect citizen's personal data and information. The media was particularly concerned as a result of personal data and information being leaked from public sector entities. As such, the concerns raised, have in large part been addressed by current data protection laws.

This Chapter discusses the data protection and privacy laws of Japan. While Japan has recognized the right to privacy dating back to the 1960s, their legislation, similar to most other Asian countries is relatively young. It was only recently that Japan's laws have made the EU's 'white list' of mutual recognition, with the European Union. Of the other countries discussed throughout the book, arguably, Japan's laws resemble something close to the EU legal framework, and appear to be providing a similar level of privacy protection to individuals over the Internet.

The Japanese Constitution has confirmed that privacy in Japan is subject to protection. Article 13 of the Japanese Constitution provides that citizens' liberty in private life shall be protected against the exercise of public authority. The Japanese courts have reinforced this Constitutional status of privacy. Nevertheless, it can be confirmed that, as one of liberties in private life, is that every individual in Japanese society has the liberty to protection of his or her own personal information from being disclosed to a third party, or made public without good reason (1965 (A) No. 1187, Judgment of the Grand Bench of the Supreme Court of December 24, 1969, Keishu Vol. 23, No. 12, at 1625, Greenleaf G 'Country Studies – B5 Japan' in Korff, D (Ed) *Comparative Study on Different Approaches to New Privacy Challenges, in Particular in the Light of Technological Developments*, European Commission, Directorate-General Justice, Freedom and Security, May 2010.).

© Springer Nature Singapore Pte Ltd. 2019

R. Walters et al., *Data Protection Law*,
https://doi.org/10.1007/978-981-13-8110-2_10

10.1 Introduction

The bases for the protection of privacy and more broadly data protection in Japan can be traced to a judgment by the Tokyo District Court on September 28, 1964.[1] The Japanese Constitution establishes the right to privacy. Article 13 of the Constitution, according to which"

> All of the people shall be respected as individuals. Their right to life, liberty, and the pursuit of happiness shall, to the extent that it does not interfere with the public welfare, be the supreme consideration in legislation and in other governmental affairs.[2]

In addition to the above, sections 709 and 710 of the Civil Code by court precedents and applied to specific cases through the general provisions of tort law in the Civil Code.[3] It was updated in 2015 and 2016 and aims to protect an individual's privacy rights and interests, while also taking into account the public utility of personal information. The updates to the law have included laws and regulations directed at the proper and effective application of personal information, and how they can effectively regulate personal data and contribute to the creation of new industries and the realization of a vibrant economic society.[4] In addition, the Civil Code also aims to enrich the quality of life for the people of Japan by setting forth the overall vision for the proper handling of personal information. It also establishes measures to protect personal information. This is particularly important to Japan as their advanced information, and communications based society evolves.[5] The laws now require small businesses handling 5,000 or less items of personal information, to also comply with the Act. This was not the case prior to May 2017.

10.1.1 Personal Data Protection

Japan's first data protection legislation came into effect in 2003. The Japanese government's goal was to support trade with Europe by providing suitably strong protection to qualify for European data-export approval.[6] Adams, Kiyoshi Murata and

---

[1] Chairman, M H (2017) *Privacy Culture and Data Protection Laws in Japan 39th International Conference of Data Protection and Privacy Commissioners Thursday*, Hong Kong Personal Information Protection Commission, Japan, https://www.privacyconference2017.org/eng/files/ppt/masao_horibe.pdf, accessed 1 June 2018.

[2] Mannocci, G, *The Public Administration and the Citizens Privacy Protection. A Comparison Between European Union and Japan,* The Italian Law Journal (2018).

[3] Ibid.

[4] Ibid.

[5] Act on the Protection of Personal Information, Ver.2 December, 2016 Personal Information Protection Commission, Japan https://www.PICP.go.jp/files/pdf/Act_on_the_Protection_of_Personal_Information.pdf, accessed 16 December 2017.

[6] Adams., A Kiyoshi Murata., Yohko Orito, K The Development of Japanese Data Protection, *Policy & Internet, Vol. 2, Iss. 2, Art. 5* (2010).

Yohko Orito argue that governmental use of personal data held on computers had been subject to regulation in Japan since 1988. The authors go onto to say that, apart from the goal of supporting European trade, they also saw the opportunity to comply with EU data protection laws. Arguably, Japan, in developing their laws largely based their policy ideas and decisions on the European Data Protection Directive 1995, along with the various voluntary guidelines the EU had established to underpin those laws.[7] There was also pressure from the community and in particular, from the media as a result of people having their personal data leaked from public sector agencies.

The legal system in Japan including the courts have been heavily influenced by the civil law codes of Germany, and to a lesser extent France.[8] Greenleaf and Shimpo note, the Japanese constitution of 1947 has also been influenced by the American common law system. This approach and adoption of law from different regions of the world, highlights how Japanese engaged in legal globalization long before globalization was endorsed by other states globally. Similar to other Asian countries in the region, Japan prides itself on arbitration and mediation to resolve disputes. While being effective at a domestic level, there are signs that this approach could also be effective in data protection, because of the varied legal and cultural approaches taken by other states. It is our view that data protection and privacy law has forced, and will continue to force, countries like Japan to adopt concepts and principles that they may otherwise not do – in other areas of the law that have a domestic rather than an international focus. It is when the law, business, technology and society collide in the international sphere that countries are forced to deviate from their historical cultural and legal backgrounds.

Nonetheless, Greenleaf and Shimpo argue that the 1980 OECD guidelines have heavily influenced the development of regulations in Japan.[9] They propose that this was particularly the case at a local government level, in adopting personal data protection regulations expeditiously and before the enactment of the OECD Guidelines.[10] In fact, in 1973, Japan introduced regulations concerning personal data protection management on computers. These were followed shortly afterwards by the privacy protection regulations adopted by Kunitachi City of the Tokyo Metropolitan area in 1975. The 2003 legislation assumed that local governments in Japan would establish their own regulations, and by April 2006, all 1,742 current local governments had done so.[11]

Notwithstanding the above, the evolution of privacy by the courts in Japan dates back to 1962. The Tokyo District Court stated the right to privacy can be found in

---

[7] Ibid.

[8] Greenleaf G., Shimpo, F *The puzzle of Japanese data privacy enforcement* International Data Privacy Law 4 (2): (2014) pp. 139–154.

[9] Ibid.

[10] Ibid.

[11] Ibid.

Article 13 of the Japanese Constitution.[12] Two years later in *S. Ct., 1965 (A) No. 1187, 23 KEISHŪ 12, 1625 (Dec. 24, 1969)*[13] the Supreme Court held that:

> [a]ll of the people shall be respected as individuals. The right to life, liberty, and the pursuit of happiness shall, to the extent that it does not interfere with the public welfare, be the supreme consideration in legislation and in other governmental affairs.[14]

That case involved a police officer who took photos of street demonstrators on the front lines of a march who were suspected of violating the conditions that the local government imposed when it issued a permit for the demonstration. The photos were submitted to the court as one piece of the evidence. The defendant claimed that taking the photos was illegal because it violated his portrait right.[15] The Court stated:

> that individuals have the right not to have their photos taken without consent. However, it also stated that this right can be restricted when it interferes with public welfare.[16]

When a police officer takes photos of suspected criminals and crime scenes in an appropriate way in a given circumstance, it does not violate someone's right to his portrait, the court said.[17] Arguably, there was no doubt that privacy was the primary consideration in the case, since it involved a level or personal information that could identify the persons photographed. However, it can be argued that the implied interpretation of personal information within these photographs would, today, fall within the current definition of personal information, even though the case was decided long before the current day personal data protection laws had been established.

The importance of personal information, and the right of data subjects to have control over that information was discussed by the Japanese Federation of Bar Associations (JFBA). The JFBA were so concerned with the lack of control afforded to data subjects over their personal data and information that they took the position that:

> it is the right to control one's personal information as a sovereign power of the people over their information.[18]

In 2003, the personal information of citizens' names, birth dates, sex, and addresses, and the assignment of an eleven-digit code to each person, were linked to a government residency registry. Community members argued that the very nature

[12] Tokyo Dist. Ct., 1962 (wa) 1882 (Sept. 28, 1964), 15 KAMINSHŪ 9, 2317, in Online Privacy Law: Japan, https://www.loc.gov/law/help/online-privacy-law/2012/japan.php, accessed 14 October 2018.

[13] *S. Ct., 1965 (A) No. 1187, 23 KEISHŪ 12, 1625* (Dec. 24, 1969), http://www.courts.go.jp/app/hanrei_en/detail?id=34, accessed 14October 2018.

[14] Ibid.

[15] Ibid.

[16] Ibid.

[17] Ibid.

[18] Online Privacy Law: Japan, https://www.loc.gov/law/help/online-privacy-law/2012/japan.php, accessed 14 October 2018.

of the registry violated their rights to privacy in accordance with Article 13 of the Constitution. The Kanazawa District Court[19] and the Osaka High Court[20] held that Jūki Net was unconstitutional. Furthermore, the Osaka High Court argued that the:

> individual's interest in determining how to deal with information concerning his/her private matters (the right to control one's own information) is guaranteed by article 13 of the Constitution, as the right is included in the right to privacy.[21]

The court further said that:

> information concerning a person's name, birth date, address, sex, and resident number is not in and of itself confidential information, but liberty in private lives can still be threatened if it is used against the data subjects' will. Therefore, this information is subject to legal protection and subject to the right to protect one's own information.[22]

The court also found there was a risk of misuse of personal information in the Jūki Net system. However, the Supreme Court reversed the Osaka High Court decision, stating that an individual's name, birth date, address and sex, and resident number are not confidential; that there is no significant system risk of leaking the information; and that misuse by people handling the information is prohibited and subject to administrative and criminal sanctions.[23] Thus, the government's acts to manage and utilize the Jūki Net did not violate the citizens' liberty in private life that was protected under article 13 of the Constitution because the disclosure of personal information to a third party or make such information public without good reason.[24] On the other hand, there was no discussion of consent or the level of control if any, that citizens had over their personal data and information. Had this case arisen in 2016, the likely argument surrounding personal data, would most likely have included the legal definition of personal information, including consent afforded to the data subject. Nevertheless, it can be seen the Japanese courts began to consider personal data and information that identified the data subject some years earlier. They had also considered regulating the use of such date sufficiently important for it to be defined by the law.

In 2010, the Japanese Federation of Bar Associations (JFBA) adopted a resolution demanding the protection of privacy in advanced information and communication networks.[25] In the resolution, the JFBA recommended legislation to protect the right to control personal information.[26] More specifically, it recommended a system whereby a data subject would be notified before his/her information was collected

[19] Ibid, *Kanazawa Dist. Ct., 2002 (wa) No. 836 and 2003 (wa) No. 114 (May 30, 2005), HANREI JIHŌ 1934, 3.*

[20] Ibid, *Osaka High Ct. (Nov. 30, 2006).*

[21] Ibid.

[22] Ibid

[23] Ibid.

[24] Ibid.

[25] Online Privacy Law: Japan, https://www.loc.gov/law/help/online-privacy-law/2012/japan.php, accessed 14 October 2018.

[26] Ibid.

for the purpose and method of collection.[27] It also recommended that the government regulate the collection of data, even if the data did not specify the identity of the data subject such as behavioral targeting advertising.[28] The JFBA and the Advanced Information and Communications Network Society advocated for guaranteeing the right to control one's personal information, and that the state should clearly provide such a mechanism as a principle of law. Additionally, the JFBA called for the legislation to establish concrete principles of law, which included the concept that personal information should be anonymized to the maximum extent possible and that personal information unnecessary for attaining a stated purpose should not be collected.[29] Due to the large quantities of data being collected and stored by organizations, the JFBA recognized the need for the impact of published personal data to be assessed, and the results of such an assessment being publicly disclosed.[30] There was also a need for secure safeguards to be established to ensure greater control over one's personal information, along with the need for the office a commissioner to be established. Doing so would ensure the independent administration of the laws from government.[31] The concepts and principles espoused by the JFBA reinforce the earlier aim of the OECD and APEC models for data protection and privacy. Moreover, these concepts can be seen today in Japan's modern day personal data protection laws.

More recently, in 2017, the Act was amended to strengthen protections and the management of personal data and privacy. The 2017 Act is supported by the 2017 Enforcement Rules for the Act on the Protection of Personal Information (Tentative translation), and the Amendment to the Cabinet Order to Enforce the Act on the Protection of Personal Information, to come into effect on May 30, 2017.[32]

The Amended Act establishes new provisions applicable to the provision of personal data to third parties in foreign countries in response to globalization of corporate activities in Japan.[33] The updates also demonstrated the readiness with which Japan developed its data protection laws to satisfy the former EU Data Protection Directive that sought to facilitate the flow of personal data within, as well as between the EU and other countries like Japan.[34]

The scope of data protection law in Japan provides that Ministries in charge of the implementation of the Act publish their commonly adopted practices and related

[27] Ibid.

[28] Ibid.

[29] Ibid.

[30] Ibid.

[31] Ibid.

[32] Supported by the Specific Personal Information Protection Assessment Guidelines (Tentative Translation), 2014, Specific Personal Information Protection Commission.

[33] Ibid.

[34] Higashizawa N, Aihara Y, *Data Privacy Protection of Personal Information versus Usage of Big Data: Introduction of the Recent Amendment to the Act on the Protection of Personal Information (Japan),* 84 Def. Counsel J. 1 (2017).

guidelines.[35] However, the guidelines are not binding on future ministry activities. Rather, they provide the relevant Ministry with directions on how to interpret and implement data protection law.[36] Moreover, the Act is underpinned by approximately forty guidelines regarding personal information protection have been issued by government agencies including the Ministry of Health, Labour and Welfare, the Japan Financial Services Agency and the Ministry of Economy, Trade and Industry.[37]

10.2 Definition of Personal Information

As reflected throughout the book, most jurisdictions have developed their own structures and definitions of data protection and privacy laws. Applied for the purpose of discussing Japan's legal structures and definitions in this chapter, a 'principal' refers to the individual who is identified by the personal data and information.

Personal information, as defined in Japanese law, like other countries in the Region, is increasingly important to the data subject and the courts. Personal information constitutes that information relating to a living individual which falls under any or each following items:

(i) those containing a name, date of birth, or other descriptions etc. (meaning any and all matters (excluding an individual identification code) stated, recorded or otherwise expressed using voice, movement or other methods in a document, drawing or electromagnetic record; and
(ii) the same requirements apply in Article 18, paragraph (2) that relates to personal information being used as part of written contract or other document (including an electromagnetic record; and
(iii) the individual identification Code.[38]

The law also provides that the handling of personal information must give special consideration to those people who have a physical or mental disability.[39] Japan, rather than define 'sensitive information', has an all or nothing approach and comprises:

- race,
- creed,

---

[35] Unsal B *Protection of Personal Data in Turkey and Japan*, 2 Turk. Com. L. Rev. 187 (2016).

[36] Ibid.

[37] Mannocci, G, *The Public Administration and the Citizens Privacy Protection. A Comparison Between European Union and Japan,* The Italian Law Journal (2018).

[38] An "individual identification code" means those prescribed by cabinet order which are any character, letter, number, symbol or other codes falling under any of each following item.

[39] Enforcement Rules for the Act on the Protection of Personal Information, to be put into full effect on May 30, 2017. Act for Welfare of Persons with Physical Disabilities Act No.283 1949, Act for the Welfare of Persons with Intellectual Disabilities Act No.37 1960, Act for the Mental Health and Welfare of the Persons with Mental Disabilities Act No.123 1950.

- social status,
- medical history,
- criminal record,
- the fact that the individual has suffered damage due to a crime, or 'other description', comparable to sensitive information protected by other countries.[40]

The reference to 'other description' provides a broad approach to allow further information, beyond the list, to be included as part of the definition. These issues could include, but are not limited to, sex or sexual orientation. This has been reinforced by the notion that the definition ensures there is no cause to unfairly discriminate against, prejudice or otherwise unfairly disadvantage the principal.[41] Supporting the above special personal information, is a Cabinet Order which was established to enforce the Act.[42] The Cabinet Order sets out what constitutes further identifiable information such as letter, number, symbol or other codes. This information can assume the form of biometric and other data, but not limited to:

- Deoxyribonucleic Acid (alias DNA);
- appearance decided by facial bone structure and skin color as well as the position and shape of eyes, nose, mouth or other facial elements;
- iris' surface undulation;
- vocal cords' vibration, glottis' closing motion as well as the shape of vocal tract and its change when uttering;
- bodily posture and both arms' movements, step size and other physical appearance when walking;
- intravenous shape decided by the junctions and endpoints of veins lying under the skin of the inner or outer surface of hands or fingers; and
- a finger or palm print.[43]

In addition, other personal information that could be used to identify a person would include passport number,[44] pension number,[45] driver's license,[46] residential record code,[47] individual number,[48] and health insurance card.[49] Japan's approach to defining personal data and information is consistent, with subtle variables, to the other jurisdictions discussed in the book. Article 36 regulates the management of

[40] Act on the Protection of Personal Information (APPI) 2016.

[41] Ibid.

[42] Cabinet Order to Enforce the Act on the Protection of Personal Information, Ver.1 December, 2016 Personal Information Protection Commission, Japan.

[43] Ibid.

[44] Passport Act, No. 267 1951.

[45] National Pension Act, No. 141 1959.

[46] Road Traffic Act No. 105 1960.

[47] Basic Resident Registration Act No. 81 1967.

[48] Administrative Procedure Act No. 27 2013.

[49] National Health Insurance Act No. 192 1958.

anonymous processed information by the information handling business operator.[50] The business handling officer must ensure that there are processes and systems in place that make it very difficult to identify an individual and restore the person's information. They are also responsible for ensuring the anonymous information is secure.[51] When disclosing anonymous information to a third party, the business handling operator must inform the public, of the categories of personal information that will be processed.[52]

In Japan, there have been several recent cases whereby the courts have had to intervene in relation to the misuse of personal information. The first case involved the city of Uji. The City negligently leaked approximately 220,000 personal records from the resident registration system. In 2001, the Kyoto District Court in *Kyoto Chiho Saibansho [Kyoto Dist. Ct.], Feb. 23, 2001, 265 Hanrei Chihoujichi 11* awarded damages of 10,000 yen for each plaintiff. However, in a later case of *Osaka Koto Saibansho [Osaka High. Ct.], Dec. 25, 2001, 265 Hanrei Chihoujichi 11 (Japan). Saiko Saibansho [Sup. Ct.], Jul. 11, 2002, 265 Hanrei Chihoujichi 11,* both the Osaka High Court and the Supreme Court dismissed Uji City's appeals.[53]

A case involving Waseda University was heard by the Supreme Court. The university invited Mr. Jiang Zemin, the former President of China, to lecture in front of a large audience. It provided a list of 1,400 student participants to the Tokyo Metropolitan Police Department for security purposes, but the participants did not consent to the provision of this information. Some students brought actions against the University. Although the Tokyo District Court and the Tokyo High Court dismissed the students' claims, the Supreme Court reversed the decision of the High Court in *Ishini Oyogu Sakana Case [Sup. Ct.], Sep. 24, 2002, 207 Shumin 289,* and awarded damages of 5,000 yen to each student.[54]

In the Yahoo! BB case, subscribers of Yahoo! BB brought an action against the Yahoo Japan Corporation and BB Technology Ltd. for leaking their data.[55] The leakage was caused by a former employee and an acquaintance of his, who stole approximately ten million records by illegally accessing the server. The Osaka District Court in *Osaka Chiho Saibansho [Osaka Dist. Ct.], May 19, 2006, 1948 Hanji 122 (Japan), Osaka Koto Saibansho [Osaka High. Ct.], Jun. 21, 2007,* granted the plaintiffs' claim, which was upheld by the Osaka High Court.[56]

In another case concerning sensitive data. A large aesthetic service provider, Tokyo Beauty Center (TBC), negligently released customers' online questionnaire results, which led to the disclosure of their bust-waist-hip measurements and interest

---

[50] Act on the Protection of Personal Information 2016, Article 36.

[51] Act on the Protection of Personal Information 2016, Article 39.

[52] Act on the Protection of Personal Information 2016, Article 37.

[53] Kreps, D., Fletcher, G., Griffiths, M *Technology and Intimacy: Choice or Coercion* 12th IFIP TC 9 International Conference on Human Choice and Computers, HCC12 Salford, UK, September 7–9, Springer (2016), pp. 88–90.

[54] Ibid.

[55] Ibid.

[56] Ibid.

in epilation services, in addition to their names, ages, addresses, phone numbers, and e-mail addresses.[57] The Tokyo District Court *Tokyo Chiho Saibansho [Tokyo Dist. Ct.], Feb. 8, 2007, 1964 Hanji 113,* granted damages of 35,000 yen to several plaintiffs and 22,000 yen to one plaintiff. Later in the same year, The Tokyo High Court in *Tokyo Koto Saibansho [Tokyo High. Ct.], Aug. 28,* 2007 upheld the decision.[58]

It wasn't until 2014 that, as a result of a court decision that the APPI would be amended. A giant education company, Benesse, leaked approximately 29 million pieces of customer data, including dates of birth, the gender of children, and the names, addresses, and telephone numbers of parents and children.[59] One of the employees of the subcontractor allegedly copied the data list from the firm's database and sold it to three data brokers. The data brokers re-sold the data to other brokers, then, finally, competitors of Benesse bought the data. Benesse sent tradable coupons worth 500 yen to each victim, which did not sufficiently compensate the victims for their damages. As of December 4, 2015, over 10,000 people had sued Benesse, claiming damages of 55,000 yen each.[60] As a result of that case, Kreps, Fletcher and Griffiths note that the APPI was amended. When a business operator handling personal information (business operator) discloses that information from a database to a third party, both parties must keep a transaction record for traceability in accordance with Article 25 of the amended Act.[61] In addition, the third party must confirm the name of the disclosing business operator and the background of such operators who have obtained that data, as required by Article 26 of the Act.[62] For criminal sanctions, where a business unlawfully benefits from the misuse of personal data they face being punished by imprisonment for not more than 1 year or by a fine of not more than 500,000 yen, as outlined in Article 83 of the amended Act.[63] The criminal penalties reflect the current day thinking that personal data is being used for criminal activities. A good example is the theft of peoples' identity, which arguably, contains important personal information that is defined by data protection law.

For more than a decade there has been debate in Japan as to how personal information is to be treated by the courts. The lower courts in the early part of this century were deciding cases in favor of the data subject, only to have their decisions overturned by higher courts. It was not until later from 2007 that the higher courts reinforced the lower court decisions.

Moreover, the amendments of 2017 introduced the concept of "Special Care-Required Personal Information", which broadly corresponds to concepts of "sensitive personal information", as seen in other jurisdictions, most notably in the

[57] Ibid.

[58] Ibid.

[59] Ibid.

[60] Ibid.

[61] Ibid.

[62] Ibid.

[63] Ibid.

EU.[64] According to Higashizawa and Aihara "Special Care-Required Personal Information" means personal information, comprising data which requires special care in handling data, so as not to cause unfair discrimination, prejudice or other disadvantage to the data holder.[65] That personal information or data includes race; creed; social status; medical history; criminal record; and history of being a victim of crime.

10.3 Business Operator [Data Controller]

The Business Operator, who is often referred to in other jurisdictions as the data controller, has extensive obligations for the management of personal data and information in Japan.[66] The Business Operator's accountability is similar to that of a data controller appointed in the EU. Business Operators are also subject to (i) the Act, (ii) the METI Guideline, and (iii) other guideline(s) issued by other Ministries that govern the field of activity of such industry, such as telecommunications, finance, and health, amongst others.[67]

Article 20 requires the Business Operator to ensure the handling of personal information and data is undertaken securely.[68] They are required to ensure that the personal data they are handling is not disseminated or used in an unauthorized manner.[69] They must also ensure that the personal data is not lost or damaged.

The Business Operator is also responsible for his or her employees that handle such personal data.[70] They must provide supervision and ensure that employees are compliant with the requirements of the Act, in making personal data secure. Article 22 adds that the Business Operator must exercise appropriate supervision over an entrusted person, so as to ensure the security control of the personal data.[71] The Business Operator also has responsibility for restricting the transfer of personal data to third parties. Consent is not required when the issues being dealt with are based on other laws and regulations. This can include, but not limited to, matters that need to protect a human life, body or wealth, and when it is difficult to obtain a principal's consent.[72] Business Operators are also responsible when the data is perceived as being of public significance and complex to obtain, such as to enhance public

---

[64] Higashizawa N, Aihara Y, *Data Privacy Protection of Personal Information versus Usage of Big Data: Introduction of the Recent Amendment to the Act on the Protection of Personal Information (Japan)*, 84 Def. Counsel J. 1 (2017).

[65] Ibid.

[66] Act on the Protection of Personal Information 2016, Article 21.

[67] Unsal B *Protection of Personal Data in Turkey and Japan*, 2 Turk. Com. L. Rev. 187 (2016).

[68] Act on the Protection of Personal Information 2016, Article 21.

[69] Ibid.

[70] Ibid.

[71] Ibid.

[72] Act on the Protection of Personal Information 2016, Article 23.

hygiene or promote the health of children. Consent is also not required when there is a need for cooperation between the central and local government organizations. This will apply when there is a possibility that obtaining a principal's consent would interfere with the performance of the task involving the collection, use, storage and dissemination of person data. However, there is no explicit right to allow a principal to withdraw consent to the use of personal data or information defined by the law. This is considered as being a major gap in Japan's laws.

A Business Operator is also responsible to ensure that personal data provided to a third party accords with the Personal Information Protection Commission's rules.[73] The purpose is to inform the principal and the Personal Information Protection Commission of the purpose, categories, method, and reason to cease providing the third party with personal data.[74] The Business Operator is required to inform the principal of the contents that were altered and to ensure they are in a state in which that principal can inform the Personal Information Protection Commission.[75] The Commission is required to disclose to the public a matter relating to the notification pursuant to rules governing the Commission. Notwithstanding the above, in circumstances where a Business Operator who alters the personal information ensure that the personal information is in a state where a principal can easily know.

The Tokyo District Court in 2007, held:

> the Act in relation to the protection of personal information did not provide the data subject with a cause of action against a data controller who withheld the data subject's personal information.[76]

The defendant operated two ophthalmology clinics in Tokyo, and each of the two plaintiffs (patients of one of the clinics) plaintiffs demanded that the defendant disclose their medical records to the defendant in accordance with Article 25-1 of [the PPI Act]. Greenleaf and Shimpo in referring to Fuse and Kosinski summarized the issue as follows:

> The plaintiffs requested the court to interpret [the PPI Act] as providing a private cause of action against the defendant for court-ordered disclosure of the data at issue and monetary compensation. In response, the defendant asserted that the legislature did not intend the PPI ACT to provide a private cause of action because the text of [the PPI Act] provides for extra-judicial conciliation methods (Article 42) and gives a clear grant of authority to the ministries to enforce the [PPI Act] (Article 34-1).[77]

The court adopted the defendant's view. However, there have been criticisms of this decision on three related grounds (i) when there is evidence from the legislative history of the PPI Act, though not from the text of the Act itself; (ii) when the legislature intended to create a civil right of action; and (iii) when the District Court did

[73] Ibid.

[74] Ibid, Article 23 (2).

[75] Ibid, Article 23 (3).

[76] Greenleaf G., Shimpo, F *The puzzle of Japanese data privacy enforcement* International Data Privacy Law 4 (2): (2014) pp. 139–154.

[77] Ibid.

not take the first two grounds into account.[78] The problem imputed by the decision, as highlighted by Greenleaf and Shimpo, is that it will compel complainants to rely on the very limited administrative remedies under the PPI Act, or upon extra-judicial mediation.

10.4 Extra Territorial Reach

The extraterritorial reach of the Act requires that three factors be satisfied. Firstly, the extraterritorial effect applies to a Business Operator in a foreign jurisdiction who has acquired personal information in the course of supplying goods or services to a person in Japan.[79] Secondly, the APPI applies to entities outside Japan if they receive personal information in connection with the provision of goods or services to individuals residing in Japan.[80] This approach taken by the Japanese, reinforces the point that other countries are looking to the EU for direction and guidance in the development of data protection law. The extraterritorial reach of data protection law is becoming an important element to ensure nation states protect the personal data of their citizens in third countries. The Business Operator must keep a record of the date, name or appellation of the third party, and other matters prescribed by rules of the Commission.[81] However, a record need not be kept for the purposes of Article 23. That is, the record must be maintained in accordance with the rules specified by the Protection Commission.

When receiving personal data from a third party, the Business Operator is to 'assess' information including the name and address of the Principal, whether it is a corporate body, and the names of its representatives. A third party is responsible for ensuring that the personal data provided to the Business Operator is in a format and accurate so as not to deceive the Business Operator. The Business Operator is required to keep a record of the personal data received from the third party.[82]

The international transfer of personal data is permitted, provided that the data subject has provided consent. The foreign country's data protection system is to provide the equivalent standard of protection to that of Japan.[83] This also extends to ensuring that the foreign country has taken steps to ensure there are adequate safe guards in place to protect the personal data of the principal and meet the Commission's rules. Noriko Higashizawa and Yuri Aihara note that a business operator must obtain prior consent of the data holder before transferring personal data to

[78] Ibid.

[79] Act on the Protection of Personal Information 2016, Article 25.

[80] Ibid.

[81] Ibid.

[82] Record must be kept in accordance with the rules of the Personal Information Protection Commission.

[83] Act on the Protection of Personal Information 2016, Article 24.

a third party in a foreign country.[84] Therefore, a legal entity located outside of Japan (including a member of the same group of companies to a Japanese company) which has a separate corporate identity from the business operator, will be considered as a third party in a foreign country.[85]

In July 2018, Japan and the EU finalized and agreed upon the recognition of each other's personal data protection systems as being equivalent.[86] This mutual adequacy finding will create the world's largest area of safe data transfers based on a high level of protection of personal data. This decision will complement and enhance the benefits of the Economic Partnership Agreement and contribute to the strategic partnership between Japan and the EU.[87] The importance of equivalency is threefold.[88] Firstly, it is argued that the EU has forced other states into adopting their model. Secondly, it goes some way to harmonizing the laws of data protection and privacy. Thirdly, in our view, it places Japan's legal framework closer to the EU than Singapore of Australia.

---

[84] Higashizawa N, Aihara Y, *Data Privacy Protection of Personal Information versus Usage of Big Data: Introduction of the Recent Amendment to the Act on the Protection of Personal Information (Japan),* 84 Def. Counsel J. 1 (2017).

[85] Ibid, for example, a subsidiary of a Japanese company which is incorporated in a foreign country will be considered as "a third party in a foreign country." In contrast, a representative office and/or branch office of a Japanese company will not be considered as a third party in a foreign country because they are part of the same corporation. A Japanese company must obtain prior consent from a data holder of personal information when it provides personal information to its subsidiaries in foreign countries except in the cases explained below.

[86] Joint Statement by Haruhi Kumazawa, Commissioner of the Personal Information Protection Commission of Japan and Věra Jourová, Commissioner for Justice, Consumers and Gender Equality of the European Commission Tokyo, 17 July 2018, https://www.PICP.go.jp/files/pdf/300717_pressstatement2.pdf, accessed 14 August 2018.

[87] Mannocci, G, *The Public Administration and the Citizens Privacy Protection. A Comparison Between European Union and Japan,* The Italian Law Journal (2018). Mannocci notes that in February 2018, the Japanese PPC reported on a plan to establish additional Guidelines being applicable to personal data transferred from the EU to process it in Japan under the mutual adequacy findings. The PPC recognizes the following major differences between the APPI and the GDPR, and plans to reflect them in the additional Guidelines: Scope of the data subject's rights on the retained personal data – the data subject's rights requesting disclosure, correction, suspension of usage, etc. shall be given to any personal data transferred from the EU regardless of the duration of the data retention period; Sensitive data – personal data regarding sex life, sexual orientation, and labor union membership transferred from the EU shall be treated as equivalent to 'special care-required personal information' under the APPI; Anonymized data – 'anonymization' of personal data transferred from the EU shall mean no one can re-identify a specific individual data subject by discarding decryption keys (different from 'pseudonymization'). Such data is treated as anonymously processed information under the APPI.At the moment, a comprehensive agreement is lacking even, though it is likely to be reached in a few months because it is a cardinal matter fundamental for both.

[88] Ibid.

10.5 Right to be Forgotten

The right to be forgotten has emerged in Japan, albeit subject to some conjecture. The right to be forgotten was recognized by Judge Hisaki Kobayashi from Saitama District court in Tokyo, in 2015.[89] The court ordered Goggle to remove information about a person's criminal record from its link. The court ruled that, depending on the nature of a crime, the right to be forgotten should be recognized with the passage of time.[90] Criminals who were exposed to the public due to media reports of their arrest are entitled to the benefit of having their private life respected and their rehabilitation unhindered. Judge Hisaki Kobayashi went further, arguing that it is extremely difficult to live a calm life once information is posted and shared on the Internet. It is this point that the court determined as critical when determining whether (the information) should be deleted.[91] This appeared to be a watershed moment in the recognition of broader rights to privacy in Japan. However, the right to be forgotten has been short-lived and in 2016, and the Tokyo High Court overturned the District court's decision.[92] The court stated that the right to be forgotten is not a privilege stated in law and its prerequisites had not been determined. As the data protection laws continue to develop in Japan, how the courts and legislature deal with and respond to the right to be forgotten will need to be watched carefully.

Moreover, in 2017, the Supreme Court presented the general criteria to be considered in judging whether it would be unlawful for search engine companies to keep providing information (URLs) containing privacy-sensitive articles. The traditional personality right under the Civil Code Article 709 may deal with issues of de-listing in Japan if the privacy harm is brought about by the original publisher.[93] The Court highlighted that this should be determined by "balancing" the legal interest for non-disclosure with the commercial and public interest rationales in support of personal information being provided via search engines.[94] Circumstances which may be considered include: the nature and details of the facts being disclosed, the extent to which facts belonging to the person's privacy is transmitted through

[89] The News Paper, http://www.tnp.sg/news/world/japan-court-rejects-mans-right-be-forgotten-google, accessed 28 December 2017.

[90] Ibid.

[91] Japan Times, https://www.japantimes.co.jp/news/2016/02/27/national/crime-legal/japanese-court-recognizes-right-to-be-forgotten-in-suit-against-google, accessed 28 December 2017.

[92] Tokyo High Court overturns man's 'right to be forgotten, https://www.japantimes.co.jp/news/2016/07/13/national/crime-legal/tokyo-high-court-overturns-mans-right-forgotten/#.W8mObvZuLcs, accessed 18 October 2018.

[93] Miyashita, H *The Right to Be Forgotten and Search Engine Liability*, Brussels Privacy Hub Working Paper, VOL. 2 (2016), A person who has intentionally or negligently infringed any right of others, or legally protected interest of others, shall be liable to compensate any damages resulting in consequence.

[94] A Right to be Forgotten Case before the Japanese Supreme Court, http://blog.renforce.eu/index.php/en/2017/02/07/a-right-to-be-forgotten-case-before-the-japanese-supreme-court/, accessed 18 October 2018.

secondary channels such as via URLs. In addition, further considerations include, the degree to which the data subject thereby suffers concrete damage; that person's social status and influence; the purpose and meaning of the said [website] articles; the social situation at the time the articles were published; social changes afterwards; and the need to include the relevant facts in the articles.[95] If the legal interest for non-disclosure clearly "outweighs" the reasons for providing that information, search engine providers can be requested to remove the relevant URLs from research results.[96] The Japanese Supreme Court did not find it necessary to oblige Google to remove information about the data subject. According to the Court, child prostitution is a penalized act subject to strong social criticisms and thus the arrest of the appellant still remained in the public interest when Google transmitted the information.[97] The Court also found that the information transmission was limited in its scope, as the search results depended on the appellant's name and his prefecture.[98] Japanese history highlights that it is generally conceived as being homogenous society, with limited multiculturalism, with Shintoism as its predominant religion, followed by Buddhism, and with Christianity having a minimal presence.[99]

10.6 Commissioner – Regulator

The Personal Information Protection Commission (PIP) came into effect in 1 January 2016. The PIP is a recent addition to the data protection regulatory framework. The PIP is responsible to protect the rights and interests of individuals, while taking into consideration the proper and effective use of personal information. It is independent of the government and has the power to monitor and oversee the implementation of the data protection laws.

Generally, the PIP can receive reports in relation to anonymized information from data users or private business operators. It has the power to conduct onsite inspections of offices and other related premises where it is believed personal data and information is being miss-handled. Similar to the Australian model, it can also provide guidance, advice and other information in relation to anonymized information. In particular, the PIP has accentuated Japan's legal framework within which it has increasingly promoted awareness of personal data protection, even though Japan's laws on the subject differ from other states in the region.

Regarding the nature and function of the PIP, Japan has not appointed a Commissioner. The Commission is headed instead by a Chairperson and eight Commission members who are appointed by the Prime Minister. These appoint-

---

[95] Ibid.

[96] Ibid.

[97] Ibid.

[98] Ibid.

[99] Religious Facts, Japan, http://www.religionfacts.com/japan, accessed 18 October 2018.

ments must also have the consent of both houses of the Diet. The PIP is an independent organ in the Japanese legal framework and functions very much within a top down organizational structure. Its central purpose is to work with the central government to regulate the proper and effective use of personal information.

In fulfilling its mandate, the PIP facilitates the development of guidelines to ensure that there is effective implementation of central government laws, notably in regulating how business operators deal with personal information. This includes developing support measures in developing and implementing central government laws that are applied by local governments.[100] This includes ensuring that complaints handled promptly;[101] and ensuring that necessary action is taken to ensure the proper handling of personal information.[102]

Primary duties of the PIP include to formulate and promote data protection and privacy policies that are consistent with the mandate of the central government.[103] These regulatory duties represent a common and important theme across all the jurisdictions studied in this book, because without organizations like the PIP performing these functions, citizens of states would not understand the complex data protection and privacy laws in their jurisdictions.[104] Typically, the PIP provides a supervisory function for government, industry and the wider community. It reports to the Diet (Bicameral Legislature) and has an important role in handling complaints and ensuing mediation.[105] In addition, the PIP is expected to promote international cooperation in the regulation of personal data, such as with partner states like Australia and Singapore.

The PIP's powers also extend to requesting reports, conducting on-site inspections, and providing recommendations to improve an organization's security measures.[106] The Specific Personal Information Protection Assessment can be undertaken to evaluate the danger and impact of leakages of personal data.[107] That assessment is aimed at preventing the ongoing impact arising from the loss or unauthorized use of personal data. The powers and non-appointment of a Commissioner in Japan are significantly different to Australia and the European Union.

The PIP can require a personal information handling business operator to submit necessary information or material relating to the handling of personal information or anonymously processed information.[108] Its officials are able to enter a business office to inquire about the handling of personal information, inspect books, documents and other property. The Commission, when there is a need to protect an

[100] Act on the Protection of Personal Information 2016, Article 8.

[101] Act on the Protection of Personal Information 2016, Article 9.

[102] Act on the Protection of Personal Information 2016, Article 10.

[103] Personal Information Protection Commission Japan, https://www.ppc.go.jp/en/aboutus/roles/mediation/, accessed 20 December 2018.

[104] Ibid.

[105] Ibid.

[106] Ibid.

[107] Ibid.

[108] Act on the Protection of Personal Information 2016, Article 40.

individual's rights can recommend the business operator to suspend current operations and to take action to address the violation in issue.[109] For more serious violations, such as not taking the required actions to redress a recurrent complaint, the PIP can request further work to be undertaken until the matter is resolved.

A corporation that handles personal information can request accreditation from the Commission in performing its functions in collecting, use and retention of personal data.[110] The accreditation helps the organization and more broadly, Japan, to promote best data management practices.

10.7 Public and Private

The APPI Act regulates both public and private sectors in Japan. It places a clear responsibility on government to develop and implement necessary measures to ensure the proper handling of personal information.[111] That responsibility also lies with local government.[112] Furthermore, Article 6 of the Act requires that the government takes necessary legislative action to protect personal information in collaboration with other countries.[113] However, there are specific exemptions that are sectorial based and that include the media and journalists, universities and other certain academic institutions. In addition, some religious groups and political parties are exempt from the Act, although they are not specifically identified. The APPI requirements also do not extend to the processing of personal data for purposes of journalism, academic research and religious and political activities.

Moreover, in accordance with Article 7 of the Act, the government has established the Basic Policy on the Protection of Personal Information (Basic Policy) which was approved by the Cabinet on 2 April 2004. The Basic Policy highlights the measures to be taken by local public bodies and other organizations, including businesses when handling personal information.[114] In accordance with Article 15(1), a business operator handling personal information must, as far as possible, specify the purpose of that use. The purpose specification is also a common requirement by other jurisdictions.[115] It restricts the use of personal data to a specific matter and ensures that businesses cannot use the data for other purposes. The Basic Policy aims to maintain society's trust of business activities. This includes ensuring that businesses announce their appropriate initiatives for processing complaints. This further entails formulating and announcing their policies (so-called privacy policies or privacy statements, etc.) and philosophies on the promotion of the personal infor-

[109] Act on the Protection of Personal Information 2016, Article 42.

[110] Act on the Protection of Personal Information 2016, Article 47.

[111] Act on the Protection of Personal Information 2016, Article 4.

[112] Act on the Protection of Personal Information 2016, Article 5.

[113] Act on the Protection of Personal Information 2016, Article 6.

[114] Act on the Protection of Personal Information 2016, Article 7.

[115] Act on the Protection of Personal Information 2016, Article 15(1).

mation protection so as not to use personal information for multiple and non-transparent uses. Consistent with this Basic Policy is the need for businesses to explain, in advance and in an easy-to-understand manner, their procedures relating to the handling of personal information, such as notification and announcement of the purpose of use and disclosure of that information, as well as complying with relevant laws and ordinances governing the use of that information.

The Basic Policy (the Policy) has become an important tool for Japan to promote and strengthen the protection of personal information. However, Japanese law does not specify what principles need to be incorporated into the Policy. Rather it is only concerned with an action to be taken by a personal information handling business operator, or in the anonymously processing of information by an accredited personal information protection organization as prescribed by Article 50.[116] The Policy specifies that the central government provide local government and incorporated administrative agency, local incorporated administrative agencies, with information to develop guidelines. The Policy also ensures that there is a process in place to promptly deal with privacy complaints and further promote the protection of personal information. The Policy is to also provide an outline of a complaints handling process and promoting measures to ensure that people and organizations understand the issues related to the handing personal information.

10.8 Retention

The retention of personal data by the Business Operator [117] must be undertaken and include the following, the name of the business operator, an outline of the intended use of the personal data. In situations in which the principal can be identified from the personal data, the Business Operator is to be informed.[118]

The principal can require the Business Operator to disclose any personal data being retained.[119] In these circumstances the Business Operator must disclose the personal data to the principal without delay as prescribed by the method outlined in a cabinet order. However, personal data that could harm a person's life, wealth, or other rights does not have to be disclosed. The personal data does not have to be disclosed where that data may interfere with Business Operators undertaking their business. Article 19 requires the Business Operator to ensure that any personal data received, processed, used and disseminated is accurate, up to date and within the parameters of the purpose underlying the use of the personal data.[120]

The Accountability and Privacy Impact Assessments (APPI) information handlers are required to take necessary and appropriate measures to ensure the security

[116] Act on the Protection of Personal Information 2016, Article 50.

[117] Act on the Protection of Personal Information 2016, Article 27.

[118] Act on the Protection of Personal Information 2016, Article 18.4.

[119] Act on the Protection of Personal Information 2016, Article 27.

[120] Act on the Protection of Personal Information 2016, Article 19.

of personal information.[121] The APPI Guidelines require each information handler to: (i) have a basic privacy policy in place; (ii) have internal rules and other internal documentary arrangements that are designed to protect personal data; (iii) have organizational structures that are designed to protect personal data; (iv) fully educate its officers and employees on data protection requirements; (v) have appropriate physical security systems; and (vi) take appropriate measures in relation to information technology systems.[122]

10.9 Collection [Acquisition] and Consent

In addition to the above, the Business Operator is charged with the responsibility not to acquire personal data by deceitful or improper means.[123] Article 23 is important because it provides the general rules that apply for multiple arrangements in which Personal Data can be transmitted, including bought and sold.[124] Furthermore, Article 24 ensures special rules apply to Personal Data provisions to a third party in overseas. That is, Article 24 requires that, in order to provide Personal Data to a third party overseas, a business must (i) obtain prior affirmative consent for cross-border provision from relevant individuals (ii) and upon identifying either the countries to which the Personal Data will be provided or situations in when the Personal Data has been provided to an overseas entity.[125] In addition, the recipient of the personal data from a third party outside of Japan, can do so, provided they are located in an overseas country designated by the PIP as having a personal data protection system equivalent to the standards of Japan, and that the provider and recipient ensure appropriate and reasonable measures have been established to protect that personal data. The recipient must also meet the relevant standards established by APEC - CBPR. This quality assurance approach to meet external country standards goes part way to strengthening the co-regulatory framework to data protection and privacy.

Moreover, the METI Privacy Guidelines 2007[126] reinforce the co-regulatory approach and provide guidance in the use of the term "consent". These represent soft law instruments aimed at directing and guiding organizations to manage the

[121] Cultures of Accountability, A cross-cultural perspective on current and future accountability mechanisms

https://www.law.kuleuven.be/citip/en/news/item/coa-workshop-report.pdf, accessed 2 September 2018.

[122] General Guidelines regarding the Act on the Protection of Personal Information, November 2017.

[123] Act on the Protection of Personal Information 2016, Article 17.

[124] Act on the Protection of Personal Information 2016, Article 23.

[125] Act on the Protection of Personal Information 2016, Article 24.

[126] Guidelines Targeting Economic and Industrial Sectors Pertaining to the Act on the Protection of Personal Information, Economy, Trade and Industry 2007.

collection, use and transaction of personal data appropriately. The Guidelines state that consent of the person means the concerned person's declaration of intent. It requires that the person agrees that personal information about the individual handling the data according to the method presented by the entity.[127] Furthermore, the phrase "obtaining the consent of the person" means that the concerned entity handling personal information recognizes the person's declaration of intent in which the person agrees, and it must be done in such a reasonable and appropriate manner that is deemed necessary for the person's to exercise judgment in providing or withholding consent.[128] This process is dependent on the nature of business and the status of handling personal information. Where a child has no ability to understand the results arisen from his or her consent to the handling of personal information, it is necessary to obtain the consent from his or her attorney. Arguably, children have become one of the most vulnerable groups in the community, and the current fragmented approach to data protection law, does not fully protect minors. This does not only apply in Japan, it also relates to the other jurisdictions covered in this book.

10.10 Notification

The Business Operator has responsibility for notifying the principal when they have acquired personal data related to them.[129] When there is a contract developed between the Business Operator and principal, the contract must state the purpose for which the personal data will be used. The Business Operator must also notify the principal of the purpose to which their personal data will be used.

Unlike other jurisdictions, there is no requirement for a breach to be reported to the Regulator. Even so, the PIPC Guidelines[130] do recommend that such notification be given, as it is considered common practice to do so. Those organizations that do not report a breach, risk the breach becoming public, posing significant reputational damage to that organization. The PIPC Guidelines recommend that companies make necessary investigations and take necessary preventive measures.[131] In addition, the company should make public the nature of the breach, as well as take the necessary steps to address it. It is also recommended that a voluntary notice be sent to the data subject of the breach or to publish the data breach, if that is necessary. The PIPC Guidelines promote self-regulation, and for companies to take responsibility for the ways in which they comply with the management of personal data.

---

[127] Ibid.

[128] Ibid.

[129] Act on the Protection of Personal Information 2016, Article 18.

[130] Personal Information Protection Commissioner, https://www.ppc.go.jp/en/legal/, accessed 20 December 2018.

[131] Ibid.

10.11 Enforcement & Breach

Greenleaf and Shimpo have noted that Japanese legislation until 2017 had not been implemented sufficiently long to fully understand the effectiveness of enforcement.[132] In 2017, the Japanese Commission released the Enforcement Rules for the Act on the Protection of Personal Information. The Business Operator could be subject to fines of up to JPY 300,000, if it does not submit the report and materials, or reports false information.[133] The unauthorized disclosure of personal data and information could lead to imprisonment for up to 1 year or a fine of up to JPY 500,000.[134] To date there is little information available regarding the enforcement of any breach. Furthermore, the Personal Information Protection Commission mediates complaints regarding the disclosure of personal information. The PIPC provides a free telephone number for any individual to make a complaint. The fines that can be imposed are considered criminal penalties, and can be imposed on an officer, employee or the organization, depending on the nature and extent of the breach.

10.12 Supporting Laws and Policy

Given that a data subject can be confused over the laws governing the protection of personal information in the Act, Japan has developed a number of legislative, and policy guidelines to support its implementation.[135] Japan, like many other countries, including the EU, have adopted a co-regulatory approach to managing and protecting personal data, information and privacy. On the other side, the Telecommunications Business Act of 1984 regulates the secrecy of communications arising from intrusion

[132] Greenleaf G., Shimpo, F *The puzzle of Japanese data privacy enforcement* International Data Privacy Law 4 (2): (2014) pp. 139–154.

[133] Ibid.

[134] Ibid.

[135] These include the Amendment to the Cabinet Order to Enforce the Act on the Protection of Personal Information;

Enforcement Rules for the Act on the Protection of Personal Information (Rules of the Personal Information Protection Commission No. 3 of 2016; Guidelines on the Act on the Protection of Personal Information (Anonymously Processed Information) (Public Notice of the Personal Information Protection Commission No. 9 of 2016); Guidelines on the Act on the Protection of Personal Information (General Rules) (Public Notice of the Personal Information Protection Commission No. 6 of 2016); Act for Partial Revision of the Act on the Protection of Personal Information and Act on the Use of Numbers to Identify a Specific Individual in Administrative Procedures (Act No. 65 of 2015); Specific Personal Information Protection Assessment Guidelines (Tentative Translation) 2014Specific Personal Information Protection Commission; and Guidelines Targeting Economic and Industrial Sectors Pertaining to the Act on the Protection of Personal Information, Economy, Trade and Industry 2007, which are based on the "Policies Concerning the Protection of Personal Information. The Ministry of Economy, Trade and Industry holds jurisdiction, and specific sectors in which the Minister of Economy, Trade and Industry is designated as a competent minister pursuant to Paragraph 1 of Article 36 of the Act.

by private organizations and institutions. Finally, marketing emails are restricted under the Act on Regulation of Transmission of Specified Electronic Mail of 2002 and the Act on Specified Commercial Transactions of 1976.

10.13 Conclusion

The general concept of privacy in Japan began to fully emerge following the second world war. However, like many other countries, Japan did not even contemplate privacy or protecting data subjects' personal identifying information until recently. In our view, Japan have taken their international obligations very seriously to ensure that they include within it national framework the core principles adopted by the OECD and APEC on data protection and privacy. This is also reflected in the fact that Japan has recently received approval from the EU having data protection laws that are equivalent to those in the EU. Furthermore, Japan has understood that government regulation alone will not provide a comprehensive legal framework that will solve all the issues related to data protection and privacy. Thus, Japan have also adopted a co-regulatory framework.

The Japanese personal protection laws began their journey in 2003. Since then, they have undergone significant reform. The first ever comprehensive changes to the legislation did not take place until 2015. They were directed at clarifying and strengthening areas of the law relating to personal protection. The changes bought the laws into the modern era, whereby it clarified cross border transfers of personal data, and went some way to harmonizing Japan's laws with that of the EU. The establishment of the independent Personal Information Protection Commission was a significant step forward – to strengthen the oversight of data protection law in the country. What followed was a significant expansion in the establishment of sector specific guidelines that now include sectors such as the medical, financial and telecommunications. The Commission and relevant governmental ministries have published sector-specific guidance providing for additional requirements, given the highly sensitive nature of personal information handled by private business operators in those sectors.

Today, the laws reflect a similar legal framework to that of the EU, Australia and other Asian nations. It is our view with Japan obtaining adequacy status from the EU, demonstrates that their laws are closer to the EU framework, than that of Singapore or Australia. The laws are based on obtaining consent from the principal to whom the personal data applies. The concept of consent, like many other jurisdictional laws, has become an important in regulating the use or personal information and its disclosure to third parties. The Business Operator is required to assume a significant level of responsibility under Japanese law, which is a similar to requirements in the EUs. However, there appears to be confusion and debate in Japan about whether the right to be forgotten exists, or should be applied.

Finally, a person can seek compensation where they can prove that a business operator has breached Japanese and has caused a loss, including but not limited to a loss of reputation. Japanese law provides that the quantum of compensation will be assessed on a case by case basis and depend on the actual loss incurred. However, the extent to which Japanese courts will recognize a data subject's loss and provide a remedy, remains uncertain.

References

Adams, A Kiyoshi Murata., Yohko Orito, K (2010) The Development of Japanese Data Protection, *Policy & Internet, Vol. 2, Iss. 2, Art.* 5

Chairman, M H (2017) *Privacy Culture and Data Protection Laws in Japan 39th International Conference of Data Protection and Privacy Commissioners Thursday*, Hong Kong Personal Information Protection Commission, Japan, https://www.privacyconference2017.org/eng/files/ppt/masao_horibe.pdf, accessed 1 June 2018.

Greanleaf G 'Country Studies – B5 Japan' in Korff, D (2010) *Comparative Study on Different Approaches to New Privacy Challenges, in Particular in the Light of Technological Developments*, European Commission, Directorate-General Justice, Freedom and Security

Greenleaf G., Shimpo, F (2014) *The puzzle of Japanese data privacy enforcement* International Data Privacy Law 4 (2), pp. 139–154.

Higashizawa N, Aihara Y, (2017) *Data Privacy Protection of Personal Information versus Usage of Big Data: Introduction of the Recent Amendment to the Act on the Protection of Personal Information (Japan),* 84 Def. Counsel J. 1

Kreps, D., Fletcher, G., Griffiths, M (2016) *Technology and Intimacy: Choice or Coercion* 12th IFIP TC 9 International Conference on Human Choice and Computers, HCC12 Salford, UK, September 7–9, Springer, pp. 88–90.

Mannocci, G, *The Public Administration and the Citizens Privacy Protection. A Comparison Between European Union and Japan,* The Italian Law Journal (2018).

Miyashita, H (2016) *The Right to Be Forgotten and Search Engine Liability*, Brussels Privacy Hub Working Paper, Vol. 2 (2016)

Unsal B (2016) *Protection of Personal Data in Turkey and Japan*, 2 Turk. Com. L. Rev. 187

Part IV

Chapter 11
Jurisdictional [Comparative] Differences

Abstract This Chapter compares the privacy and data protection laws of Australia, India, Indonesia, Japan, Malaysia, Singapore, Thailand and the European Union. It does not provide a comprehensive discussion of the policy or legal gaps between the respective jurisdictions, which is addressed in the country specific Chapters. This Chapter will focus on comparing the key principles and concepts outlined in each of the jurisdictions. Due to the breadth and depth of the privacy law, it will address and compare the following legal and policy concepts and principles:

- Introduction - Privacy and Data Protection Laws;
- Definitions;
- Application to Public and Private Sectors;
- Control & Enforcement;
- Consent & Collection;
- Storage [Retention & Localisation];
- International Transfer;
- Code of Practice;
- Data Portability; and
- Right to be Forgotten.

The Chapter also highlights that the data and privacy laws of Australia, India, Indonesia, Japan, Malaysia, Singapore, Thailand and the European Union differ, and that these divergences need further explanation. The divergence of approach taken by each jurisdiction makes for a complex legal environment, when having to address a problem that is not limited to a single nation state.

11.1 Introduction

Jurisdictions such as the EU, Australia, Singapore, Japan and Malaysia have established comprehensive data protection laws. To a lesser extent, Indonesia and India have begun to develop their privacy and data protection laws. However, at the time of writing this book, Thailand was a way off from fully implementing their similar laws. More importantly, there are significant differences between Australia, India,

© Springer Nature Singapore Pte Ltd. 2019

R. Walters et al., *Data Protection Law*,
https://doi.org/10.1007/978-981-13-8110-2_11

Indonesia, Japan, Malaysia, Singapore, Thailand and the EU privacy and data protection laws. Arguably, the EU has and continues to lead the way in data protection law and policy. The EU have not only expanded their data protection laws in recent times, they have elevated the laws from a Directive to a Regulation. This is a significant step, because in the EU context, a Regulation has greater authority than a Directive.

This Chapter will repeat some of the material outlined in earlier chapters, so as to highlight the comparative differences. One of the core reasons for this varied approach across the countries studies is, on the one hand the acceptance and treatment of privacy as a fundamental right. On the other hand, there are differences in the manner in which that right is construed and protected including in relation to personal data. One of the main drivers is that different states have different social, economic and cultural traditions and sovereign needs that, have had a significant influence on the development of data protection law. Importantly, there is a lack of consensus of what privacy constitutes over the Internet. However, it is argued that the fundamentals of data protection and privacy law explored in this book indicate a beginning in the development of a level of consensus in relation to personal data.

Dating back to 2006, Bennett and Raab developed the idea of dividing data protection and privacy laws into regions.[1] Their idea of policy blocs can be best described as the OECD and European Union Council being one entity in regard to the data protection and privacy policy they have adopted. For Bennet and Raab, the Asia-Pacific Region constitutes the United States, Australia and Southeast Asian countries of Japan, China, and Korea, as a new bloc.[2] Their idea of grouping the Asia-Pacific countries the way they did, is somewhat different to the way we conceive the make up of this Region. However, Bennet and Raab wrote during a period in which most of South East Asian countries, within ASEAN, had no specific data protection laws.[3] Today, this has changed significantly, with Singapore and Malaysia having introduced specific data protection laws, although Indonesia's data protections law remains in its early development. Nonetheless, grouping data protection and privacy law into blocs is a viable option for comparing data protection, and subsequently privacy law, because these blocs of countries have very different economic and social policy needs.

11.2 The Definition of Personal Data and Personal Information

It is recognised that a direct comparison of data protection policies and laws across the countries studied can be misleading, as the respective laws in these countries often use different terms to mean essentially the same thing. Some countries refer to personal data, while others refer to personal information. In addition, the term "sensitive information" is also dependent on what information is actually treated as

[1] Bennett, C., Raab, C *The Governance of Privacy*. Cambridge, MA: MIT Press (2006).

[2] Ibid.

[3] Ibid.

sensitive and if so, whether to accord greater protection to that information compared to other information. Again, the answers to these questions diverge across the jurisdictions studied.

Generally, personal information and personal data includes but is not limited to: the name, date of birth, and residential address of the data subject. This common approach is not new and has been used as identifying information since modern records of personal information began. The contemporary view of the definition of personal data is derived from the OECD. From the mid-1970s, the OECD sought to provide guidelines to its member states on respecting privacy as a right. These guidelines have influenced the definition of personal data in legislation in many jurisdictions across the world.[4] However, the OECD highlights how the definition of personal data in the twenty-first century is a moving target, due to the new dimensions added by advanced information communication technologies and including intrusive devices, use of biometrics, social media, powerful search engines and maintenance of transnational databases.[5] The OECD[6] defines personal data as any information related to an identified or identifiable person. This includes, but is not restricted to, a data subject's full name, address, occupation, affiliations, physical and mental health, sexual orientation, and even his or her opinions.

The EU via the GDPR has responded to these rapid technological developments by seeking to protect personal data of "natural" persons processed by "automated means", including online identifiers such as Internet Protocol addresses and cookie identifiers that create profiles on individuals and identify them.[7] These basic concepts are also not new and were commonly found in a passports and other identity documents issued by states.

Australia and Singapore specifically state what and how personal data and information is to be defined. Australia defines general personal data and information to be a person's full name, alias or previous name, date of birth, sex, current or last known address, and driver's license. Interestingly, and as detailed in Chap. 5 an important identifying information under Australian law also includes a person's current and last employer. Unlike Singapore and Japan, Australia does not have a national identification card. However, once a person has begun working or undertaking business, no matter what age, that person does have a Tax File Number. But,

[4] Ibid.

[5] Ibid.

[6] The Organization for Economic Cooperation and Development (OECD) Guideline, governing the Protection of Privacy and Transborder Flows of Personal Data' ('OECD Guidelines. The OECD member countries are: Australia, Austria, Belgium, Canada, Chile, the Czech Republic, Denmark, Estonia, Finland, France, Germany, Greece, Hungary, Iceland, Ireland, Israel, Italy, Japan, Korea, Luxembourg, Mexico, the Netherlands, New Zealand, Norway, Poland, Portugal, the Slovak Republic, Slovenia, Spain, Sweden, Switzerland, Turkey, the United Kingdom and the United States of America. The Commission of the European Communities takes part in the work of the OECD, http://www.oecd.org/sti/ieconomy/49710223.pdf, accessed 15 June 2018. A notable absentee from this list is Singapore.

[7] Ibid.

unlike Singapore, that tax file number does not capture every person, because it only applies to those people that are registered to pay tax.[8]

The use of the term 'personal data' in the EU may have some significance, as it was the advent of new technology in the 1970s that resulted in easily accessible datasets that served and the catalyst for the establishment of a data protection framework.[9] The EU GDPR does not apply to non-automated processing of personal data which is not intended to be part of a filing system.[10] India, through its proposed Privacy Bill is considering using the term personal data that will include any form of information. This approach, if adopted, will expand the extent of information that will be used to describe personal information. However, it remains to be seen whether this broad approach will be adopted, and to what extent any exemptions might apply.

Indonesia, along with Malaysia, has adopted a broad approach and defines personal data as individual data that is stored, maintained and kept for correctness[11] and protected for confidentiality.[12] Additionally, 'individual data' means any correct and actual information that relates to any individual and is identifiable directly or indirectly. As being subject to adaptation and implementation under the applicable laws and regulations. The definition could be either viewed restrictively or very broadly and could include those other elements of personal data that other countries have defined as sensitive personal data. Malaysia, having taken a commercial approach to transactions of data and information, does not define what that data or information might be. Rather, Malaysia describes the circumstances in which the data or information might be processed (recorded). Japan also considers personal information to be stated, recorded or otherwise expressed, using voice, movement or other methods in a document, drawing, or an electromagnetic record (see Chap. 10).

11.2.1 Sensitive Information [Data]

Some jurisdictions have begun to identify personal data as being sensitive. The starting point is the fact that some data is perceived as being more sensitive than other data.[13] However, the level of sensitivity differs in two aspects. At the basic

[8] Privacy Act 1988.

[9] Article 29 Data Protection Working Party, Opinion 4/2007 on the Concept of Personal Data', European Commission (20 June 2007), available at: http://ec.europa.eu/justice/data-protection/article-29/documentation/opinion-recommendation/files/2007/wp136_en.pdf, accessed 17 December 2017.

[10] General Data Protection Regulation, Official Journal of the European Union 2016/679. Article 2

[11] Electronic Information and Transaction 20/2016.

[12] Implementation of the Electronic System and Transaction 82/2012.

[13] Etzioni, A (2015) *A cyber age privacy doctrine: More coherent, less subjective, and operational,* Brooklyn Law Review, 80(4), pp. 1263–1308. Pesciotta, D. T. (2012) *I'm not dead yet: Katz, Jones, and the Fourth Amendment in the 21st century* Case Western Reserve Law Review, 63, p. 187.

level, sensitive data is separate from non-personalized data, such as is attributed to "pseudonymized data" and "anonymized data".[14] Secondly, sensitive personal data is distinguished from other personal data that is deemed to be less private.[15] The sensitivity of personal data is arguably one of the most important factors in determining an individual's perception of privacy, and the gradation of sensitivity could decide the security level that controls access to such data.[16] The loss of sensitive data is a significant concern for individuals whose data may be at risk of being publicized or otherwise disclosed.[17] Furthermore, sensitive data is often considered as the core[18] of both privacy and data protection law and requires stricter protection in legislation.

Australia has specified sensitive personal data to include" racial or ethnic origin; political opinions; membership of a political association; religious beliefs or affiliations; philosophical beliefs; membership of a professional or trade association; membership of a trade union; sexual orientation or practices, or criminal record; health information about an individual; genetic information (that is not otherwise health information). In addition, biometric information can be used for the purpose of automated biometric verification or biometric identification, which is being treated as sensitive information [data].[19]

Japan defines sensitive information to be race; creed; social status; medical history; criminal record; fact of having suffered damage by a crime, or other descriptions, which is similar to sensitive information that has been described by other countries.[20] In Malaysia sensitive information has been defined in a similar way to Australia and Japan. However, sensitive data there has been limited to political, religious or other belief, the commission or alleged commission of any offence or any other personal data as the Minister may determine (see Chap. 8). Singapore has generally grouped sensitive and personal data together in a similar manner to the EU. Apart from the full name of the person, Singapore has been for decades consciously implementing mechanisms that can identify people easily such as the National Registration Identity Card (NRIC). Singapore is also the only jurisdiction to mention the passport and mobile phone number as identifiable information

[14] Zuiderveen, B. F. J., Van, E. M., & Gray, J (2015). *Open data, privacy, and fair information principles: Towards a balancing framework,* Berkeley Technology Law Journal, 30(3), p. 2077.

[15] Taddicken, M (2014) *The "privacy paradox" in the social Web: The impact of privacy concerns, individual characteristics, and the perceived social relevance on different forms of self-disclosure,* Journal of Computer, p. 270–272.

[16] Al-Fedaghi, S (2007) *How sensitive is your personal information?* In Proceedings of the 2007 ACM Symposium on Applied Computing (pp. 165–169).

[17] Photopoulos, C (2011) *Managing catastrophic loss of sensitive data: A guide for IT and security professionals.* Rockland, MA: Syngress, p. 3.

[18] Ojanen, T (2014) *Privacy is more than just a seven-letter word: The Court of Justice of the European Union sets constitutional limits on mass surveillance:* Court of Justice of the European Union Decision of 8 April 2014 in Joined Cases C-293/12 and C-594/12, digital rights Ireland and Seitlinger and others. European Constitutional Law Review, 10(3), pp. 528–541.

[19] Privacy Act 1988, section 6.

[20] Act on the Protection of Personal Information (APPI) 2016.

(see Chap. 4). Whatever other jurisdictions describe as sensitive data and information, Singapore has generalised this to include: facial images, voice recordings, fingerprints, iris images and DNA profiling. The EU, rather than define sensitive data, has also adopted a broad approach that is a combination of Malaysia and Singapore.[21] However, neither India, Indonesia nor Thailand provide a clear definition of sensitive information or data.

The definition of personal data or personal information will provide ongoing challenges for governments, regulators, legal profession and entities operating across international borders. Arguably, this area alone needs further work, especially in light of uncertainty over how technology has developed and how it has and will capture personal identifying information in the future. Importantly, in responding to these questions, a continuing concern is whether the definitions of personal data and information, as well as sensitive data in particular, will remain adequate.

11.2.2 Anonymization and Pseudonymization

The EU, Australian and Singapore are the only jurisdictions to identify anonymization or pseudonymization data, but to differing degrees (see Chaps. 3, 4, and 5). Pseudonymization occurs by anonymizing data subject's name and other identifying information, so that the individual cannot be identified. In the same way as anonymization, pseudonmization does not identify the person, or that person's information (name address, biometrics, amongst other personal attributes).

Anonymization constitutes the process used to remove personal data and information so that the data subject is no longer identifiable. This is an important addition to the legal framework in recent years, and strengthens the control and ownership of personal data by the data subject. The terminology varies across jurisdictions and is closely associated with pseudonymization. It relates to reducing the risk to the data of individuals and as a method of meeting data protection obligations. It also prescribes the applicable technical and organizational safeguards.[22] These concepts are becoming an important addition to the overall legal and policy framework for data protection, and the ability for data subjects to have some level of control over their privacy – online.

11.3 Private and Public

The approach to the application of privacy and data protection laws to both public and private organisations varies greatly. The EU, Australia and Japan's current day laws apply to both the public and private sectors. However, Australia has limited the application to the private sector only to those entities with a turnover of more than

---

[21] General Data Protection Regulation 2016/679, Article 9.

[22] Ibid, Recitals 26, 28 and 29.

$3 million. Even though Australia's approach is somewhat limited to a health service provider, amongst others, this, arguably, needs to be reconciled. In part, the sectorial approach taken by India and Indonesia also applies to various public and private organizations. Singapore, along with Malaysia, has restricted the law to only the private sector and exempts the public sector. Thailand's current laws only apply to state agencies. The reason for such a varied approach can be seen in the size, culture, history and sovereign needs of each jurisdiction. For instance, Singapore has created a business friendly environment, and subsequently treated human rights as secondary (see country Chapters). This is the case for most jurisdictions, except the EU. However, it is argued that, as the world becomes increasingly interconnected and where government's interaction with the private sector becomes even more blurred through the use of technology, any exemptions afforded to government agencies are likely to be reviewed at some stage. The areas likely to be untouched are those that relate to the domestic interest such as national security. Other areas of public policy that states share, however, are likely to be reviewed and conceivably, might override domestic public policy concerns. This is likely to be the case in relation to the OECD. Even though some of the jurisdictions discussed in this book are not members of the OECD, concepts espoused by the OECD, such as transparency and accountability, can only be fully implemented when data protection laws are applied to both public and private sectors (see Chap. 16). It is our view that there are areas of the public sector whereby, the public interest test would override narrower domestic conceptions, even those identified with national security.

11.4 Controllers & Enforcement

The establishment of controllers or business operators within the legal framework, arguably establishes a single point of responsibility for the collection, use and storage of personal data. The inclusion of such responsibilities within the legal framework is ground-breaking, as it holds both individuals and entities accountable, and provides a single point of reference for data subjects. This section will discuss whether the different jurisdictions appoint data users who have such responsibility for the management of personal data. This section does not compare their specific functions and is intentionally descriptive in nature.

The multi-layered approach taken by the EU has created four core appointments: (1) Data Controller, (2) Joint Controllers, (3) Processor and (4) Data Protection Officer. However, other jurisdictions do not have the same multi-layered approach because, unlike the EU, they are not part of a supranational polity (see Chap. 3). Their roles also vary, depending on where – within an organization – they are located. On the other hand, Australia makes an organization accountable rather than appoint a particular person responsible for data protection. Australia does not distinguish between a data controller or data processor (see Chap. 5). Data controllers are not appointed in India. Instead, to date, India refer to Grievance Officers who have a limited role and are not accountable for extensive areas of privacy or data protection (see Chap. 6). Similarly,

Thailand does not have a framework that requires data users to be appointed as controllers, joint controllers, processor or protection officers. However, Thailand's proposed data protection law proposes to appoint a controller (see Chap. 9).

Under the laws of Indonesia, there is no legal requirement for a Data Protection Officer to be appointed. Indonesia refers to an Electronic System User (ESU). The ESU is any person, state administrator, or business entity, and the public that uses the benefit of goods, services, facilities, or information that are made available by an Electronic System Provider as stated in Chap. 7, or, country chapter reference Japan, while adopting a similar approach to the other jurisdictions, appoints data users as business operators. The business operator is also obliged to follow the rules of the Commission. There is no requirement for data users to appoint a data protection officer, controller or processor in Malaysia. Although, a data user use may be registered there, this is normally reserved for organizations, for example the airline industry. Singapore, only requires an organisation to designate a person to be responsible for the data protection laws. Singapore does not specify the title of the designated individual within an organisation. Arguably, the appointment of a dedicated controller or some other officer within and organisation will go some way to address the risks associated with the collection and use of personal data. The challenge, which this book has not addressed, is to ensure their functions are the same.

11.4.1 *Notification of Breach*

Data breaches can take many forms including; hackers gaining access to data through a malicious attack; lost, stolen, or temporarily misplaced equipment; employee negligence; and policy and/or system failure. This section is limited to determining whether each jurisdiction requires an organization to notify a relevant authority whether there has been a breach of the law. It does not compare the process or timeframes for notification to be undertaken.

As highlighted in Chap. 5, the Australian *Privacy Amendment (Notifiable Data Breaches) Act 2017* came into effect in February 2018, and requires organizations to notify the Information Privacy Commissioner where there has been a breach of the law. The EU GDPR requires that, in the case of a breach, the controller shall provide notification of the breach without undue delay and, where feasible, not later than 72 hours after having become aware of it (see Chap. 3). Neither do Indonesia, India, Japan, Malaysia, Singapore or Thailand, require that a breach be reported to the Regulator (see Chaps. 4, 5, 6, 7, 8, and 9). Malaysia and Singapore encourage organizations to notify the regulators of breaches. In Malaysia, the Personal Protection Commission Guidelines recommends that notification be undertaken. To date, in Singapore, the Personal Data Protection Commissioner considers that organisations should implement the Managing Data Breaches guide, which only requires an organisation to notify data subjects of a breach, and not the Commission. This varied approach, exemplifies the growing value and need for legal convergence and harmonisation. Doing so, sends a message to the broader community, no matter what country they are located in, that the same data protection procedures will apply there. More importantly, the notification of a breach adds a further but important layer to the overall management of personal data. It is our view that having this regulatory oversight strengthens the deterrence for breaches of the law, and provides a strong platform for entities to improve the management of personal data.

11.4.2 Complaints Mechanism

The enforcement process becomes even more effective by having a complaints mechanism in place. It, in part, goes some way to underpinning the notion of notification of breach. Australia allows a person to make a complaint with the Office of the Australian Information Commissioner. The Commissioner can investigate the complaint to determine whether the organization has established processes to ensure that there is no repeat of the conduct (see Chap. 5). India has established adjudicating officers who are responsible for hearing and deciding cases when there have been breaches of the IT Act (see Chap. 6).

Indonesia has not established a formal complaints mechanism. Furthermore, each country chapter highlights whether the country has adopted such a mechanism. Japan, on the other hand, has through the Personal Information Protection Commission, provided for the mediation of complaints in regard to personal information. Similarly, Malaysia through the Department of Personal Data Protection has established a formal complaints handling mechanism. Singapore, through the Personal Data Protection Commission, has an online complaints and review system. Thailand does not currently have a formal complaints mechanism. However, like Indonesia, there are sector specific systems in place for the banking and other industries. Finally, EU member states supervisory authorities have wide powers to manage complaints made by data subjects.

11.4.3 Penalties

The penalties imposed by each jurisdiction vary significantly. This section only compares the level of penalty that can be imposed in each jurisdiction. This section retains the currency of each jurisdiction, and will not convert the currencies in different jurisdictions into a single currency. As highlighted in Chap. 5, in Australia, the Commissioner must apply for a court order and for civil penalty of up to 2000 penalty units (AU $420,000). The extent to how a penalty will be imposed will depend on the breach, and by whom it is made. If the court is satisfied that the entity has contravened certain provisions of the Privacy Act, it may order the entity to pay a pecuniary penalty as it determines.[23] The Indian IT Act has limited civil penalty to

[23] Privacy Act 1988, section 80(W), body corporate can be fined 5 times the amount of the pecuniary penalty specified for the civil penalty.

a maximum of Rs.25,000 (see Chap. 6). Within EU, the GDPR provides two tiers of administrative fines. Firstly, organisations can be subject to administrative fine up to €10,000,000 million EURO, or in the case of undertakings, 2% of global turnover, whichever is higher. Secondly, an administrative fine can be imposed of up to €20,000,000, or in the case of undertakings, 4% of global turnover, whichever is the higher (see Chap. 3). The Personal Data Protection Commissioner of Singapore, which has been discussed in Chap. 4, can impose fines of up to SG$1,000,000, for breach of the data protection laws. In Malaysia the courts can impose fine of between RM100 to 500,000 to an organisation. The Law Concerning the Electronic Information Technology specifies penalties of fines between Rp600 million to 12 billion. This will vary, depending on the crime and breach. Furthermore, country Chapter discuss at length the level of penalties that countries have adopted. The Japanese Personal Information Protection Commission can issue corrective orders and fines for non-compliance with the laws (see Chap. 10). A breach of a corrective order can result in six-month imprisonment or a fine of up to Y300,000 to the individual and or the business operator. Thailand has no specific penalties, but rather, it is spread across tort and criminal law. The fines across these countries are uneven, but in some cases like the EU, the potential fines are considerable. However, given the wealth generated by organisations such as Google and Facebook, it is questionable whether these levels of penalties that exist today are in fact adequate to deter and seriously penalise breaches of data protection and privacy laws. What is not explored and requires further research is the application of arbitration and mediation in data protection law, particularly where there are disputes between entities that are trading in personal data.

11.4.4 Compensation

Compensation also underpins the enforcement framework underlying the protection of law. The basis for such compensation is that a data subject has incurred harm (loss or damage) as a result of a data controller's failure to comply with the applicable data protection laws, which has been discussed in each country Chapter. In Australia, the OAIC can decide whether an individual is entitled to compensation for any loss or damage.[24] That loss or damage will be assessed according to the injury of humiliation suffered by the individual. However, a decision by the OAIC is not binding and therefore not conclusive.

The GDPR imposes compensation for a person's material and non-material loss. That individual can bring court proceedings before a competent court of the member state in order to protect his or her right to receive compensation for any ensuing loss or harm (see Chap. 3).[25]

---

[24] Privacy Act 1988, section 52.

[25] Regulation 2016/679, Article 82.

Indonesia has provided a very broad approach to compensation, but without providing any specific guidance. Article 26 (2) of the Law Concerning Electronic Information and Transactions states that, unless provided otherwise by Laws and Regulations, use of any information through electronic media that involves personal data of a Person must be made with the consent of the Person concerned. Any Person whose rights are infringed as intended by section (1), may lodge a claim for damages incurred under this Law.

In Japan, a person can claim compensation where they can prove that the business operator has breached the law, which has resulted in a loss to that person. The level of compensation will be assessed on a case by case basis and depend on the actual loss incurred. However, in Malaysia and Singapore, claims of compensation cannot be brought directly to the Regulator, but can be brought to the courts. Thailand regulates that compensation that can be awarded for the loss suffered as a result of a data breach or non-compliance with the laws. Moreover, any compensation is limited to a breach of contract or non-compliance of data protection provisions, and is on a sectorial basis. The difficulty in determining compensation is in regulators and courts to measure the level harm to the data subject (see Chap. 17).

11.5 Consent & Collection

Consent is arguably now one of the most important principles of data protection and privacy law (see country Chapters). Consent is broad and has been used as a tool to verify and confirm that data subjects approve of their personal data and information being collected and processed. The OECD Guidelines regards consent as an important step to the lawful collection and processing of personal data (see Chap. 16).[26] The extent of the consent from the data subject that is required differs from jurisdiction to jurisdiction. Consent relates to both children and adults and the nature and extent of notice that must be given in securing consent from the data subject.

In Chap. 5, it was highlighted how consent in Australia, can be in the form of express or inferred (implied),[27] written, verbal or by silence (implied). The definition of consent requires that an individual be adequately informed of the issues and obligations before giving consent (express or implied).[28] Consent must be current and specific, or voluntary. Importantly, the person must have the capacity to understand and communicate that consent. Generally, India does not require consent for the processing of general personal data (see Chap. 6). However, the rules surrounding sensitive personal data there are somewhat different and do require consent.

[26] Organization for the Economic Co-operation and Development, Guidelines on the Protection of Privacy and Transborder Flows of Personal Data 2013. http://www.oecd.org/sti/ieconomy/oecd-guidelinesontheprotectionofprivacyandtransborderflowsofpersonaldata.htm, accessed 20 February 2018.

[27] *Giller v Procopets* (2008) 24 VR 1.

[28] Privacy Act 1988, section 6.

Such consent can be obtained in a number of forms, and includes letter, fax or email. Electronic consent via tick box such as an 'I Agree' tab is also permitted.

Consent in Indonesia requires prior consent of the person to whom the personal data applies. The process for obtaining consent by an electronic system provider is through a Standard form in Bahasa Indonesia, and agreement sought by the personal data owner. However, because of the sectorial approach taken by Indonesia, the law of consent is not as advanced as in most other jurisdictions studied. This is likely to change when Indonesia establishes specific data protection laws.

The Business Operator in Japan can acquire personal data upon the principal's consent where there is a need to protect a human life, body or fortune (see Chap. 10). The Business Operator must also acquire the principal's consent for the collection of personal data from a successor business or the business operator of another organization arising from a merger. In order to provide Personal Data to a third party overseas, a business must: (i) obtain prior affirmative consent for cross-border provision from relevant individuals (ii) and upon identifying either of the countries to which the data will be transferred (see Chap. 10).

In Malaysia, personal data cannot be processed unless it is processed for a lawful purpose. However, a notable exemption is that the data user is able to process data without any consent in relation to the performance of a contract to which that individual is party. Consent is also required to ensure the best interest of the person is protected, such as that person's life, liberty or security. The general principles also prohibit the processing of any data, unless it is for a lawful purpose that is directly related to the activity of the data user. For sensitive personal data in relation to physical, mental health or condition, political opinions or religious beliefs, explicit consent has to be obtained.

In Singapore consent assumes a similar form to most other jurisdictions studied (See Chap. 4). An organization is required to obtain consent that is in writing or recorded in a manner that is accessible for future reference. The PDPA deal with a number of issues relating to the Consent Obligation. In particular, an individual is deemed not to have given consent, unless the individual has been notified of the purposes for which his or her personal data will be collected, used or disclosed and that individual has provided consent for those purposes. An organization may also obtain consent verbally, although it may correspondingly be more difficult for an organization to prove that it had obtained consent. Deemed consent has similarities to implied consent. However, in Singapore deemed consent arguably operates more broadly that implied consent because, when data subjects hand over their personal data, they have automatically done so voluntarily and therefore are deemed to have consented.

Within the EU (see Chap. 3), consent can be freely given. Even so, consent should be given by a clear affirmative act that establishes a freely given, specific, informed and unambiguous indication of the data subject's agreement to the processing of personal data. Active consent is to be provided by an organization. Consent can also be obtained in writing.

Most of the examined jurisdictions provide for the right of an individual to withdraw consent. However, the timing of that withdrawal varies. Consent is not provided if the individual has no genuine or free choice, or is unable to refuse or withdraw consent at any time. It is also important for an organization to establish systems and processes to monitor and record whether actual consent has been provided or not. Significantly, the withdrawal of consent also does not ordinarily affect the lawfulness of processing based on consent before its withdrawal.

These requirements do not pertain to all the jurisdictions studied. Given such divergent approaches to the provision and application of consent, the cross border legal issues that will arise in this area of law are likely to be significant in the future. It is our view that along with the definition of personal data, there must be a greater level of harmonisation of the law in regards to consent. Failing to do so, will only continue to complicate the understanding and implementation of data protection law for data subjects and individual entities that operate internationally in one or more jurisdictions. This will only further add to the compliance costs to industry.

11.6 Storage & Localisation

The development of data protection law has also resulted in countries retaining their sovereign control over the place where data is stored. There are two components discussed in this section. The first highlights how states have developed laws to ensure storage is retained in the state. The second explores the concept of storage limitation in the EU. For further discuss see specific country Chapters.

Since the Edward Snowden incident, countries have been realigning their data protection laws and policy, as Livingston and Greenleaf state, within the borders of the country concerned.[29] That is, countries are ensuring that certain personal data is stored and processed within the country in which it is collected. Russia was one of the first countries to impose localisation laws in 2015.[30] Courtney Brown points out that Russia was very quick to maintain sovereign data protection laws, requiring all personal data collected from Russians to be only stored and processed within the Russian state. Arguably, this approach was created on policy grounds (national security), whereby Russia could avert or limit the likelihood of state secrets being stored and processed outside the sovereign state, and also to limit access to stored data to key people in senior political and security positions in the Russian state.

Australia has, to a limited extent, localised the laws on storage and use of personal data, in relation to health records only. Section 77 of the *My Health Records Act 2012* requires that the System Operator cannot hold, take, process or handle

---

[29] Livingston, S Greenleaf, G (2017) *Data Localisation in China and other APEC Jurisdictions*, 143 *Privacy Laws & Business International Report*, 22---26, October 2016 [2017] *UNSWLRS* 11, UNSW Law UNSW Sydney NSW 2052 Australia.

[30] Bowman, C *Data Localization Laws: an Emerging Global Trend*, JURIST – Hotline, Jan. 6, 2017, http://jurist.org/hotline/2017/01/data-localization-laws-an-emerging-global-trend.php, accessed 3 January 2017.

records outside of the Australian territory. Singapore and India have not localized sectors of the law in the same way as Australia, for example in regard to the storage of health records. Japan have resisted data localization laws. Indonesia, on the other hand have adopted a different approach. Indonesia require the systems administrators for public services to be located in the country. Moreover, Regulation 20/2016[31] provides that electronic system providers are required to process protected private data only in data centers and disaster recovery centers located in Indonesia. The EU have taken a very hands off approach to localizing data protection laws. Likewise, Malaysia has taken a minimalist approach, only requiring the personal data of Malaysian citizens to be stored on servers based in that country. The current status of storage of personal data in Thailand is, at the time of writing this book in early 2018, that no laws had been established to regulate the storage or processing of personal data by the private sector (companies and businesses).

The localization of data laws allows countries, not only to protect their national, economical and social interest, but also to assert their own cultural priorities over their citizens' personal information. States will also want to continue to strengthen the protection of their citizens interests. This is particularly important where the country believes that another country has placed less value on the management and governance of personal data. As data protection law continues to be diverse in its approach and application, it is likely that data localisation laws and practice will only strengthen. This area of the law will only be repealed in and when nation states, and their citizens no longer view that their personal data requires protection. It is argued that this is unlikely to occur, as states treat personal data very differently (economically and socially).

11.6.1 Storage Limitation

A key feature of the EU's framework for the retention or storage of personal data is the storage limitation principle. The principle transfers responsibility to the data controller to ensure that data is deleted and not stored for an unnecessary length of time. Australia has adopted a much broader approach. Data is to be destroyed if it is no longer needed for the purpose to which is was collected. Singapore's approach is similar to Australia, although slightly broader as they apply the 'standard of reasonableness', having regard to the purpose for which the data needs to be retained. Again there is no time limit placed on any organisation. The standard of reasonableness is left to the judiciary to determine on a case by case basis. Throughout India, the Privacy Rules specify that sensitive personal data cannot be retained for longer

[31] Regulation 20/2016 on Personal Data Protection in Electronic Systems. Due to the sectorial approach taken by Indonesia, they have to date established a number of laws. These include, Government Regulation 82 (Reg 82/2012), the Minister of Communication and Informatics (MOCI) Regulation 20 of 2016 regarding Protection of Personal Data in Electronic Systems (Reg 20/2016), and the MOCI Circular Letter No. 3 (2016).

than is required for the purpose for which it was lawfully collected, or as otherwise required under another law. Under the data retention provisions set out in various laws, companies are generally required to retain data for eight financial years. Malaysia, adopts a similar approach whereby the retention of data should not be kept any longer than is necessary. The broad and open ended approach requires the judiciary to set the direction on the length of time that is appropriate. Japan's business operators must endeavor to delete personal data without delay when its use is no longer required. It does not specify a timeframe. Thailand nor Indonesia on the other hand does not impose a limited period for storage of data. A further discussion in relation to storage limitation can be found in country specific Chapters.

This area of the law can be best described as a watching brief, as no person knows how individual countries will act in the future. It is complex and will vary from country to country. The impact of such laws have the potential to restrict or deter businesses to expand into other regions and countries because of the high costs involved in setting up, storing and processing personal data in separate countries. On the other hand, it appears that those countries that have embraced this concept, have seen the wider impact to their own sovereignty for not maintaining some sort of control.

11.7 International – Transfer

The international transfer of data is unexceptional today, and will only increase. Each jurisdiction has different requirements for the transfer of data to another country (see country Chapters). These differences range from, but are not limited to, requiring consent from the data subject, assessing the receiving countries' data management system, purpose, or data management plan.

An entity in Australia will be responsible not to mishandle the data, once it is received by the recipient. Australia has adopted what could be considered a broad approach, whereby an APP entity that discloses personal information about an individual to an overseas recipient, must take reasonable steps to ensure that the recipient does not breach the APPs in relation to that information. Taking reasonable step means that the judiciary or Commissioner will determine whether the steps taken were adequate to protect the personal data.

In India, the IT Rules are more specific and predominantly regulate the transfer of sensitive data by a body corporate. The transfer may be allowed only if it is necessary for the performance of a lawful contract between the body corporate or any person acting on its behalf, and the provider of information or where data subject has consented to data transfer. Moreover, the transfer of sensitive data may require that it be transferred to a foreign jurisdiction if specific conditions must be met. The transfer can only be undertaken if the transferee entity has standards in place that are not lower than those set by IS/ISO/IEC 27001. No other country specified standards, such as ISO, must be used when transferring data to another country (see Chap. 6).

Indonesia, to date, requires that the transmission of personal data must be in the form of a personal data transmission plan that outlines the country of destination, that is, where the personal data will be disseminated. Furthermore, as part of the plan, the organization is required to report on the transmission activity. This process would, if it is also adopted under the proposed data protection laws, strengthen the management of personal data being transferred outside the country (see Chap. 7).

Japan does not require a plan or standard to be met, but requires only that the business operator document the recipient's name and personal data that is to be transferred. The data cannot be transferred without obtaining consent. Furthermore, the transfer of personal data outside of Japan can only take place when the receiving country has demonstrated they have the same level of safety and controls in place. Even though a plan of transmission is not needed, it does not preclude organizations from establishing any such requirements through contracts or memoranda of understanding. With Japan recently receiving EU adequacy approval, the transfer of personal data from Japan, can arguably be accepted as being protected to an equivalent standard to that of the EU.

In Malaysia, the processes is different again. The transfer of personal data can only be undertaken after the Minister of Communication and Multimedia has published the name of the receiving country in the Government Gazette Order.[32] However, the Order has not been approved or confirmed by the Malaysian Government. Unless and until the proposed Order takes effect, transferring personal data outside of Malaysia requires an organization to obtain consent and the transfer must also meet contractual obligations. The organization is not required to have a transmission or transfer plan. As discussed in Chap. 8, and unlike EU law, Malaysian does not require transfer contracts to be made for the benefit of third parties.

Generally, transferring data from Singapore requires that the legislation be complied with and that the receiving country has a comparable standard. However, the standards set by Singapore are different to other jurisdictions. The recipient will be bound by the enforceable obligations specified by Singapore law. The 2015 Key Concepts Guideline also direct organizations on what and how the international transfer of data should be undertaken. The Key Concepts Guidelines also sets out the scope of contractual clauses[33] with which recipients must comply in protecting personal data received in accordance with Singapore's laws (see Chap. 4). Finally, as highlighted in Chap. 9, Thailand does not have a comparable framework to the other countries or the EU for managing the transfer of personal data outside of the country.

It can be assumed that there are likely to be laws in place to regulate and assist industry sectors such as banking, education and telecommunications when transferring personal data around the world. At issue, is the wide and varied approach to meet this control mechanism. It is understandable that the state of origin would want the recipient state to have processes in place to ensure they meet the state of origin's standards. This issue of managing the transfer of personal data alone calls for some level of consistency. With the large quantities of data being transferred daily, the varied approaches taken range from minimal to quite comprehensive.

---

[32] Government Gazette Order 201732.

[33] The Key Concept Guidelines 2015, section 19.

11.7.1 Adequacy Test and Privacy Shield

The EU was the first to establish laws that require an assessment (test) for the transfer of data to third countries.[34] The adequacy decision is complex and involves a proposal from the European commission, an opinion from the European Data Protection Board, approval from EU countries, and the adoption of the decision by European commissioners. To date the 'white list', those third countries outside the European Economic Area, only include Andorra, Argentina, Canada, Faroe Islands, Guernsey, Israel, Isle of Man, Japan, Jersey, New Zealand, Switzerland, Uruguay and the United States as providing adequate protection. Currently, Australia, India, Indonesia, Malaysia, Singapore or Thailand also have made the 'white list'.

The adequacy assessment is a game changer for modern day legal framework for data protection. The impact of this assessment, is the requirement that a foreign jurisdiction assess another nation state's legal framework, to ensure that it meets the former's standards. Even though the adequacy test is beginning to appear in other nation states' legal frameworks for data protection and privacy, this has put the EU in a very powerful position. It has allowed the EU to effectively determine the playing field in which data protection law throughout the world will be determined. In other words, countries that do not meet the EU adequacy test will restrict companies and organizations from doing business in the EU (see Chap. 17).

The European Commission has established the Privacy Shield Framework[35] which enables the transfer of data between EU and US. The US have also established a similar process with Switzerland. However, to date there are no other similar schemes established between the EU and Australia, India, Indonesia, Japan, Malaysia, Singapore or Thailand. Malaysia is the only other country to have prepared a list that is published by the government detailing the countries that have comparable systems of data protection.

The Privacy Shield program provides companies operating in the respective jurisdictions with a mechanism to comply with data protection requirements when transferring personal data. The Privacy Shield[36] program, is administered by the International Trade Administration (ITA), U.S. Department of Commerce. It is a voluntary program; however, once an organization makes the public commitment to comply with the framework's requirements, the commitment will become enforceable under U.S. law. Voluntary programs become important, and form part of the current co-regulatory approach taken towards data protection. Another layer to this co-regulatory approach is the application of codes of practice.

---

[34] Regulation (EU) 2016/679, Article 45.

[35] US Department of Commerce, Fact-Sheet: Overview of EU-US Privacy Shield Framework 2016, https://www.commerce.gov/sites/commerce.gov/files/media/files/2016/fact_sheet-_eu-us_privacy_shield_7-16_sc_cmts.pdf, accessed 30 December 2017.

[36] Commission Implementing Decision (EU) 2016/1250, 12 July 2016, pursuant to Directive 95/46/EC of the European Parliament and of the Council on the adequacy of the protection provided by the EU-U.S. Privacy Shield, Official Journal of the European Union L 207/2.

11.8 Codes of Practice

A number of different approaches have been adopted across each jurisdiction to enhance the enforcement of data protection and privacy laws. They can be best described as a combination of co-regulation or the command approach. In other words, the co-regulatory approach is specified in the law and requires entities to establish soft standards that are enforceable outside of the stand-alone legislation. The use of Codes of Practice, Guidelines, Policies and Procedures have, to varying degrees, been established by the relevant information Commission, Commissioner, Agency, or Department. The use of Codes of Practice as a regulatory tool have been very effective in many other industries. Australia, EU and Malaysia require Codes to be prepared to strengthen the co-regulatory approach to managing personal data. Singapore, rather than requiring Codes to be formally established, require an organization to adopt Guidelines. India, along with Indonesia and Japan, do not require codes of practice to be developed. Currently, Thailand does not require Codes to be implemented or approved. Although, sectorial laws may require industry sectors to develop codes, procedures or guidelines for the management and co-regulation of data, co-regulation is alive and well in all jurisdictions that underpin government regulation of personal data and information. They also enhance the risk-based regulatory approach to data protection and privacy. However, the level and extent of co-regulation varies greatly, although they all reflect the core principles of their respective data protection and privacy laws. That is, they generally reflect the need for an organization and data user to consider and adopt the principles of consent, accuracy, collection, disclosure, transfer and retention.

11.9 Data Portability

The right to data portability is a recent addition to the data protection legal framework, particularly the EU (see Chap. 3). However, that right is most important by 1). both in granting control rights to data subjects and 2). operating at the intersection between data protection and other fields of law (competition law and intellectual property consumer protection). The right to data protection itself provides another valuable layer to protect an individual's personal data over the Internet. Moreover, the ability to freely port personal data from one controller to another, is undoubtedly developing as a tool that allows data subjects to create competition between digital services and the interoperability of platforms. This in turn, strengthens the ownership and control of personal data by the data subject. The rationale for the right to portability is the specific role that the controller has throughout the life cycle of data protection.

The right to data portability allows data subjects to receive the personal data that they have provided to a controller in a structured, commonly used and machine-readable format. The data portability right also provides the data subject with the right to request that a controller transfer their data to another controller. The exercise of that right is arguably becoming one of the most important additions to the EU GDPR framework. In Australia, the right to access data is restricted to allowing for the correction of personal information relating only to credit information.[37] In Singapore, sections 21 and 22 of the PDPA provide data subjects with the right to request access to their personal data for the purpose of correction (see Chap. 4). However, at the time of writing this book, neither Indonesia, India, Japan, Malaysia, nor Thailand have established the right to portability.

11.10 Right to Be Forgotten

The right to be forgotten or otherwise known as the right to erasure has quickly become an important principle of data protection and privacy law, and has been discussed in each country Chapter. It allows persons to request that their personal data and information be deleted or removed. Jeffery Rosen asserts that the intellectual roots of the right to be forgotten can be found in French law, which recognizes *le droit à l'oubli*, or the right of oblivion which is a right that allows a convicted criminal who has served his time and been rehabilitated to object to the publication of the facts of his conviction and incarceration.[38]

The modern right to be forgotten has grown largely out of the EU (see Chap. 3). The concept allows an individual to request a search engine to remove personal data and information pertaining to them. The request can come in different forms, but is essentially the right of an individual to seek to have their personal data removed or deleted (delisted) so that they cannot be located when someone else conducts a search on the Internet. The drafting of the GDPR resulted in the right to be forgotten specifying that natural persons would obtain the right to have publicly available personal data and information erased.[39] According to the European Commission, this right would help people better manage data protection risks online by enabling them to delete their personal data and information if there was no legitimate grounds for retaining that data. The draft GDPR also elucidated that such a protection had to be reconciled with the right to freedom of expression.[40]

The right to be forgotten has not gone unnoticed by the Court of Justice of the European Union (see Chap. 3). In *Google Inc. v. Agencia Espanola de Proteccion de Datos, Mario Consteja González*[41] the Court of Justice of the European Union ruled that an internet search engine operator is responsible for the processing that it

---

[37] Privacy Act 1988, Sub-Division 3.

[38] Rosen J *The Right to Be Forgotten* Stanford Law Review Online 64, (2012) http://www.stanfordlawreview.org/online/privacy-paradox/right-to-beforgotten, accessed 5 May 2018.

[39] Draft General Data Protection Regulation, European Commission, European Commission 2012 European Commission (2014), Article 17, Memo: Data Protection Day 2014, Full Speed on EU Data Protection Reform. http://europa.eu/rapid/press-release_MEMO-14-60_en.htm, accessed 13 April 2018.

[40] Draft General Data Protection Regulation, European Commission, Article 80.

[41] Case C-131/12 *Google Inc. v. Agencia Espanola de Proteccion de Datos, Mario Consteja González*, 95–96.

carries out on personal data which appear on web pages published by third parties, and hence upholding a right of erasure and in essence the right to be forgotten. The Spanish newspaper *La Vanguardia* had published two announcements regarding the forced sale of properties arising from social security debt. The announcements were published on the order of the Ministry of Labour and Social Affairs website and the purpose was to attract as many bidders as possible. One of the properties described in the newspaper as belonging to Mario Costeja González. Mario Costeja González, alleged that, when an Internet user entered his name in Google's search engine, the user would be provided with links to the 1998 *La Vanguardia* newspaper, which continued to announce the foreclosure auction on Mr. Gonzalez's home. Arguably, this case began to set the foundations of what the right to be forgotten would constitute. The ruling, while only enforcing the right to erasure on search engines that are operating in Europe does not extend to those search engines operating in Australia, India, Indonesia, Japan, Malaysia, Thailand or Singapore.

More recently the balancing of competing and conflicting interest was no more evident than in the case of in *NT1 and NT2 v Google and The Information Commissioner* the High Court of England and Wales.[42] This case is important because it is arguably one of the most high profile cases regarding the right to be forgotten in common law jurisdictions.[43] What arose out of this case was the need to balance the right to be forgotten and a person who had received a criminal conviction, and wanted the information regarding the conviction removed from the Internet. In other words, the claimants sought the removal by the defendant, Google, of search results concerning their previous convictions on the basis that the results conveyed inaccurate, out of date and irrelevant information, failed to attach sufficient public interest and/or otherwise constituted an illegitimate interference with their right to be forgotten as established in *Google Spain*.[44] Costello argues that the decision in *NT1/NT2* is particularly relevant given the traditional hostility of common law jurisdictions to rights of privacy that extend to historical criminal convictions.[45] Common law jurisdictions have traditionally privileged principles of open justice in contrast to the approach of many civil law jurisdictions which, in general, opposes punitive shaming and presumes criminal records to be confidential.[46] The civil law approach is reflected in the 1981 Council of Europe Convention for the

---

[42] *NT1 and NT2 v Google and The Information Commissioner* [2018] EWHC 799 (QB).

[43] Costello R, *The Right to be Forgotten in Cases Involving Criminal Convictions*, NT1 and NT2 v Google and The Information Commissioner [2018] EWHC 799 (QB), European Human Law Rights Review (2018).

[44] Case C-131/12 *Google Spain SL & another v Agencia Espanola de Proteccion de Datos (AEPD) and another*. Henceforth *Google Spain*.

[45] For a comparative analysis as between a common law and civil law jurisdiction see JB Jacobs and E Larrauri, "Are criminal convictions a public matter? The USA and Spain" (2012) 14(1) *Punishment and Society* 3. On the still recent change in the Irish position, see TJ McIntyre, "Criminals, Data Protection and the Right to a Second Chance" (2017) 58 *The Irish Jurist* 27.

[46] JB Jacobs and E Larrauri, "European Criminal Records & Ex-Offender Employment" *New York University Public Law and Legal Theory*, http://lsr.nellco.org/nyu_plltwp/532/, accessed 9 December 18.

Protection of Individuals with regard to Automatic Processing of Personal Data as well as Data Protection Directive Article 8(5) the General Data Protection Regulation.[47] *NT1/ NT2* thus represents an explicit departure from traditional common law attitudes toward criminal histories.[48] Moreover, Costello points out that the court confined to its facts due to the emphasis placed by the judge on a subjective assessment of credibility and remorse.[49] Despite this, the case offers a tentative first step towards clarifying the criteria for a delisting order in cases involving criminal convictions and offers a significant endorsement of the right to be forgotten in such cases.[50] Thus, the balance of rights, competing and conflicting interests will unlikely never be set in concrete because there are far too many variables.

The EU, unlike Asia and Australia, has preferred that its data protection laws, including the right to be forgotten, have extra territorial reach (see Chap. 3). Before the GDPR came into effect, Article 4 of Directive No. 95/46/EC introduced the extraterritorial scope of the right to be forgotten. Article 4 of the Directive required that each member state was to apply the national provisions it adopted pursuant to that Directive to the processing of personal data in which the location of an establishment of the controller and the location of equipment being used.[51] That is, member states must apply their national data protection laws where the processing is conducted by an entity of the controller. Where the processing is being undertaken in another state, the national laws of that state apply. Article 46 of the GDPR has retained this extraterritorial reach embodied in Directive No. 95/46/EC. Furthermore, where a data controller or processor has, through a code or agreement, established with the European Commission that the third country in which it is located can receive data from the EU, then that controller or processor will be obligated to remove personal data in response to data subject's application.

---

[47] Article 6 provides that criminal convictions "may not be processed automatically unless domestic law provides adequate safeguards."

[48] *Supra note 41.*

[49] Ibid.

[50] Ibid. The judge also referred in his decision regarding NT2 to the fact that the crime at issue was not one of 'dishonesty.' However, there was no discussion of whether the differentiation as between a crime of dishonesty and other crimes was a determinative factor. Again, the implication from the judgment is that, as with a spent conviction, this will be a consideration rather than determinative factor. Focusing on the question of what is in the public interest, emphasized differential impacts on the public in its discussion of the offenses of both claimants it muddied the waters by introducing dishonesty as a factor. The result is an unclear mélange of a public interest test with a categorical sliding scale of offenses defined in relation to their relative degrees of deception. The implication that a conviction for a violent crime committed without deception, would be more favourably treated than a non-violent offense of dishonesty is problematic on a public policy basis. As criminal acts generally involve an individual recklessly, or knowingly breaking the law, invariably in a manner which seeks to avoid detection, the merits of using honesty as a distinguishing metric is of questionable merit. The most substantively consideration what that treatment of self-help. Both claimants, on the advice of reputation management professionals, had generated content with the express aim of influencing Google's list of returned results prior to the decision in *Google Spain.*

[51] Article 29 Data Protection Working Party, 'Opinion 8/2010 on applicable law' (2010) WP 179, 8, http://ec.europa.eu/justice/policies/privacy/docs/wpdocs/2010/wp179_en.pdf, accessed 25 January 2018.

The Australian privacy laws do not provide a direct right to be forgotten. To date the right has received little consideration by the courts of the commonwealth or any state in Australia. The court in the Australian state of South Australia (see Chap. 5), in *Duffy v Google Inc*[52] found that, after a reasonable time had passed following the removal requests, Google became a secondary publisher of the defamatory material.[53] The court argued that, even continuing to make a URL to the offending content available after a take-down request had been received, could make Google responsible as a secondary publisher.[54]

The right to be forgotten was recently considered in India. However, it has been restricted to sensitive cases involving sexual assault. *Justice K.S. Puttaswamy (Retd.)& Anr. v. Union of India & Ors*[55] focused on whether the right to be forgotten exists in India. The court there also reinforced that the right to be forgotten was fundamentally important to the citizens of the country.[56] It ruled further, that the impact of the digital age results in information on the Internet becoming permanent. Moreover, any endeavour to remove information from the Internet may not result in its absolute obliteration. It is argued that in the digital world preservation is the norm and forgetting a struggle.[57] Citizens are entitled to re-invent themselves and correct their past actions. It is privacy which nurtures their ability and removes the shackles of things which they may have been done in the past.[58]

The Karanatak High Court of India in *Sri Vasunathan v. The Registrar General*, 2017 SCC *Online*, had to determine whether the right to be forgotten existed (see Chap. 4).[59] In this case a woman, hereinafter called X, had filed a First Information Report (FIR) against a man, Y, involving crimes of a grave nature, such as forgery, compelling to get married, and extortion. She also filed a civil suit for annulment of her marriage with him. She requested an injunction to restrain Y from claiming any marital rights. Later, both reached an out-of-court settlement and the cases were closed. Subsequently, X got married. However, her father filed another petition, realizing that an online search would reveal his daughters' connections to all the legal disputes. This could result in affecting X's personal life and her public image. The father pleaded for the court to mask X's name in cause title of the cases and prayed the same for any other copy available at online portals.[60] The court upheld the petitioners' claim and recognized the 'right to be forgotten. Justice Byraredy concluded the judgment in the following terms this would be in line with the trend in Western countries that enforce the 'right to be forgotten' in sensitive cases involv-

[52] *Duffy v Google Inc* [2015] SASC 170.

[53] Ibid.

[54] Ibid.

[55] *Justice K.S. Puttaswamy (Retd.)& Anr. v. Union of India & Ors*, (2017) 10 SCALE 1.

[56] Ibid.

[57] Ibid, paragraph 65.

[58] Ibid.

[59] *Sri Vasunathan v. The Registrar General*, 2017 SCC Online Karr 424.

[60] Ibid.

ing women in general and highly sensitive cases involving rape or affecting the modesty and reputation of the person concerned.[61]

The issue for India, unlike other countries, is that currently their data protection laws do not provide clear direction on the deletion or removal of an individual's personal data and information. Relying on the court to expand the jurisprudence in this area is a positive sign and may force legislative change. Conversely, it could arguably result in a very ad hoc and inconsistent approach to determining when and how the right would be applied. Nevertheless, the move by these countries to adopt western thought, traditions and rights continues to be significant, in a region of the world that, while being heavily influenced by law from the United Kingdom, has very different cultural and religious values when compared to Australia and Singapore.

Indonesia is moving very slowly towards recognising the right to be forgotten. To date the right to be forgotten has not been considered by any court in Indonesia.[62] In October 2017, the Parliament of Indonesia passed the revised EIT[63] law that enables a data subject to request that their personal data be deleted. Article 26 section 3 requires that Each Operator Electronic System delete irrelevant information and records of an individual's personal data and information under its control, but only upon the direction and request issued through a court order (see Chap. 6). However, it is likely that the application of the principle right to be forgotten in the revised EIT Act will pose practical problems to its effective implementation. This is because of the gaps in the law, whereby the EIT does not clearly define personal data.[64]

In Japan, the right to be forgotten has also been recognised. However, this has not gone without controversy. The right was first recognised by the courts in Japan in 2014, but in 2017 the courts took a very different view. In 2014, Google was ordered to remove links to personal data that identified peoples' criminal past. Unfortunately, that court did not set a precedent, and a higher court dismissed a decision from a lower level court's ruling in 2015 in which the right to be forgotten had been cited. The higher court's decision was based on the traditional legal frameworks governing privacy rights. However, the right to be forgotten was short-lived in Japan; and in 2016, the Tokyo High Court overturned the District Court's decision. The court stated that the right to be forgotten is not a privilege provided for in Japanese law and that its prerequisites had not been determined.[65]

Across Singapore and Thailand, the right to be forgotten has not gained recognition accorded it in other parts of the world. Thailand does not recognise that right,

---

[61] Ibid.

[62] Hak untuk dilupakan Direvisi UU ITE Masih Belum Berlaku, https://tekno.kompas.com/read/2016/11/29/09250047/, accessed 10 July 2018.

[63] Electronic Information and Transactions Law No. 19/2016.

[64] Zeller, B,. Dr. Trakman, L., Walters, R., Dewl Rosadi, S, *The Right to be Forgotten – the European Union and Asia Pacific Experience* (Australia, Indonesia and Singapore), European Human Rights Law Review (2018), under review.

[65] Japan Times, https://www.japantimes.co.jp/news/2016/07/13/national/crime-legal/tokyo-high-court-overturns-mans-right-forgotten/#.W62Yw2VeL-Y, accessed 2 September 2018.

and has not implemented significant data protection laws. Singapore, on the other hand grants a restricted right to a person to withdraw consent for the collection and disclosure of personal data. This restricted version of the right to be forgotten, does not meet the full right to requesting personal data be deleted. However, one could explain Singapore's failure to focus on the right to be forgotten, given that Singapore has concentrated primarily on economic activity and growth since its independence (see Chap. 8).[66] On the one side is the accusation that Singapore unjustifiably prioritizes economics as the expense of personal freedoms. On the other side is the recognition that Singapore continues to be an economic powerhouse throughout the Region.[67] Whatever the view, Singapore has restricted the right of a person to withdraw consent for the collection and disclosure of personal data. This restricted right falls somewhat short of the right to request that personal data be deleted.[68] The authors are not advising or directing Singapore as to the policy position in regards to the right to be forgotten. It is up to Singapore to decide. Nonetheless, it will be a watching brief, particularly if citizens pressure government to provide greater privacy protection of personal data over the Internet. The notion of a right to be forgotten could be contained in section 25 of the Personal Data Protection Act, which requires the destruction or de-identification of personal data when there are no longer any legal or business and any other purpose for the retention of that data.[69] However, it is argued that this is inconclusive, and further research and jurisprudence is required to confirm the effectiveness or otherwise of the right to be forgotten in Singapore.

11.10.1 Adoption of the Right to Be Forgotten

In summary, there is a varying recognition and adoption of the right to be forgotten. The right is likely to be continually debated, but countries are characterising that right based on local national needs. Some countries have dismissed the idea of promoting such a right altogether, while others are taking a cautious approach towards it. The varied approach today would be as a result of how rights are balanced within each jurisdiction. The legal and policy debate regarding the right to be forgotten, on its own, and when balancing it against the freedom of expression, will continue for some time. That is, until courts have clearly identified when, where and how it will be definitively applied, the right to be forgotten, like every other area of data protection and privacy law, will continue to transmute and be transformed. This may only

[66] Ibid.

[67] Ibid.

[68] Ibid.

[69] Personal Data Protection Act 2012, section 25, An organisation shall cease to retain its documents containing personal data, or remove the means by which the personal data can be associated with particular individuals, as soon as it is reasonable to assume that—the purpose for which that personal data was collected is no longer being served by retention of the personal data; and retention is no longer necessary for legal or business purposes.

ever occur when the laws of each jurisdiction place an obligation on an organisation or its data controller – processor that is located in a third country, to be fully responsible for the deletion or removal of personal data.

11.11 Conclusion

The historical beginnings of the European Union, Central Asia and the Asia Pacific are very different. Arguably, the EU framework has evolved to provide a high level of privacy protection over the Internet. Singapore's model meets their current sovereign and policy needs in order to retain themselves as a business hub for the region. Australia's model arguably falls somewhere between the two, while the remaining countries are either combination of all, or, specific data protection laws do not exist. The current day data protection and privacy laws reflect a risk based approach in each jurisdiction. However, this approach diverges from one jurisdiction to the next. The EU serves as the high-end benchmark and authority in data protection and privacy law, extending its model around the world. That is, if one views the laws through the lenses of protecting personal data and providing a level or privacy over the Internet. However, if one looks at the law through the lens of being business friendly, Singapore arguably takes the lead. This is a complex ideological question as to what is the right way forward. Is it the model presented by the EU or Singapore, or does the Australian model occupy the more fitting middle ground? The authors do not attempt to answer this question because there are too many policy and political variables. Additionally, there are other models in existence, and neither the United States, Canada nor China's model has been examined as part of this book.

The brief comparison of data protection regimes highlights some remarkable findings. It has revealed which countries are frontrunners and which lag behind in developing their data protection laws, whatever the ideological basis they adopt. Arguably, the changing landscape of technology and the Internet has challenged the very notion of what is the best, or even the right, legal framework for data protection into the future. Nevertheless, the tension between governments, civil rights organizations, and data protection authorities vary from country to country in which they exist. This is expressed by and in the law, because laws are so fragmented that they lack any meaningful convergence or harmonization. It must also be noted that the political and policy discourse among states also differs according to the intensity and scope of public debate, public awareness and understanding that an issue actually exists.

Even though the EU model is not being universally adopted, many of the principles and concepts that have been adopted by other jurisdictions have evolved from the GDPR or the OECD. As highlighted, upon comparing the laws of each jurisdiction, they appear similar conceptually. However, on closer inspection those laws have been developed primarily to meet national needs. The acceptance of international concepts and principles has only served to be a starting point for consideration at the national level.

The internationalization of the Internet and its infrastructure calls for greater legal convergence and harmonization across countries, including in the region studied. The current regulatory environment not only creates uncertainty and confusion; it allows entities to move their operations around the world, so as to minimize their exposure to laws that are an obstacle to their operations. In other words, where the laws are the strongest, for example in the EU, organizations are unlikely to establish bases in that jurisdiction or to move their bases outside of its jurisdiction. This is arguably the case as Google considers moving a significant part of its base out of Ireland and the jurisdiction of the EU. On the other hand, Singapore is likely to continue to provide a greater balance between data protection and accommodating commercial entities through their business friendly approach. Moreover, countries that have no data protection laws could be considered very welcoming and favorable for business to locate their operations there. This tension, conflict and confusion does not bode well for addressing future challenges in the law or policy governing data protection. Ultimately, the most difficult question to answer is – what model does the international community want to see for the future of data protection, and to a lesser extent, privacy law over the Internet?

References

Al-Fedaghi, S (2007) *How sensitive is your personal information?* In Proceedings of the 2007 ACM Symposium on Applied Computing (pp. 165–169)

Courtney Bowman, *Data Localization Laws: an Emerging Global Trend*, JURIST – Hotline, Jan. 6, 2017, http://jurist.org/hotline/2017/01/data-localization-laws-an-emerging-global-trend.php, accessed 3 January 2017.

Etzioni, A (2015) *A cyber age privacy doctrine: More coherent, less subjective, and operational,* Brooklyn Law Review, 80(4), pp. 1263–1308

Pesciotta, D. T. (2012) *I'm not dead yet: Katz, Jones, and the Fourth Amendment in the 21st century* Case Western Reserve Law Review, 63, p. 187.

Livingston, S, Greenleaf, G (2017) *Data Localisation in China and other APEC Jurisdictions*, 143 *Privacy Laws & Business International Report, UNSWLRS* 11, UNSW Law UNSW Sydney NSW 2052 Australia.

Ojanen, T (2014) *Privacy is more than just a seven-letter word: The Court of Justice of the European Union sets constitutional limits on mass surveillance:* Court of Justice of the European Union Decision of 8 April 2014 in Joined Cases C-293/12 and C-594/12, digital rights Ireland and Seitlinger and others. European Constitutional Law Review, 10(3), pp. 528–541.

Photopoulos, C (2011) *Managing catastrophic loss of sensitive data: A guide for IT and security professionals.* Rockland, MA: Syngress, p. 3

Rosen J *The Right to Be Forgotten* Stanford Law Review Online 64, (2012). http://www.stanfordlawreview.org/online/privacy-paradox/right-to-beforgotten, accessed 5 May 2018.

Taddicken, M (2014) *The "privacy paradox" in the social Web: The impact of privacy concerns, individual characteristics, and the perceived social relevance on different forms of self-disclosure,* Journal of Computer, p. 270–272

Zuiderveen, B. F. J., Van, E. M., & Gray, J (2015). *Open data, privacy, and fair information principles: Towards a balancing framework,* Berkeley Technology Law Journal, 30(3), p. 2077.

Zeller, B, Dr. Trakman, L, Walters, R, Dewl Rosadi, S, *The Right to be Forgotten - the European Union and Asia Pacific Experience* (Australia, Indonesia and Singapore), European Human Rights Law Review (2018), under review.

Part V

Chapter 12
Intellectual Property

Abstract Efforts to reconcile the tension between personal data (protection) law and intellectual property law remains contentious. This Chapter explores the nature of personal data in intellectual property. The growing value of personal data has, over the past decade, begun to lend itself to question whether this data contains a level of intellectual property. However, to demonstrate whether intellectual property exists in personal data, there are key principles and concepts within the current day data protection laws that also need to be considered. The concept of consent and the definition of personal data within the law, become very important because they determine the level of control and subsequently the level over ownership data subject wills have of their personal data. More recently, the European Union have arguably strengthened the possibility that personal data contains intellectual property as a result of the introduction of the right to data portability.

This Chapter explores whether personal data, defined by current day national data protection and privacy laws, constitute intellectual property, or ought to do so. It also considers whether the personal data of an individual ought to be subject to a privacy right, and justifiably extended to being an intellectual property right. The Chapter will show how the current definition of personal data, the concept of consent, the right to data portability, and the right to access personal data in national and supranational laws, is steering personal data towards having an intellectual property right. The Chapter will also highlight how neither individual states nor the EU have excluded intellectual property from personal data within their respective data protection laws. Along with the remainder of the book, the Chapter also highlights that the internationalization of the Internet and its supporting infrastructures requires an international response to data protection and privacy rights through legal harmonization across jurisdictions. It stresses how recent case law in different jurisdictions have afforded greater protection to personal data, including through intellectual property rights.

© Springer Nature Singapore Pte Ltd. 2019

R. Walters et al., *Data Protection Law*,
https://doi.org/10.1007/978-981-13-8110-2_12

12.1 Introduction

A data subject can be identified in a number of ways over the Internet. On the one side, the traditional notion of personal identification has been a person's name, date of birth and place of residence. On the other side, with the rise of the Internet, and the supporting technology infrastructure, systems and platforms can now identify a data subject through biometric characteristics like appearance, height, weight, fingerprints, DNA (Deoxyrebonucleic Acid) and retinal patterns, passport, bank account, social security number, taxation permanent account number and in some cases national identification cards. These and other identifying information can now be found in modern day data protection law.

Conceptualizing personal data as intellectual property (IP) is an evolving concept in the law. IP and the rights afforded to a data subject in relation to their personal data rights is driven by the common desire to control the distribution of personal information as it is defined by law.[1] It is well understood and accepted that commercial data can be the subject of an intellectual property right. However, extending legal rights to data subjects over their personal information is an important source of new thought. The rationale behind this new thought is that the introduction of data protection laws has significantly promoted the protection of an individual's privacy over the Internet. In effect, the protection of privacy over the Internet is considered as being a by-product of data protection law that has evolved to regulate the collection and use of personal data.[2] However, there is a policy tension between protecting personal data and privacy based on their respective economic and social goals. On the one side, personal data provides an economic benefit to entities. On the other side, protecting the personal data promotes the social goal of protecting privacy over the Internet.

Mark Lemley takes a conservative approach in addressing these issues. He warns that, from a privacy perspective, intellectual property is regularly signed away. Importantly, he has concerns that the information revolution may reduce the protection that individuals ought to have over their personal data.[3] Lemley's viewed are espoused, in part, by Omer Tene and Jules Polonetsky. They argue that personal information should be regarded as neither an exclusive asset of individuals (the treatment of which may impinge on business trade secrets and intellectual property rights), nor as the exclusive property of businesses that exclude individuals from

---

[1] Zittrain, J *What the Publisher Can Teach the Patient: Intellectual Property and Privacy in an Era of Trusted Privication*, 52 Stanford Law Review (2000) p. 1203. Zittrain is discussing how the music industry moved from being vulnerable to developing technological systems that could the intellectual property of those that make music.

[2] Trakman., L, Walters., R Bruno, Zeller B *Is Privacy and Personal Data set to become the new Intellectual Property*? International Review of Intellectual Property and Competition Law, forthcoming 2018

[3] Lemley, M *Private Property: A Comment on Professor Samuelson's Contribution,* 52 Stanford Law Review (2000).

benefiting from that property.[4] The authors reject both propositions, arguing instead that personal information should be treated as a valuable joint and shared resource, that can promote value, creation and innovation. The position put forward by Tene and Polonestky contests the very notion that an intellectual property right can be extended to personal data.

12.1.1 Internet Systems, Platforms and Infrastructure

A starting point in determining the perceived economic and social value of protecting data as intellectual property is to look at the supporting infrastructure that supports the Internet. The systems, platforms and technology that supports the Internet have become a lucrative economic activity. Atul Singh notes that Black's Law Dictionary defines a database as a compilation of information arranged in a systematic way and offering a means of finding specific elements it contains, often by electronic means.[5] Singh argues that the definition of a database being an organized collection of information held on a computer under the Oxford Dictionary of Law, also relates closely to automated processing of data. He stresses that the ability to discover astonishing co-relations between data unrelated *per se*, using techniques such as data mining and big data analytics, reveals the knowledge power contained in large databases and underlines the need to protect such databases.[6]

However, Singh argues that the United Kingdom, may provide a way forward to better understand the connection between computer systems, platforms and infrastructure, personal data and intellectual property rights. To begin with, one must look to copyright. He further argues that in the United Kingdom, copyright was addressed under the Copyright, Designs and Patents Act, 1988 which had no specific provision for a database as it stood originally, though it could be considered a compilation of data. The European Parliament and the Council was of the opinion that either database was not sufficiently protected, or, if protected, the protection varied with national legislations across the EU.[7]

[4] Tene O., Polonetsky., J *Big Data for All: Privacy and User Control in the Age of Analytics*, 11 Nw. J. Tech. & Intell. Prop. 239 (2013).

[5] Singh, *Protecting Personal Data as a Property Right*, ILI Law Review (2016).

[6] Ibid.

[7] Ibid, in 1996, the European Parliament and Council adopted the Directive 96/9/EC19 for legal protection of databases. To implement the provisions of this Council Directive, Statutory Instrument 1997 No. 3032, the Copyright and Rights in Databases Regulations, 1997, was approved by a resolution of the Houses of the Parliament. These regulations give effect to the Directive 96/9/EC recognizing *sui generis* right protecting databases in England and Wales. A database right exists in a database if there has been a substantial investment in obtaining, verifying or presenting the contents of the database even if the work fails to satisfy the threshold of originality. A database right is, therefore, separate from, and in addition to, a copyright which may exist in a database. Regulation 16 makes extraction or reutilization of all or substantial part of a database, without the consent of the owner thereof as an infringement of database right.

In *Flogas Britain Ltd.* v. *Calor Gas Ltd*[8] the plaintiff sought damages from the defendant for use of a database maintained by the plaintiff, containing information on its customers, their name, address, contact details, contract dates, pricing and other information. The defendant made commercial communications to the customers of the plaintiff. The High Court of England and Wales held that:

> the information such as the names and addresses of the customers was protected by a database right and transfer of all or a substantial part of the contents of the database to another medium by any means or in any form amounted to such extraction as to constitute infringement of a database right.[9]

Arguably, the above case paved the way for some level of right to be afforded to the information (personal data or otherwise) contained within data base. It stopped short of providing a property or intellectual property right. However, the case also highlights that, where there has been an abuse of the information, it would extend to include an infringement of the data base itself; and therefore would likely result in the right afforded to the information also being breached. Thus, it is our view that indirectly, a level of right is afforded to the information within a data base, because it is the data base itself that, in this case, had a copyright.

However, Singh argues that there are competing views over ownership of a person's name. One view is that the data subject has a right, as the original "owner" of personal data, to control the use and dissemination of that information. Another view is that such information should belong to the data collectors who have gathered personal information by expending time, money and effort.[10] He goes onto extend this second view is the ownership claim of data processors who have aggregated such personal information into meaningful databases. Contradicting this view, however, is the contention that such processors and database creators merely hold the personal information as trustees.[11] Given their trust duties, they would not be able to exploit the rights of data subjects whose information is considered as intellectual property.[12]

Kenneth Laudon expounds on the economic value of personal data, while also recognizing right of data subjects in that data. He favours the commoditization of personal information, with a property right vested in data subjects in respect of their personal data.[13] The position is that data subjects have a right to deal in their personal data for a value, and that this affords them a level of intellectual or property right. Laudon takes the position that a National Information Market and a National Information Exchange would aggregate personal information and lease it on a

[8] [2013] EWHC 3060 (Ch).

[9] Ibid.

[10] Singh, A *Protecting Personal Data as a Property Right*, ILI Law Review (2016).

[11] Ibid.

[12] Ibid.

[13] Laudon, K *"Markets and Privacy"*, 39 (9) Communications of the ACM, 92–104 (1996).

regulated information market thus creating economic stakes for data processors or data controllers and data subjects.[14]

Nevertheless, Singh adds a note of caution. He stresses that Indian Courts have been grappling with the concept of property at a personal level for some time. In *Vikas Sales Corporation v Commissioner of Commercial Taxes*[15] the Supreme Court of India made an elaborate analysis of the meaning of the expression 'property', and noted that:

> In the strict legal sense, property is an aggregate of rights which are guaranteed and protected by the government. ... The term is said to extend to every species of valuable right and interest. More specifically, ownership, the unrestricted and exclusive right to a thing; the right to dispose of a thing in every legal way, to possess it; to use it, and to exclude everyone else from interfering with it. That dominion or indefinite right of use or disposition which one may lawfully exercise over particular things or subjects. The exclusive right of possessing, enjoying, and disposing of a thing.

From this case, it can be concluded that, India is considering the broader area of that which a thing or subject might be, this does not rule out personal data being a thing or subject. While that court associated a thing as being land and chattel, it did not narrow, confine or exclude any such rights to personal information. What the court did conclude is that the conception of property constituted rights that have an economic value.[16] Singh believes that a property based approach to the protection of personal data is fraught with uncertainty, including, but not limited to, issues arising from costs of acquisition of data, alienability and onward transfer of property rights in data. The position taken by Singh has been reinforced by Lennart Chrobak[17] who believes that copyright law confers intellectual property rights with regard to digital data to end users generating the data.

Nevertheless, it is our view that personal data, today more than ever, has an economic value because it is being traded – at a price for which entities make a profit. This becomes evident when understanding the current day data protection laws. In other words, as this Chapter will show how data subjects has a level of ownership over their personal data, which is treaded by organizations that make a profit, and thus in turn, have created an economic and commercial value in that personal, defined as by the law.

[14] Ibid.

[15] AIR 1996 SC 2082.

[16] Ibid

[17] Chrobak, L, Propietary Rights in Digital Data? Nromative Perspectives and Principles of Civil Law, in Mor BakhoumBeatriz Conde GallegoMark-Oliver MackenrodtGintarė Surblytė-Namavičienė, *Personal Data in Competition, Consumer Protection and Intellectual Property Law Towards a Holistic Approach?* MPI Studies on Intellectual Property and Competition Law, Springer (2018).

12.1.2 Economic Value Personal Data

The economic rationale for intellectual property law arises significantly from the public interest in regulating markets in information-based products. This, in turn, has given rise to laws that regulate intellectual property.[18] The economic rationale behind intellectual property law therefore is to overcome the abuse of information.[19] Moreover, in the absence of intellectual property rights, there may be little incentive to induce an optimal level of private investments in the production and dissemination of intellectual products, notably including personal data on the Internet.[20]

Dorothy Glancy[21] seeks to protect personal rights from being subject to economic exploitation. She argues that personal information is at risk as it moves from the person who creates the personal information into files and databases of personal information controlled by others.[22] However, Glancy argues that, before devising a speculative intellectual property regime to protect personal information, it is essential to consider that the value of protecting personal information varies, based on cultural, religious, political and economic differences across and within countries. Even the definition of personal data diverges, not only according to competing perceptions of the value of protecting personal information, but because of a failure to consider the multitude of software systems that capture someone's personal data, however minimal that impact may be. Glancy argues that personal data, at inception, does constitute property – it has a level of value. In particular, personal information is initially the intangible intellectual property of the person who creates it. This personal information is frequently mixed thereafter with the intellectual property rights of others in what amounts to a co-ownership arrangement. Her suggestion supports the development of appropriate intellectual property rules to address the protection of personal information as intellectual property, while recognizing its economic value to others.[23] The issue for Glancy is whether existing protections can be generalized into a more comprehensive intellectual property right in personal information.[24] Her response is in the negative, on grounds that other more effective public and private regulatory measures displace the need for intellectual property protection.[25]

One response to divergence over how to protect data is by developing an experimental model of data protection. Such a model can help to determine how intellectual

[18] Merges, P, Menell, P, Lemley, M Jorde, T, *Intellectual Property in the new Technological Age*, New York: Aspen Law & Business (1997), pp. 11–20

[19] Ibid.

[20] Ibid.

[21] Glancy, D *Santa Clara Personal Information as Intellectual Property*, https://www.law.berkeley.edu/files/bclt_IPSC2010_Glancy2.pdf, accessed 14 May 2018. The article recounts a thought experiment into what recognition of personal information as intellectual property might look like.

[22] Ibid.

[23] Ibid.

[24] Ibid.

[25] Ibid.

property and general property protection can be extended to personal data. Some advances in this experimental direction are already reflected in existing literature, such as in evaluating when it may be appropriate to treat personal information and privacy as intellectual property, based on when and how such information is most likely to be mishandled by others.

Economists and privacy advocates, too, have proposed giving individuals property rights in their personal data.[26] Pamela Samuelson argues that, providing greater protection to personal data in cyberspace and elsewhere, is based on the incentives of the company that acquires private information.[27] That incentive includes the full benefit of using that information in its marketing efforts, or the benefit from the fee it receives when it sells the information to third parties. That company, however, does not suffer losses from the disclosure of the private information being transferred because the data subjects often will not learn of the disclosure of their information; nor will the data subject be able to discipline the misuse of that information due to that subject's lack of economic and other resources to control that company's misuse of it. In economic terms, the company internalizes its gains from using their personal information; but it can externalize some of its losses, giving it a systematic incentive to overuse, not limited to selling, that information.[28] Nevertheless, Samuelson warns against propertizing personal information as a way of achieving information privacy goals.

To some extent, the position put forward by Samuelson is supported by Gianclaudio Malgieri who argues that personal data has an intrinsic value that makes it eligible for treatment as an intellectual property right.[29] However, Malgieri points out that the intersection between personal data and intellectual property is blurred.[30] Nevertheless and despite these blurs, personal data has been the subject of *sui generis* regulation.[31] In particular, when the protection of personal data is in the public interest, such as for national security or law enforcement purposes, propertization of that personal information may be stipulated for in law. Schwartz argues that regulation is also needed to control the commodification of information, even personal information, in recognition of the additional uses and prospective transfers of personal data.[32]

---

[26] Samuelson, P *Privacy As Intellectual Property?* Stanford Law Review Vol. 52:1125 (2000).

[27] Ibid.

[28] Ibid.

[29] Malgieri, G *User-provided personal content' in the EU: digital currency between data protection and intellectual property*, International Review of Law, Computers & Technology, 32:1, (2018) pp. 118–140.

[30] Ibid.

[31] Ibid, user-provided data are the only piece of information that is explicitly recognized as 'commodifiable' as a kind of digital good of individuals. Indeed, on the one hand, it is the only set of personal data that can be 'ported' from one platform to another. It is the only kind of (personal) data that the (proposed) law would consider a legitimate counter-performance other than money for the provision of digital content.

[32] Schwartz, P *Property, Privacy, and Personal Data*, Harvard Law Review, vol 117, No 7 (2004). Susan Rose-Ackerman's definition, an "inalienability" is "any restriction on the transferability,

Similarly, Karki while taking a more cautious approach, argues that the property rights model offers two principal benefits. Firstly, it would establish a right in individuals to sell their personal data and thereby capture some of the value their data has in the marketplace. Secondly, a property rights model would force companies to internalize certain social costs now borne by others from the widespread collection and use of personal data.[33]

Lawrence Lessig's instrumentalist theory of propertization provides an economic argument for recognizing property rights in personal data.[34] Lessig argues that property rules allow individuals to decide what information to disclose and what information to protect for privacy reasons. He is of the view that information privacy constitutes control over personal information.[35] In support of this view, he identifies the gradual shift towards tighter controls exerted over the collector and user of personal information. Lessig also believes that, having property rights in personal data, has increasingly forced businesses to negotiate with those individuals whose personal data is in issue.

Litman maintains that the position advanced by Lessig corresponds to the current legal position.[36] In effect, propertizing personal information[37] requires the inalienability of property in a system that protects privacy.[38] This leads to the hybrid inalienability of personal information, consisting of a use-transfer restriction, plus an opt-in default.[39] According to Litman, the hybrid inalienability of personal data permits an initial transfer of that data from data subject to data collector, but only if the data subject is granted an opportunity to block further transfers or uses by unaffiliated entities.[40]

Importantly, property is defined as opposite to the liability rule.[41] If any entitlement, is protected by a property rule, whether or not that right extends to personal data, that property is inalienable and cannot be taken away.[42] While the property rule protects the entitlement of the data subject, the liability rule allows that subject to

---

ownership, or use of an entitlement. Susan Rose-Ackerman, *Inalienability*, in The New Palgrave Dictionary of Economics and the Law, Yale Law School Faculty Scholarship (1985) p. 268.

[33] Karki, M, *Personal Data and Privacy and Intellectual Property*, Journal of Intellectual Property Rights, Vol 10, (2005) pp. 58–64.

[34] Lessig, L *Code and Other Laws of Cyberspace*, New York, Basic Books (1999).

[35] Ibid.

[36] Litman, J *"Information Privacy/Information Property"*, Stanford Law Review, No 52, (2000) p. 1295.

[37] Ibid, p. 1283.

[38] Schwartz, P *Property, Privacy, and Personal Data*, Harvard Law Review, vol 117, No 7 (2004). Susan Rose-Ackerman's definition, an "inalienability" is "any restriction on the transferability, ownership, or use of an entitlement. Susan Rose-Ackerman, *Inalienability*, in The New Palgrave Dictionary of Economics and the Law, Yale Law School Faculty Scholarship (1985) p. 268.

[39] Ibid.

[40] Litman, J *"Information Privacy/Information Property"*, Stanford Law Review, No 52, (2000).

[41] Calabrese, G., Melamed, A *"Property rules, liability rules, and inalienability: one view of the cathedral"*, *Harvard Law Review*, 1972, No 85.

[42] Ibid.

transfer the liability or responsibility for the use of that right to a third party, whether of that that right is conceived as protecting property. Today, to some degree, data protection laws have followed the liability rule. That is, the current model allows the transfer of responsibility for personal data to a third party, notably to a data controller. However, this approach has not been adopted by every country under study. Scholars, since that 1970s, have been grappling with the idea of introducing property rights in personal data.[43] Some have argued that introducing property rights in personal data would help individuals to gain control over their personal data; or that it would establish a sustainable connection between individuals and the protection of their personal data under theories of information privacy.[44]

A further complicating issue in treating personal information as property is, not only the alienability of property, but the manner in which property rights are bought, sold, and otherwise exchanged. In effect, the purpose of property law is to prescribe the conditions for the transfer of that property.[45] The concern is that property gives the owner of that property control over it, including the right to sell or license it. That control includes the right to exclude the person whose private information is being sold to third parties. The misuse of that person's control over that property is therefore necessarily subject to regulation. That regulation is not provided directly through the law of property, but by governments protecting privacy rights, and individuals who protect their rights by private law means, such as through tort law.[46]

A significant challenge ahead is the multilayered approach to personal data regulation does not lend itself easily to affording intellectual property, or, a general property right in personal data. While data protection is often viewed as a right that ought to be legally protected, personal data is also increasingly recognized as being valuable (particularly to data collectors, data miners, and personal data merchants). As a result, it is necessary to consider both quantitative and normative factors in determining when and how to protect that information.

The result of these different scholarly views is disquiet over why, when and where intellectual property rights in personal data ought to exist, if at all. In contention is when such an intellectual property right commences and concludes through the data use lifecycle; and how far policy makers can extend the cycle of regulation in protecting the data subject's intellectual property. Regulators who purport to regulate users along the full data use cycle are likely to protract the regulatory regime, rendering it too costly and ineffective to apply in a complex data environment. Conversely, restricting regulation to the relationship between the data subject and the immediate data collector is likely to encourage the misuse of the data subject's personal information further down the data user cycle. Noteworthy, too, is that

---

[43] Agre, P., Rotenberg., M, *Technology and Privacy: The New Landscape*, Cambridge, MIT Press, (1997).

[44] Solove, D "*Privacy and Power: Computer Databases and Metaphors for Information Privacy.*", *Stanford Law Review*, No 53, (2001) pp. 1440 - 1446

[45] Litman, J "*Information Privacy/Information Property*", Stanford Law Review, (2000), No 52, pp. 1295–1296.

[46] Ibid.

protecting personal data as intellectual property is likely to attenuate regulation over the full lifespan of data usage.

Ultimately, there is no easy response to choosing among these competing regulatory measures outlined above. One option is to make normative assumptions about the value of each in isolation; or to value each in relation to the others, such as establishing a continuum between private and public regulatory measures. This approach is well entrenched in those countries that have specific data protection laws, such as Singapore, Australia, Japan, Malaysia, and the EU. However, Indonesia and India have a way to go, to replicate the data protection laws in other jurisdictions. Another option is to attach costs and benefits of each measure, such as to protecting the human rights of the data subject, the economic rights of the data user, and the right of the public to be informed. More importantly, the law in supporting different notion, has arguably been assisted, whether by accident or design, through the establishment of key legal concepts such as consent to the use of personal data. That consent is discussed immediately below.

12.2 Consent & Personal Data

Paul Schwartz argues that, under the consent-based model of decentralization, regulatory commissions should be established to oversee compliance with decentralized laws that regulate the collection and use of personal data.[47] Schwartz's model is reflected to some degree in the establishment of regulatory structures that have established Commissions, Commissioners and dedicated agencies that are responsible to regulate the collection and use of personal data. Even though Schwartz proceeds with caution over the propertization of personal data, the elements he espouses identifies a bundle of interests that are subject to regulation. These interests encompass inalienabilities, defaults, and a right to exit through the principle of consent, damages, and institutions.[48] However, some commentators have expressed concern that regulating the use of personal data is defective in providing that persons never fully own their personal information once that information has a footprint on the Internet or network. The feared result is that data controllers and processors will have the most control over personal data through laws that protect their database rights. In effect, they will be the primary beneficiaries of the economic value extracted from that data.[49] Schwartz further argues that the protection of personal data is accomplished by combining privacy and property to enhance the inalienable use and transfer of personal data.[50] That is, data subjects are provided

[47] Schwartz, P *Property, Privacy, and Personal Data*, Harvard Law Review, vol 117, No 7 (2004).

[48] Ibid.

[49] Karanasiou A, Douilhet, E *Never Mind the Data: The Legal Quest over Control of Information & the Networked Self*, http://eprints.bournemouth.ac.uk/23392/1/PID4084429%20%285%29.pdf, accessed 24 April 2018.

[50] Ibid.

with a level of control and ownership over their personal data which enables them to restrict the use and transfer of that data.

The growing concern in response to this realization, is that a person does not truly own his/her personal data throughout the cycle of its use, commencing with that person accessing the Internet or network. The related worry is that the ownership of personal data is likely to be lost once that person has provided consent to the controller or processor to use that data. The inferred contractual result is that the data subject loses any proprietary (and contractual) interest thereafter in that data, regardless of whether the data subject consented to the manner of its use, or was properly informed about the consequences arising from its use.

Centrally in issue are several pervasive socio-economic, technological and political obstacles in regulators deciding over whether, and if so, when and how to protect the privacy of personal data, including through intellectual property or general property rights in that data. These obstacles are protracted by the ever-expanding potential uses to which personal data can be placed in our constantly evolving technological revolution.

Notwithstanding these obstacles arising from data collectors and processors protracting the consent of data subjects, the concept of consent is also becoming a cornerstone of data protection and privacy law.[51] The OECD Guidelines regards 'consent' as an important step to the lawful collection and processing of personal data.[52] However, the level and extent of consent required by data subjects differs from jurisdictions to jurisdiction. Consent is also influenced by the person allegedly so consenting, varying from 1) consent by an adult, 2) consent relating to children and 3) consent by notice. At the national level and across them, consent takes the form of actual (verbal and written), whether it is deemed or implied. In addition, consent can be withdrawn. These issues are discussed below in light of the nature of consent to use personal data. That discussion also encompasses the extent to which the legal construction of that consent diverges according to whether or not the data subject is deemed to have a property right in that data.

Consent that complies with legal requirements provides data subjects with greater control over and ownership of their personal data and information.[53] Arguably, such consent is given at that moment at which personal information is exchanged. However, personal data is often captured, stored and used by many other platforms beyond those consenting parties, such as by companies with which the data subject is unlikely ever to identify or interact. Therefore, the ability to consent to the use of personal data in such circumstances has been somewhat limited in

---

[51] Ibid.

[52] Organization for the Economic Co-operation and Development, Guidelines on the Protection of Privacy and Transborder Flows of Personal Data 2013. http://www.oecd.org/sti/ieconomy/oecd-guidelinesontheprotectionofprivacyandtransborderflowsofpersonaldata.ht m, accessed 20 February 2018.

[53] Trakman., L, Walters., R Bruno, Zeller B *Is Privacy and Personal Data set to become the new Intellectual Property*? International Review of Intellectual Property and Competition Law, forthcoming 2018

law as in practice, given that the party using the data is unknown to the data subject. This limitation is attributable to the fact that the data subject ordinarily only provides consent one step down on the supply chain, namely, to the data controller or data processor that sits within an organization that trades in that subject's personal data. Once traded, that data subject appears to lose any further control over the applicable data.

Consent in data protection law as a concept has become very important in determining whether to treat privacy as an intellectual property right. The concept of implied or deemed consent is significant because it adds another layer of complexity to the notion that personal data is the subject of an intellectual property right, while not negating that right. Indeed, it is plausible that the right of individuals to both expressly and impliedly consent to the disclosure of their personal information presupposes that they have some form of intellectual property right in that data. This will be discussed further in the sub-sections below.

12.2.1 Withdrawal of Consent

It is arguable that the ability of data subjects to withdraw their consent to the use and processing of their personal data, further strengthens their control and ownership of that data. Once that individual has withdrawn consent, an entity can no longer rely on past consent for any future use or disclosure of that individual's personal information.[54] However, in practice the withdrawal of consent is not necessarily clear, unless the entity has provided information to the data subject that he/she has the option to deny consent to further trade in that personal data. Otherwise, in most circumstance the data subject is likely to be clueless as to whether withdrawing consent is an accessible option.

Having the legal power to withdrawal one's consent, arguably would provide the data subject with a greater level of control and ownership over personal data. It is arguable that such consent is conceived as being given at the moment at which personal information is first exchanged, namely, when the data subject consents to transfer that data to the first user. However, the ability to consent to the use of personal data in such circumstances is limited by the fact the data subject only ever provides consent to the first data controller or processor, not to downstream users who are unknown to the data subject.[55]

[54] Ibid.

[55] See Council Regulation (EU) 2016/679, General Data Protection Regulation, Article 7(4) affirms that the consent is not freely given if it is conditional. Article 6 requires that processing of personal data is to be lawful only if and to the extent that at least one of the following criteria applies: the data subject has given consent to the processing of his or her personal data for one or more specific purposes. Consent in Australia is conceived broadly. There is no direct requirement or pre-requisite for collecting personal data or information from a data subject. However, for 'sensitive information' a person's consent must be provided. The Australian Privacy Principles (APPs) require that personal information should be collected directly from the individual, unless the individual has

It is our view that consent reinforces the proposition that an intellectual property or general property right can exist in personal data. This is because a data subject who has intellectual property in that data acquires a legally supportable level of control over their personal data beyond a personal right. Were a data subject to have intellectual property in that data, it would be difficult to conclude that, by consenting to its use by the data collector, the data subject had impliedly consented to its use by a host of further downstream users.

While it is out of scope of this Chapter to examine the different definitions of personal data, what can be confirmed is that each jurisdiction that has defined such data has adopted a slightly different approach,[56] as was discussed in Chap. 11. What can be confirmed is the unlikelihood that personal data defined by law includes all personal data that is captured, mined and harvested by Internet platforms. Generally, all the jurisdictions define personal information and data as constituting the name, date of birth, residential address, but differ in defining the attributes of personal data.

12.2.2 Sensitive – Personal Data

Importantly, some jurisdictions have begun to identify personal data based on whether it is conceived as being "sensitive". Sensitive data is considered as being a higher level of personal data and information about an individual which must be protected, including through ownership by that data subject. Arguably, sensitive data warrants being accorded a more stringent level of propertization, whether as intellectual property or property in general; and also requires stricter protection mechanisms. This propertization of sensitive data has begun to be developed in current day data protection laws in some jurisdictions. However, other jurisdictions have included sensitive data as part of general personal data which has not been accorded any attributes of property. Sensitive data is dealt with briefly below to highlight the prospective propertization of that data.

Generally, sensitive personal data constitutes racial or ethnic origin; political opinions; membership of a political association; religious beliefs or affiliations; philosophical beliefs; membership of a professional or trade association; membership of a trade union; sexual orientation or practices, or criminal record; health information about an individual; genetic information (that is not otherwise health information). In addition, sensitive information can include biometric information

consented to collection from other sources, or if it is authorized by law. The APPs define consent as 'express consent or implied consent. Section 13 of Singapore's Personal Data Protection Act 2012, provides for a form of implied (deemed) consent, and prohibits organizations from collecting, using or disclosing an individual's personal data unless the individual gives, or is deemed to have given, his consent for the collection, use or disclosure of personal data.

[56] See Regulation (EU) 2016/679 on the protection of natural persons with regard to the processing of personal data and on the free movement of such data, and repealing Directive 95/46/EC (General Data Protection Regulation) [2016] OJ L 119/1, Article 4, sub (1).

that can be used for the purpose of automated biometric verification or biometric identification.[57] Some jurisdictions go further and include facial images, voice recordings, fingerprints, iris images and DNA profiling. Other jurisdictions such as the EU, rather than define sensitive data, include sensitive data in their general definitions of personal data.

Notwithstanding the differences across jurisdictions identified above, both sensitive and less sensitive data are tradeable, portable, and therefore able to be transported from one jurisdiction to another.[58] To do so under the legal framework that currently prevails, requires the consent of the data subject (individual), at least to the point at which the data controller or processor is in full control of that personal data. In other words, it is arguable that there is nothing in the definition, or the laws of these jurisdictions, that would preclude this personal information from having a proprietary right. Nevertheless, the definition of personal data alone has created confusion because no data is personal from the outset, and all data can become personal.[59] This reinforces the argument that the clash between privacy and property is blurred.[60]

Björn Lundqvist[61] asserts that the definition of personal data is extensive because such information that is non-personal in nature might also indirectly, be combined with other information that identifies a natural person, and therefore than non-personal may become personal data.[62] He believes that it can be wise to calibrate the collection mechanism, to transfer and collect non-personal data, such as in an industrial-Internet setting, when that non-personal data can be based on personal data. He proposes keeping such personal and non-personal data sets intact and to maintain their separation. Lundquist further elaborates that data information, irrespective of how private and how valuable it is, is not currently covered by a property right.; and that no one owns personal data. However, Lundquist also acknowledges that a'data subject'in the EU holds some rights to it, in accordance with the GDPR. Thus, the position put forward by Lundquist, in our view, reinforces the notion that personal data is afforded a level of intellectual property rights because the definition of personal data in some national laws, notably within the EU, is wide enough to deal with some, if not most, of the identifying information whether directly or indirectly.

---

[57] Ibid.

[58] Ibid.

[59] Janeček, V *Ownership of personal data in the Internet of Things* Computer Law & Security Review (2018).

[60] Ibid.

[61] Lundqvist, B Big Data, Open Data, Privacy Regulations, Intellectual Property and Competition Law in an Internet-of-Things World: The Issue of Accessing Data, in Mor BakhoumBeatriz Conde GallegoMark-Oliver MackenrodtGintarė Surblytė-Namavičienė, *Personal Data in Competition, Consumer Protection and Intellectual Property Law Towards a Holistic Approach?* MPI Studies on Intellectual Property and Competition Law, Springer (2018).

[62] Janeček, V *Ownership of personal data in the Internet of Things* Computer Law & Security Review (2018).

Regardless of how personal data is defined, if that data is coupled with consent, data subjects do have a level of control over their personal data. Janeček believes that the EU law defines personal data reversely. In effect, he argues that data is the source of information which, if personal, leads to the reverse implication, namely, that the original data is also personal. His definition of personal data, however, leads to a seemingly paradoxical situation in which no data is personal from the outset and all data can become personal from the outset. The added perception of a clash between privacy and property then assumes the form of a chicken/egg problem in which it is unclear what comes first. Do the information-centered privacy arguments prioritize the personal chicken? Do data-centered property arguments fall on the side of the data egg? The problem in determining the source of personal information and data is a different one. The trick is that an egg that is made of data does not need to reveal or contain the chicken's personal information in every single case and can still be considered valuable and worth protecting. The egg can also be valued at different levels of abstraction beyond being valued only at the level of personal information. For example, the egg contains precious albumen as well as information about resistant constructions – you may try to crack the egg in your fist yourself. Data and information simply cannot be compared to each other at the same level of analysis because they are fundamentally different in kind. On this account, Janeček contends, it is clear that personal and non-personal data are not conceptually incompatible categories. In large part they are compatible, because while some personal data is defined by the law, and nonpersonal data may not be captured by the same definition, both personal and nonpersonal data provides identifying information, which identifies the data subject.

Janeček highlights further, that a clash arises between information-centered privacy which cannot be owned, and personal data that, conceivably, can be controlled. He then asserts that there is a need to restrict the potential scope of ownership and control over personal data "from an opposite direction", asserting that, in that case … the key question must be whether some data contain personal information intrinsically and therefore *cannot* be defined as non-personal data from the outset. Janeček concludes cautiously that, since ownership of personal data still cannot be satisfactorily explained and justified, such initiatives should remain investigatory, analytic and descriptive.

Trakman, Walters and Zeller express the view that Janeček's line of reasoning is somewhat circular, primarily because control over personal data is provided for by law.[63] Current data protection laws strengthen the idea of information-centered privacy, although they do not fully remedy deficiencies in promoting such information-centered privacy. The view they have adopted is that this strength in current data protection laws is further reinforced by renewed emphasis given to the definition of personal data and information, although the precise definition varies across jurisdictions. In particular, most jurisdictions capture the same data, even though some

[63] Trakman., L, Walters., R, Bruno, Zeller, B *Is Privacy and Personal Data set to become the new Intellectual Property*? International Review of Intellectual Property and Competition Law, forthcoming 2018

jurisdictions have gone into more depth than others. The problem will arise when there is no clear definition, because ownership will vary at different stages of the defined personal data.

Janeček argues for a strict conceptual separation of personal data from personal information; and focuses on whether the concept of ownership/property can be applied to personal data in the context of the Internet of Things. He maintains that, given that personal data protection originated with the fundamental right to respect for private life as is reflected in GDPR, the starting point ought *not* to be when that data can be protected by the law because personal information can never be protected by the law because it would violate free access to information. Identifying the GDPR as the root of the problem, he asserts that EU law defines personal data reversely: data are the source of information which, if personal, reversely implies that the original data are also personal. Janeček notes that this definition leads into a seemingly paradoxical situation in which no data are personal from the outset and all data can become personal from the outset. He responds by proposing to restrict the scope of the potentially or controlled personal data from an opposite direction. In that case, he argues that the key question must be whether some data contains personal information intrinsically and therefore *cannot* be defined as non-personal data from the outset. He also identifies examples of such data in the jurisprudence of the European Court of Human Rights (ECHR); and concludes that, intrinsically, personal data must be excluded from my [his] definition of personal data for the purposes of ownership issues. In effect, the [o]wnership of such data would thus conceptually imply ownership of people's identities; and the owner of the intrinsically personal data cannot exclude the individual's demands on these data unless he/she neglects the individual's identity in the first place.[64]

However, it is arguable that Janeček reasoning does not adequately address the primary fact, that control over personal data is provided for by law.[65] Not only does the definition of personal data provide a clear pathway to guiding people as to what constitutes personal data, the concept of consent is also an important determinant. Without a definition of personal data, consent cannot be provided by a person or a data subject. In other words, where there is no definition of personal data – the data subject cannot consent to anything. Put another way, without personal data being defined, a data subject has nothing to which to consent.

It is arguable that property rights are included in the definition of both personal data and privacy within national and supranational laws. It is also reasonable to hold that individuals have a certain level of control over their personal data arising from their consent to its use. In fact, based on Litman's conception, the consent of individuals to the use of their personal data is based on their rights to privacy, which arguendo, is also the source of their property rights. This proposition is further enhanced on grounds that personal data that is protected by privacy, is also capable of being traded as property.

[64] Ibid.

[65] Ibid.

12.3 Data Portability

The right to data portability enables data subjects to receive their personal data that they have provided to a controller. More importantly, data portability provides the data subject with the right to request that a controller transfer their data to another controller who sits in another organization. In recognizing that the right of data subjects to portability enables them to move their personal data from one controller to another, the result is a more formidable and propertied right in personal data. They is, arguably, now the case in the EU, in which personal data is accorded a higher level of property rights.[66] The portable right of the data subject also goes some way in redressing the tension between the protection personal data and the alleged intellectual property right of data subjects in that data.

In Australia, the right to access data is restricted to allowing for the correction of personal information relating only to credit information.[67] In Singapore, sections 21 and 22 of the PDPA provide data subjects with the right to request access to their personal data for the purpose of correction. However, at the time of writing this book, neither Indonesia, India, Japan, Malaysia or Thailand had established the right to portability. Importantly, the right to portability allows data subjects to move their personal data, both legally and functionally, from one organization to another.[68] While this right has implication for competition law, it does raises questions as to who has the intellectual property, or other property right in that data. The two most likely persons or entities with rights in that data are the data subject and the first receiver of that personal data. However, the GDPR makes it clear that the right to data portability should not have an impact on any other rights, which arguably includes intellectual property rights. Reconciling the tension among different kinds of rights, including property rights, is likely to be complex because the collection of data, in the first instance, likely falls within the scope of the GDPR. However, once that data (the raw data) has been transformed into and mixed with other data, it is yet to be confirmed whether the GDPR applies to it. Nonetheless, there is case law that points to personal data being protected as an intellectual property right. These developments are discussed below.

---

[66] Ibid, Article 15. Recital 63 provides some protection for controllers concerned about revealing trade secrets or intellectual property, which may be particularly relevant in relation to profiling. It says that the right of access 'should not adversely affect the rights or freedoms of others'. However, only under rare circumstances should these rights outweigh individuals' rights of access; controllers should not use this as an excuse to deny access or refuse to provide any information to the data subject. These rights should be considered in context and balanced against individuals' rights to have information. Recital 63 also specifies that where possible, the controller should be able to provide remote access to a secure system which would provide the data subject with direct access to his or her personal data.

[67] Privacy Act 1988, Sub- Division 3

[68] Trakman., L, Walters., R Bruno, Zeller B *Is Privacy and Personal Data set to become the new Intellectual Property*? International Review of Intellectual Property and Competition Law, forthcoming 2018

12.4 Emerging Case Law

The courts in the United Kingdom have provided some direction on whether privacy can constitute intellectual property. In 2012, the Court of Appeal (England Wales) in *Coogan v News Group Newspapers Ltd* & Anor [*2012*] 2 All ER 74 ruled that confidential personal information is intellectual property under section 72 of the *Senior Courts Act 1981*.[69] In examining the construction of section 72 (2),[70] the court went to some lengths to explain intellectual property and the meaning of commercial information. It also maintained that the meaning of the expression 'technical or commercial information' has to be assessed by reference to the purpose of section 72, the immediate context in which that information is used, and the natural meaning of the words used.[71] Section 72 states:

> (1) In any proceedings to which this subsection applies a person shall not be excused, by reason that to do so would tend to expose that person to proceedings for a related offence: (a) from answering any question put to that person in the first mentioned proceedings; or (b) from complying with any order made in those proceedings. (2) Subsection (1) applies to the following civil proceedings in the High Court, namely: (a) proceedings for infringement of rights pertaining to any intellectual property or for passing off.[72]

The Court further argued that intellectual property covers confidential information. It maintained that it is unsatisfactory to place undue weight on a single generic term that covers all intellectual property rights. In quoting the earlier case of *Price v Hal Roach Studios* Inc., 400 F. Supp. 836 (S.D.N.Y. 1975),[73] the court highlighted that:

> [t]here is no single generic term that satisfactorily covers' all rights which comprise intellectual property. However, the courts went on to say intellectual property protects information and ideas that are of commercial value. The position taken by the court is that as long as the personal information is confidential and of commercial value it has a property right and can be treated as intellectual property.[74]

This quotation refers to a dispute over the ownership of the commercial rights to use the names and likenesses of Stanley Laurel and Oliver Hardy, two famous but long deceased comedians. The Complaint was filed on January 29, 1971 by plaintiff Larry Harmon Pictures Corporation ("Harmon"), a California corporation, against defendants Hal Roach Studios, Inc. ("Roach"), a Delaware corporation with its principal place of business in New York, and Richard Feiner & Co. ("Feiner"), a New York partnership. Jurisdiction in the case was predicated upon diversity of citizenship. The plaintiffs, widows of Laurel and Hardy, and sole beneficiaries under the comedian's wills, claimed to have exclusive rights to their late husband's names

[69] *Coogan v News Group Newspapers Ltd* [2012] EWCA Civ 48, [2012] 2 WLR 84, [2012] EMLR 14, [2012] 2 All ER 74.

[70] Ibid.

[71] Ibid.

[72] Ibid.

[73] *Price v. Hal Roach Studios*, Inc., 400 F. Supp. 836 (S.D.N.Y. 1975).

[74] Ibid, para 1-01.

and likenesses. The defendant asserted that, because the comedians were now dead, their names and likenesses were part of the public domain. They argued further, that their rights in their names terminated on their deaths; that these were property rights which were assignable in the public domain including for commercial purposes.[75]

The court decided that:

> it might have been possible for the performers to have waived some of their rights to privacy while they were alive, in them being so well known by their names, but that the property rights in one's own name was not waivable.[76]

The significance of this historical case lies in the fact that the current day data protection and privacy laws of the countries studied in this article, have defined a person's name as personal data that is subject to protection through such laws.[77] That protection, in turn, conceivably includes a property right in one's name.

12.5 Moving Forward

Intellectual property in personal data is evolving into another part of overall jigsaw puzzle that data protection law has become. Trakman, Walters and Zeller note that regulators and courts have also begun to accord intellectual property protection through the conception of the data subject's consent to the use of personal data, most notably in the EU. Regulators and courts have also extended such protection through the conception of implied consent in jurisdictions beyond the EU.[78] In doing so, they have enlivened a concept of implied or "deemed" consent that is complex, not only in nature, but also in its scope of application.[79] Regulators and courts have also extended such protection to include a notion of consent, although that extension is limited to the first receivers of such personal data (the controller or processor) that have acquired from data subjects, ownership rights in, or control over, personal data.

However, meeting the challenge in protecting personal data an intellectual property extends beyond the boundaries of express or implied consent. The obstacle in protecting such data is also attributable to the fact that protecting personal data as property resides at the very divide between private and public rights, in which disparate conceptions of the public good, conceivably, extend beyond private rights. On the private side, there is the virtue of allowing, and indeed requiring, that individuals protect their own rights by private law means, of which consent is central to

---

[75] Ibid.

[76] Ibid.

[77] Trakman., L, Walters., R Bruno, Zeller B *Is Privacy and Personal Data set to become the new Intellectual Property*? International Review of Intellectual Property and Competition Law, forthcoming 2018.

[78] Ibid.

[79] Ibid.

such protection. On the public side is the virtue of governments imposing requirements on the use of personal information beyond the consent of the individual, on grounds of a fathomable and not overly expansive public good. Regulators and courts have already, albeit limitedly, entertained such protection of personal data on public policy grounds. They have done so in protecting "sensitive" personal data.[80] However, they have yet to redress effectively conflicting conceptions of the public good which encourage the free flow of personal data for the benefit of informing society at large.

The comparative analysis adopted in Chap. 11 highlights the further tension between the common and civil law, namely, in the control and ownership of that data. Arguably, the common law provides greater flexibility in the types of ownership that can be created and protected as property rights.[81] However, the civil law of the EU has a restricted number of property rights and a limited number of legal objects that can be subject to these property rights (*numerus clausus*).[82] Janeček argues that the civilian idea of ownership is an absolute dominion that encompasses all the listed rights (*numerus clausus*) over the relevant object. That differs from the common law tradition in which ownership includes a variety of different rights over the same property.[83] Therefore, unlike in civil law, acquiring ownership of personal data in the common law can be gradual. An individual or entity can also have more, or less, ownership, depending on the size of that person's bundle of property rights in the data object.[84]

It is our view that a pathway forward includes the need for greater legal convergence and harmonization, not only across jurisdictions, but also legal systems. Such harmonization extends beyond narrow rules of law to key concepts and principles, such as the definition of personal data and consent to its collection, processing and other use. This approach, arguably, has been successful in other areas of private international law, such as, in international trade law. However, the challenge for policy makers lies in the multi-layered approach and direction that data protection laws have assumed to date.

Nevertheless, these developments are likely to be arduous and inevitably, strained. The reality is that attempts to arrive at a pervasive regulatory framework flies in the face of the somewhat fragmented and *ad hoc* manner in which nation states seek to develop their domestic laws to meet their internal needs in response to their localized public policies. This reality is likely to sublimate the global need to protect the transmission of personal data across states. A cohesive regulatory need, at the least, is for nation states to recognize the value in subscribing to transnational public policies in protecting against abuse in the transmission of personal data, such as in

---

[80] Ibid.

[81] Gordley, J *Foundations of Private Law: Property, Tort, Contract, Unjust Enrichment* (OUP 2006) 49.

[82] Akkermans, B *The Principle of Numerus Clausus in European Property Law* Intersentia (2008).

[83] Janeček, V *Ownership of personal data in the Internet of Things* Computer Law & Security Review (2018).

[84] Ibid.

modes of data delivery. The benefit in such protections is that, while they may differ in precise detail, they can ideally be shared in principle across state boundaries, rather than be subordinated by divergent laws based on divergent legal principles.

12.6 Conclusion

An important part of expansion of data protection law is to address the extent to which personal information and data ought to be treated as property, and ought to be the subject of more pervasive intellectual property rights. The ability to balance the rights afforded to individuals. while maintaining levels of controls by the data subject over the use of that data, is an unavoidable and ongoing challenge for regulators and policy makers both nationally and internationally. How far or how much control is afforded to data subjects through the law is debatable. Whether that protection extends to the first, second, third, fourth, fifth, or some further point in the cycle of collection and use of personal data remains an open question. While the principles espoused by the OECD, particularly regulating transparency and accountability, are sound, the issue is to determine how far those principles ought to be extended (see Chap. 16). These raise policy challenges that need to be addressed on a continuing basis, in an area of the law that needs clarification as technologies develop over time, place and space.

The growth in technology, its infrastructure, systems and platforms now enables large quantities of personal data to be traded across the world. The infrastructure, systems and platforms supporting the Internet, have arguably created a level of copyright which, in turn has gone part of the way to protecting personal data on those systems and platforms. It is the view of the authors that the evolving data protection laws increasingly provide a level of intellectual property right to personal data, such as through their definitions of personal data. This growth of intellectual property rights in personal data is supported, in part, by the consent which a data subject provides an organization to collect and use that subject's personal data as circumscribed by law. In our view, consent provides data subjects with a level of control over their personal data. It is this control, that arguably provides them with level of personal ownership in and over their personal data. Their right to data portability and to provide access to their personal data has further strengthened their control and ownership. These developments, evaluated in tandem, have enhanced the contention that personal data is gaining intellectual property protection. Providing an intellectual property to personal data will also have many other economic benefits, and may go some way to addressing some of the issues associated with consent, beyond the first consent provided by the data subject.

Nevertheless, the nature and scope of intellectual property in personal data is unlikely to determined coherently, cohesively, or effectively in the immediate future. States diverge over the nature and extent of such protection – locally and regionally, based on dissimilar common, civil and customary law roots, and influenced by socio-economic ideologies.

However, the international community faces immediate pressures, calling for shared solutions to shared problems. What is seminal, is how states can balance the need for data protection and privacy with the need for economic activity in a developing global digital economy. Finding that balance is becoming increasingly more complex as more people are exposed to the misuse of personal data. That consternation is likely to grow further as data collectors and processors secure access to ever expanding sources of personal data in global markets. In issue is their capacity to use such information commercially so as to place themselves as a competitive advantage over other data collectors and users across vast business sectors.

What remains contestable is whether the current regulatory and policy tools have gone far enough to provide a sufficiently solid legal framework to protect personal data both fairly and effectively. That contest is particularly acute in determining whether protecting personal data intellectual property will enhance a fragile global regulatory framework that historically treated the Internet as a predominantly free market in data. In doubt is whether local and regional governments believe that further regulatory and policy intervention is needed to ensure that a continual balance is maintained between innovation, economic activity and privacy protection. Should the answer to this unresolved question be "yes", the challenge will be for governments to regulate personal data that has growing economic and commercial value to data users, offset by growing exposure to data subjects to that profitable use. On the one side of this tryst is intellectual property in personal data as a compelling means of protecting data subjects from divestiture of their "personal" property. On the other side is the threat that intellectual property will privilege data subjects, enabling them to sell their propertied data for extortionate prices, or otherwise stifle the transmission of data – not limited to personal data – over the Internet.

References

Agre, P., Rotenberg, M, *Technology and Privacy: The New Landscape*, Cambridge, MIT Press, (1997)

Akkermans, B *The Principle of Numerus Clausus in European Property Law* Intersentia (2008)

Calabrese, G., Melamed, A *"Property rules, liability rules, and inalienability: one view of the cathedral", Harvard Law Review*, 1972, No 85.

Chrobak, L, Propietary Rights in Digital Data? Nromative Perspectives and Principles of Civil Law, in Mor BakhoumBeatriz Conde GallegoMark-Oliver MackenrodtGintarė Surblytė-Namavičienė, *Personal Data in Competition, Consumer Protection and Intellectual Property Law Towards a Holistic Approach?* MPI Studies on Intellectual Property and Competition Law, Springer (2018)

Cohen, J *Examined Lives: Informational Privacy and the Subject as Object*, 52 Stanford Law Review, 1373, (2000), pp. 1423–28

Glancy, D *Santa Clara Personal Information as Intellectual Property*, https://www.law.berkeley.edu/files/bclt_IPSC2010_Glancy2.pdf, accessed 14 May 2018.

Gordley, J *Foundations of Private Law: Property, Tort, Contract, Unjust Enrichment* (OUP 2006) 49

Janeček, V *Ownership of personal data in the Internet of Things* Computer Law & Security Review (2018)

Karki, M, *Personal Data and Privacy and Intellectual Property*, Journal of Intellectual Property Rights, Vol 10, (2005) pp. 58–64

Laudon, K*"Markets and Privacy"*, 39 (9) Communications of the ACM, 92–104 (1996)

Lemley, M *Private Property: A Comment on Professor Samuelson's Contribution,* 52 Stanford Law Review (2000)

Lessig, L *Code and Other Laws of Cyberspace*, New York, Basic Books (1999)

Lundqvist, B Big Data, Open Data, Privacy Regulations, Intellectual Property and Competition Law in an Internet-of-Things World: The Issue of Accessing Data, in Mor BakhoumBeatriz Conde GallegoMark-Oliver MackenrodtGintarė Surblytė-Namavičienė, *Personal Data in Competition, Consumer Protection and Intellectual Property Law Towards a Holistic Approach?* MPI Studies on Intellectual Property and Competition Law, Springer (2018)

Litman, J *"Information Privacy/Information Property"*, Stanford Law Review, No 52, (2000)

Malgieri, G *User-provided personal content' in the EU: digital currency between data protection and intellectual property*, International Review of Law, Computers & Technology, 32:1, (2018) pp. 118–140

Merges, P, Menell, P, Lemley, M Jorde, T, *Intellectual Property in the new Technological Age*, New York: Aspen Law & Business (1997), pp. 11–20

Samuelson, P *Privacy As Intellectual Property?* Stanford Law Review Vol. 52:1125 (2000)

Schwartz, P *Property, Privacy, and Personal Data*, Harvard Law Review, vol 117, No 7 (2004).

Singh, A *Protecting Personal Data as a Property Right*, ILI Law Review (2016)

Solove, D "*Privacy and Power: Computer Databases and Metaphors for Information Privacy.*", *Stanford Law Review*, No 53, (2001) pp. 1440–1446

Tene O., Polonetsky, J *Big Data for All: Privacy and User Control in the Age of Analytics*, 11 Nw. J. Tech. & Intell. Prop. 239 (2013)

Trakman, L, Walters, R Bruno, Zeller B *Is Privacy and Personal Data set to become the new Intellectual Property*? International Review of Intellectual Property and Competition Law, forthcoming (2019)

Zittrain, J *What the Publisher Can Teach the Patient: Intellectual Property and Privacy in an Era of Trusted Privication*, 52 Stanford Law Review (2000) p. 1203

Chapter 13
Competition Law and Personal Data

Abstract For decades, competition law has been effective when intervention is warranted of companies cause or may cause harm to competition within a single market or across several market and diminish consumer welfare. Today more than ever competition law, along with many other areas of the law is being challenged by the introduction of data protection laws around the world. Competition law has played an important role in protecting the consumer. With the recent introduction of data protection laws, it has emerged that they are also playing a role in consumer protection.

This Chapter explores some of the issues in competition law, from the collection and use of personal data. Moreover, this Chapter will briefly demonstrate whether the current regulatory approach is adequate, or, requires non – regulation such as Internet platforms to intervene. This Chapter draws on earlier work by the authors, which proposes a possible solution to the potential problems faced by the intersection between data protection and anti-competitive behaviour. This Chapter demonstrates that personal data, which has been stolen or used without the data subjects consent, and provided that data has been defined by the law – it is automatically protected. Although, personal data not defined by law has very little to no protection, which provides the basis for individuals and entities to acquire the data, so as they can gain a competitive edge in the market.

13.1 Introduction

The developments in digital technology have, and will continue to make it easier to process large amounts of commercial and personal data. This Chapter will briefly highlight the competition issues in regard to the abuse of power, consumer, web browser, mergers and acquisitions, and predatory pricing.

© Springer Nature Singapore Pte Ltd. 2019

R. Walters et al., *Data Protection Law*,
https://doi.org/10.1007/978-981-13-8110-2_13

There have been increasingly calls for competition regulators to incorporate the possession of personal data into their analyses of anticompetitive[1] practices and behaviour. The point the authors make is that the control of large amounts of both commercial and personal data will give companies an unfair advantage over competitors.[2] It is well settled that in the past commercial data has been used to create anticompetitive practices, which has allowed companies employing such practices to capture a dominant position.[3] However, it has only recently emerged that personal data is also being used as a tradable commodity that is placing entities in a position whereby they can use the data to bargain for a stronger position in the market, because they have exclusive access to personal data.

One of the most pressing issues in the new digital economy is the lack of knowledge and understanding about the competing forces between competition and personal data protection law. That is, competition law regulates the behavior of individuals and organizations with regard to products, choice and price. However, data protection laws focus on protecting the privacy of the individual person's – personal data that has been defined under national or supranational law. This protection has become necessary because more importantly, large amounts of personal data are being used in anticompetitive behavior[4] and because organizations use their market power to take advantage of consumers and competitors.[5] Nevertheless, competition and data protection laws both converge to provide a level of consumer protection.

Anti-competitive behavior, from the collection, use and application of personal data can be traced to predominantly two different forms. Firstly, personal data defined by law that is stolen or used without the consent of the data subject to enhance market power by corporations. The second corresponds to situations in which personal data, which is defined by law, and also captured by Internet systems and platforms, is used in a way that causes harm, resulting in economic inefficiency. Anti-competitive behavior can be defined as personal data being harvested or mined, whether illegally or legally, to gain a dominant position in the market.

One of the problems is that the price effectively paid by consumers for Internet services now extends far beyond punctual advertising breaks (such as when using the music-streaming service, Spotify) or banner as flashing next to a search entry.[6]

---

[1] Walters, R., Zeller, B., Trakman, L, *Personal Data Law and Competition Law – where is it heading?* European Competition Law Review (2018).

[2] Ibid.

[3] Ibid.

[4] Stucke, M., Grunes, A *Big Data and Competition Policy* New York: Oxford University Press, (2016).

[5] Bernasek, A., Mongan, D *Our Massive New Monopolies: Amazon, Google and Facebook Have the Power to Move Entire Economies,* Salon, (2015) https://www.salon.com/2015/06/07/our_massive_new_monopolies_amazon_google_and_facebook_have_the_power_to_move_entire_economies, accessed 22 June 2018.

[6] Organisation for Economic Co-operation and Development, Big Data: Bringing Competition Policy to the Digital Era, (2016), https://one.oecd.org/document/DAF/COMP(2016)14/en/pdf, accessed 5 August 2018.

Data and search entries are often analyzed by data mining software that can result in various levels of intrusiveness, which in turn can create a system and environment whereby entities gain a competitive edge. The process of data mining provides the entity with specific information that a competitor who does not have the access or the systems and infrastructure to undertake the same activity – is at a disadvantage. It must also be noted that the collection, mining or harvesting of data may also provide many benefits to the consumer, for instance, improved services,[7] recommending certain products to the market or providing content that is free to the end user.[8]

Nevertheless, the data protection concerns specifically in relation to personal data are likely to remain because one of the most significant concerns arising from this behavior and practices is the rise in privacy breaches. Secondly, the privacy debate has also extended to difficulties for Internet users to be able to cope with privacy due to information problems and behavioral biases that have developed. For instance, it is argued that users are intentionally kept uninformed or misled about the extent of the tracking of their behavior over the Internet. That tracking, to some degree, provides identifying data and information about the person. Moreover, people do not feel as though they have enough control over how their data is collected and specifically used by online platforms, systems and infrastructure.[9] When the data subject does not know how their data is collected and how the data holders may use that data, even the sophisticated consumer cannot protect themselves against these breaches.[10]

Wolfgang Kerber questions the extent to which secret collecting of data (through tracking with cookies and web bugs) should be prohibited, and if allowed, whether there should be a duty to inform users of a service or a website relating to the data collection?[11] By prohibiting the secret collecting, mining or harvesting of data, it is acknowledging that this activity amounts to data being stolen. However, the answer

---

[7] Alessandro, A., Varian, H *Conditioning prices on purchase history,* Marketing Science (2005) 24(3): pp. 367–381.

[8] Avi, A., Tucker, C *"Online Advertising."* In The Internet and Mobile Technology Advances in Computing, (2011) 81, pp. 290–337.

[9] Stucke, M., Grunes, A *Big Data and Competition Policy* New York: Oxford University Press, (2016).

[10] Kerber, W *Digital Markets, Data, and Privacy: Competition Law, Consumer Law, and Data Protection*, No. 14–2016. Marburg Centre for Institutional Economics (MACIE), School of Business & Economics, Philipps-University Marburg (2018).

[11] Ibid. See also: EU ePrivacy Directive (2002/58/EC) and EU Cookie Directive (2009/136/EC) which permit the use of cookies if the users give their opt-in consent, whereas in the US the Do-not-track proposal of the FTC (Federal Trade Commission) in 2012 follows an opt-out approach. *FTC*, Protecting Consumer Privacy in an Era of Rapid Change, FTC Report March 2012, and for the EU *Luzak*, Privacy Notice for Dummies? Towards European Guidelines on How to Give "Clear and Comprehensive Information" on the Cookies' Use in Order to Protect the Internet Users' *Right to Online Privacy, Journal of Consumer Policy* (2014) p. 547.

may lie in what Kerber highlights is the effectiveness of the concept of consent. Kerber argues that currently individuals, particularly across EU member states, are informed about the "privacy policies", and implicitly consent to them by using the service, website or Internet platform. Effectively, data subjects provide a level of consent that their data can be collected, harvested or mined. Therefore, reinforcing the point that, where consent has not been obtained or granted for such an activity, the data is simply stolen.

Despite the traditional regulatory approach 'one group of solutions try to solve the problem of weak competition among Internet platforms in order to increase the incentives of the firms to offer their services in a more privacy-friendly way. For example, by being more responsive to the heterogeneous privacy preferences of their customers.[12] Therefore, the option of granting access to the already accumulated data of a dominant platform (as an essential facility) to other competitors for eliminating a huge entry barrier might admittedly help competition, but can be viewed critically from a privacy protection perspective due to further spreading private data.[13] To alleviate these competition concerns, it is argued that the right for data portability reduces switching costs that, in turn, lead to more competition between platforms, particularly in regard to social networks.[14] Furthermore, it is arguable that there is a link because of the singular power that corporations like Facebook and Google have, even though people can choose not to use them. However, with more and more of our daily lives being conducted over the Internet, the choice not to use these platforms continues to diminish. In effect everyone is slowly being directed to eventually use the Internet, and consequently these types of platforms. Therefore, the challenge is, not only to determine the need for anti-competitive regulation, but also to combine that regulation with other regulation and non-regulatory mechanism to secure privacy, and ensure the right balance between these two regulatory regimes.

It is not within the scope of this Chapter to fully examine the theoretical concept of consent, except to acknowledge that it is a key concept in the data protection law. It is argued that consent provides individual data subjects with a level of control over their personal data that has been defined in law. They argue further that consent is conceived as being given at that moment at which personal information is exchanged. The ability to consent to the use of personal data in such circumstances is limited, given that the party using the data is unknown to the data subject.[15] This limitation is attributable to the fact that the data subject only ever provides consent

[12] Ibid.

[13] Ibid.

[14] Ibid.

[15] Ibid.

to the data controller or data processor that sits within an entity.[16] The OECD[17] has identified this conception of consent as the key to strengthening the management, governance and regulation of data and privacy across all areas of law (see Chap. 16). Coupled with competition law, the concept of consent arguably has its challenges. Consent can come in the form of actual or implied consent, depending on the national or supranational laws. However, the question arises as to what actual personal data or personal information to which an individual is consenting. That can only be found in data protection laws. However, and as already highlighted throughout this book, consent varies from jurisdiction to jurisdiction (see country-jurisdictional Chapters and Chap. 11).

13.2 Data Protection and Competition

Data protection and competition did not suddenly develop from nowhere. Arguably, the collection and use of personal data and data generally is now fast becoming an important part of the economy. The balance between data protection, particularly personal data which is defined the law and anti-competitive behavior, walks a thin line. That line becomes even more blurred when, on the one hand governments do not want to stifle innovation, while on the other hand data (commercial and personal) needs to be protected. Subsequently, there has been considerable debate as to whether a problem actually exists in determining the relationship between personal data and anti-competitive behavior. That is, the nature of this relationship has never been clear, even in Europe which arguably is the leader in the development of data

[16] Council Regulation (EU) 2016/679, General Data Protection Regulation, Article 7(4) affirms that the consent is not freely given if it is conditional. Article 6 requires that processing of personal data is to be lawful only if and to the extent that at least one of the following criteria applies: the data subject has given consent to the processing of his or her personal data for one or more specific purposes. Consent in Australia is conceived broadly. There is no direct requirement or pre-requisite for collecting personal data or information from a data subject. However, for 'sensitive information' a person's consent must be provided. The Australian Privacy Principles (APPs) require that personal information should be collected directly from the individual, unless the individual has consented to collection from other sources, or if it is authorized by law. The APPs define consent as 'express consent or implied consent. Section 13 of Singapore's Personal Data Protection Act 2012, provides for a form of implied consent, and prohibits organizations from collecting, using or disclosing an individual's personal data unless the individual gives, or is deemed to have given, his consent for the collection, use or disclosure of personal data.

[17] Organization for the Economic Co-operation and Development, Guidelines on the Protection of Privacy and Transborder Flows of Personal Data 2013. http://www.oecd.org/sti/ieconomy/oecd-guidelinesontheprotectionofprivacyandtransborderflowsofpersonaldata.htm, accessed 20 February 2018. Organization for the Economic Co-operation and Development, Guidelines on the Protection of Privacy and Transborder Flows of Personal Data (2013), http://www.oecd.org/sti/ieconomy/oecdguidelinesontheprotectionofprivacyandtransborderflowsofpersonaldata.htm, accessed 20 February 2018.

protection law. In 2006, the Court of Justice of the European Union made reference to the possible intersection between competition law and personal data, concluding that personal data, "*as such*", was not a matter for competition law. At an early stage, the European Commission (EC) took the position that it refused to assess data protection in competition law cases. In *Case No COMP/M7217 - Facebook/ Whatsapp*[18] it was stated that:

> Any privacy-related concerns flowing from the increased concentration of data within the control of Facebook as a result of the Transaction do not fall within the scope of the European Union competition law rules but within the scope of the EU data protection rules.[19]

The position taken in this case arguably demonstrates the thinking at a time when technology was not significantly advanced. Thus, it took another 5–7 years before the general thinking in this area of law began to change. In 2013, the German and French competition authorities (*Bundeskartellamt* and the *Autorité de la Concurrenc*) published a joint paper on *Competition Law and Data*' clearly acknowledged that, despite personal data concerns having specific laws at a supranational and national level, data protection laws did not preclude competition law from intervening. It was stated that the "fact that some specific legal instruments serve to resolve sensitive issues on personal data, it does not entail that competition law is irrelevant to personal data".[20] Moreover, it was so stated in 2015, when Alec Burnside summarized the interrelationship between data protection and antitrust (competition), from a privacy perspective. Burnside stated that:

> (...) It is hardly a blanket assertion that privacy is irrelevant to antitrust, or that antitrust must not address facts to which privacy laws may also be relevant. Rather, it indicates that antitrust rules should be applied in pursuit of antitrust goals. And indeed that is what the Court did in the case before it: apply the antitrust rules to a set of facts to which privacy disciplines had a parallel application.[21]

Furthermore, in referring to the former European Commission for Competition, Margrethe Vestager, Burnside described personal data as the new currency of the Internet.[22] Privacy is also viewed as a by-product of this new currency when traded according to applicable rules and laws. However, privacy is becoming ever more important when the data is harvested or mined illegally, heightening the potential to establish anti-competitive practices. Burnside believes that there is a need for anti-

---

[18] *Case No COMP/M7217 - Facebook/ Whatsapp* [2014] European Commission Decision, para.165.

[19] Ibid.

[20] Autorité de la Concurrence and Bundeskartellamt, *'Competition Law and Data'*, p.23. *Case C-32/11 Allianz Hungária* [2013] Court of Justice of the European Union, ECLI:EU:C:2013:160, para. 46–47.

[21] Burnside, A *'No Such Thing As A Free Search: Antitrust And The Pursuit Of Privacy Goals'* (2015) Competition Policy International, https://www.competitionpolicyinternational.com/assets/Uploads/BurnsideCPI-May-15.pdf, accessed 4 August 2018.

[22] Ibid.

trust law to evaluate the role of datasets when they arise in the factual matrix of any assessment, such as dominance, restrictive practices, or a merger review.[23] The rationale is that competition law cannot be set aside when a data set, of any size, contains personal data defined by the law, and is used to establish a dominant market position. This assessment is similar to that which has been espoused by Peter Wire[24] and Robert Lande.[25] Both seek to promote the need to undertake assessments of the potential or actual harm, choice and quality of the data that is being used to create an environment that would exclude any competitor.

Notwithstanding the above, if data about ourselves really is the price we pay for content and access to the Internet, why should competition law not limit a company's ability to collect and analyze that data?[26] At one level, there appears to be no issue with this concept, provided the data subject agrees to the collection and use of that data. On another level, this become very problematic because of the privacy issues related to the data obtained when there has been no agreement (consent) by the data subject. The resulting effect is likely to deter data collection,[27] and which has mutual benefits for innovation and the economy more generally. James Cooper makes an important point that, in understanding data from a privacy perspective within the competition sphere, is not easy. Copper states:

> We live in a world where a large portion of online content is free. We do not pay to search on Google or Bing, post our photos on Facebook or MySpace, or read the latest news on CNN.com or Foxnews.com. Apps like Angry Birds are available for free in Apple's and Google's app stores. Why does everyone give away things online? The answer, in some ways, is that they do not. These businesses ("publishers") monetize the content they provide for free by selling access to our attention. By collecting more data about their users, publishers can improve their products and target ads more precisely to the consumers who are most likely to respond.[28]

---

[23] Ibid.

[24] Swire, P *Submitted Testimony to the Federal Trade Commission Behavioral Advertising Town Hall*, (2007), http://ftc.gov/os/comments/behavioraladvertising/071018peterswire.pdf, accessed 12 August 2018. Peter Swire argues that the combination of deep and broad tracking resulting from the Google-DoubleClick merger is one example which goes some way to strengthening the protection of personal data. According to Swire, "this sort of quality reduction is a logical component of antitrust analysis [A]ntitrust regulators should expect to assess this sort of quality reduction as part of their overall analysis of a merger or dominant firm behavior.

[25] Lande, R *The Microsoft- Yahoo Merger: Yes, Privacy is an Antitrust Concern*, FTC: WATCH, (2008) p. 1. Lande argues that consumers also want an optimal level of variety, innovation, quality, and other forms of nonprice competition, including data protection.

[26] Cooper, J *Privacy and Antitrust: Underpants Gnomes, the First Amendment, and Subjectivity*, George Mason University School of Law, (2015).

[27] Ibid.

[28] Ibid. Copper goes onto say that by doing more searches on Google - Google learns more about you. Combine your search data with what Google knows from your Gmail and other interactions with Google properties, as well as reports from tracking cookies placed by its display advertising network, and Google has a pretty good idea of what you like. Google can use this information to provide you with better search and map results, as well as more relevant ads, both of which will help Google's bottom line. First, better content makes for a more attractive product, encouraging greater use of Google's services, increasing both ad revenue and Google's database of consumer

Similar issues have arisen in Japan, and in July 2012, a former store manager of an agent company of a mobile phone company was arrested for disclosing customer personal information of the mobile phone company to a research company in respect of violation of the Unfair Competition Prevention Act.[29] Consequently, the Nagoya District Court in November 2012 gave the defendant a sentence of one year and eight months' imprisonment with a four-year stay of execution and a fine of ¥1 million.[30] Two years later in Japan, it was revealed that the customer information of an educational company (Benesse Corporation) had been stolen and sold to third parties by employees of an outsourcing contractor of the educational company.[31]

In September 2014, Japan's Ministry of Economy, Trade and Industry promulgated an administrative guidance requesting that the educational company reform its security control measures and supervision of outsourcing contractors in respect of violations of the duty regarding security control measures under Article 20 APPI.[32] The organisation was also found to be a violation of the duty to supervise an outsourcing contractor under Article 22 of the Act on the Protection of Personal Information. Subsequently, the organization distributed a premium ticket (value of ¥500) to its customers, to compensate for the damage incurred. Currently, however, a lawsuit is pending before the Supreme Court of Japan brought by a customer requesting damages of ¥100,000 following the Osaka High Court's dismissal of the customer's claim.[33] It is anticipated that the Supreme Court will deliver an opinion clarifying the liability of businesses handling personal information and the calculation of personal damages as a result of the leaking of that person's personal information. Nonetheless, the interconnectedness between the use of personal data and anti-competitive behaviour is not limited to a few countries, and the practice will continue to transcend international borders and other organizations transmit personal data, as the growth in trade in personal data expands in the future.

Competition law may be appropriate to regulate the use of such personal data where the potential harm is actual, or, potentially undermines future economic efficiency. In other words, anti-competitive behavior can be identified by the very use of data or by the technology created to harvest the data, rather than by relying on consumer laws that are ineffective as means of regulating the use of such data. Attempting to unify competition and consumer protection laws creates needless risks for the Internet economy. In particular, it could destabilize the assessment of

---

information. Second, the expansion of Google's database also allows Google to earn more revenue by facilitating targeted ads that are more likely to elicit consumer responses. See also Howard Beales, *The Value of Behavioral Targeting*, 1–2, 2010, http://www.networkadvertising.org/pdfs/Beales_NAI_Study.pdf, accessed 29 August 2018.

[29] Ishiara, T *The Privacy, Data Protection and Cybersecurity Law Review*, (2017) https://thelawreviews.co.uk/edition/the-privacy-data-protection-and-cybersecurity-law-review-edition-4/1151289/japan, accessed 5 October 2018.

[30] Ibid.

[31] Ibid.

[32] Ibid.

[33] Ibid.

the anticompetitive use of personal data, pulling it away from rigorous, scientific and allegedly objective methods of such assessment developed in the last few decades, and reverting back to the influence of subjective noncompetitive factors. Indeed, trying to expand competition law, as some have proposed, better reflects legal thinking in 1915, not 2015. However, privacy can be (and is today) a dimension of competition, whereby the more direct route to protecting privacy as a norm lies in consumer protection laws.[34]

Maureen Ohlhausen and Alexander Okuliar in 2015 argued that privacy, as a result of entities obtaining personal data outside of the current legal framework, is part of a non-price dimension of competition that can hurt individuals in general arising from some companies abusing their market power.[35] What these authors contend is that, where there is too much market power, the possible result is a total reduction in data protection and subsequently privacy – in the absence of any regulation. Today, there are a number of countries that have either no or a limited regulatory framework for personal data. The authors argue that competition law should look at data protection and subsequently privacy issues, even if no competitive implications exist. They go onto say that, by rejecting attempts to incorporate data protection and privacy concerns into competition policy, three major problems have arisen: (1) competition deals with harm to competition, not to privacy harms; (2) competition is concerned with market-wide effects, whereas privacy policy focuses on the individual relationship between the company and the consumer; and (3) competition remedies are inadequate to handle privacy concerns, specifically because companies can accomplish the same outcome through private contracts than through mergers.[36] However, when looking at their propositions more closely, the authors are referring to data that has been obtained within the context of the law. That is, they have assumed that the data subjects have provided a level of consent for their data to form part of a contract within the confines of a merger. This assumption does not account for the data that has been illegally obtained (stolen), or where no contract or adequate level of consent has been provided.

In 2016, Germany and France released a white paper arising from concerns raised in relation to market power and data. Firstly, the three broad areas of concern included, but were not limited to, the fact that the collection and exploitation of data may raise barriers to entry and may be a source of market power. Secondly, these barriers may reinforce market transparency, which may impact upon the functioning of the market. Thirdly, different types of data-related conduct relating to an undertaking may raise competition concerns.[37] Data may be obtained without consent to the user, through search engines and services including social networks that use

---

[34] Ohlhausen, M., Okuliar, *A Competition, Consumer Protection, and The Right [Approach] to Privacy*, 80 *Antitrust Law Journal* 121 (2015).

[35] Ibid, 134–36.

[36] Ibid.

[37] Competition Law and Data, Germany and France, White Paper, http://www.autoritedelaconcurrence.fr/doc/reportcompetitionlawanddatafinal.pdf, accessed 17 December 2017

cookies and sensor data to track web surfing. The European Commissioner for Competition, in 2016, highlighted that:

> It's possible that in other cases, data could be an important factor in how a merger affects competition. A company might even buy up a rival just to get hold of its data, even though it hasn't yet managed to turn that data into money. We are therefore exploring whether we need to start looking at mergers with valuable data involved, even though the company that owns it doesn't have a large turnover.[38]

In 2017, Inge Graef argued that both competition and data protection law are interlinked, even though they perform different functions.[39] Graef maintained that, ultimately, both sets of laws aim to protect consumer welfare,[40] by (1) regulating anticompetitive behavior and (2) by ensuring an individual's privacy has a level of protection. Furthermore, data analytics also poses challenges in attempting to apply both competition and data protection law. Simply put, data analytics enables the use of analytics to predict an individual's behavior over the Internet. The analytics is capable of capturing personal data that is both defined and not defined by the law, in order to identify an individual. The data analysis can extend beyond personal data identified with online shopping practices, to encompass a person's health, education, recreational activities, sport preferences and even political or religious preferences. Not only will existing businesses be impacted by this behavior, but purported new entrants into global markets may find themselves shut out because they cannot get access to the systems and data that creates this information.

More recently, Giuseppe Colangelo and Mariateresa Maggiolino have explored the interface between data protection and competition (anti-trust) law.[41] Firstly, in an economy in which data is collected in exchange for free services, low levels of privacy could be indicative of high levels of market power, including the harvesting of large amounts of data that are concentrated amongst a few dominant market entities. Secondly, the authors contend that antitrust law can make up for the pitfalls of data protection law. They highlight that such pitfalls arise, for example, in considering whether a practice that renders a product less privacy-friendly could be considered to be anticompetitive, or as a basis to allow antitrust law to intervene to protect privacy-enhancing technologies.[42] Colangelo and Maggiolino add that it is increasingly more difficult today to identify a competitive quantity of consumer data (the

---

[38] Vestager, M European Commissioner for Competition, *'Big Data and* Competition' (Speech at the EDPS-BEUC Conference on Big Data, Brussels) (2016), http://ec.europa.eu/commission/2014-2019/vestager/announcements/big-data-and-competition, accessed 29 July 2018.

[39] Graef, I Beyond Compliance: How Privacy And Competition Can Be Mutually Reinforcing', *Computers, Privacy & Data Protection Conference* (2017), https://www.youtube.com/watch?v=Af1qLye_-Ok, accessed July 2018.

[40] Ibid.

[41] Colangelo, G., Maggiolino, M Data Accumulation and the Privacy- Antitrust Interface: Insights from the *Facebook* case for the EU and the U.S. Transatlantic Technology Law Forum, Stanford Law School and the University of Vienna School of Law, (2018).

[42] Ibid.

quantity of personal data that firms would naturally collect in competitive markets).[43] Whereas, in the analogue economy, the competitive level of the market price can be approximated by looking at marginal costs (or other measures of costs), in the fast growing digital economy, no one has as yet quantified the benchmark for assessing the competitive quantity of personal data. Even data protection laws cannot help in this regard because current market analytics only regulates the way in which personal data is collected, without addressing the quantities of personal data that individuals may transfer to entities.[44] Additionally, the value of personal data varies according to the data considered, which is very hard to measure. Valuing data also does not lend itself to any form of inter-personal comparison, and cannot become a tool for measuring aggregated, or market, phenomena.[45]

In 2018, The Australian Competition and Consumer Commission released its Digital Platforms Inquiry – Issues Paper,[46] The paper raises concerns in relation to big data, including whether data platforms are able to provide consumers with adequate levels of privacy and data protection. One of the major concerns has been the use of large sets of personal data for commercial purposes to enhance an entity's competitive position in the market.[47] The Issues Paper went on to say that, using accumulating consumer behaviour data to expand targeted advertising may improve services provided to advertisers (and potentially be of greater interest to their audiences), but also represents a cost to consumers in the form of a loss of privacy.[48] The Issue Paper also highlights how an increase in the level of personal data obtained from users or the supply of more data to third parties, is being viewed as an effective source of increasing the market price or decreasing the quality of the 'free' service (e.g. social media interaction or search functionality) supplied to consumers.[49] Moreover, another potential source of concern is the extent to which consumers are aware of the amount of data they provide to digital platforms, the value of the data provided, and how that data is used.[50] This concern has arisen in Australia because consumers are required to provide wide-ranging consent regarding the collection and use of their data across a number of Internet platforms to ensure that they are

---

[43] Ibid.

[44] Ibid. It must be observed that the above reasoning and the resulting link between market power and personal data has been elaborated, as previously stated, in relation to multi-sided media platforms, with the ultimate purpose of appreciating their market power. However, other tools and variables can be used to this end, such as: (i) the price of advertising space; (ii) the amount of advertising space imposed on users (i.e., the amount of users' attention required); and (iii) the quality of the "free" products and services.

[45] Ibid.

[46] Australian Competition and Consumer Commission, https://www.accc.gov.au/system/files/DPI%20-%20Issues%20Paper%20-%20Vers%20for%20Release%20-%2025%20F.._%20%28006%29.pdf, accessed 10 September 2018.

[47] Ibid, 9.

[48] Ibid.

[49] Ibid.

[50] Ibid.

supplied with adequate information on the data collection and in order to be able to secure informed consent in order to use that data.[51]

Notwithstanding the above, jurisdictions such as Australia, the EU, Malaysia and Singapore begun to pave the way to addressing competition related issues within their respective current day data protection and privacy laws. Australia and Singapore, for example, have established Do Not Call Registers. It is argued that these registers, to a limited extent, provide restrict direct marketing from organizations from contacting individuals by using their personal data to make the contact. In restricting this contact, these registers help to minimize the ability for organizations to obtain certain information that could be used to gain a dominant position in the market and develop anti-competitive practices. The EU has taken this one step further by providing the right to object, which can be used by data subjects to restrict direct marketers from using personal data for marketing purposes. The right afforded to a data subject to object to the specific use of personal data, is arguably far reaching, and in the case of the EU, does go some way in allowing data subjects to restrict the use of their data for marketing purposes. The resulting effect also limits the ability of organization to use that data to enhance their position in the market. Japan, Indonesia, India, Thailand are far from establishing even a minimal approach to addressing this concern. Thus, there is a need to better understand the issues and potential solutions to the tension between data protection and competition law.

13.3 Issue & Solution

Walters, Zeller and Trakman argue that the debate in relation to data protection and competition law is, arguably, complex and requires that a balance be struck between the broader public benefit arising from a competitive market in personal data compared to the risk of breaching the privacy of a single individual or group. It is a global issue that requires a global response, beyond simply looking at individual national and regional responses.[52] This global setting also serves as the backdrop for determining the public benefit derived from the Internet and its supporting systems, platforms and infrastructure. Furthermore, further legal development is key to striking a lasting balance between data protection and competition.

The authors go onto say that it is well understood that economic scholars argue that restricting competition stifles innovation and change, which has broader economic impacts on the economy and society. Firstly, companies such as Google have provided a public benefit, by ensuring greater access to information, whether that is medical, personal, entertainment (sport and music), legal or business. Google and other Internet platforms have also enhanced and changed the way people shop, interact and have access to justice, arguing making these processes more efficient

[51] Ibid.

[52] Walters, R., Zeller, B., Trakman, L, *Personal Data Law and Competition Law – where is it heading?* European Competition Law Review (2018).

and more contentiously, user friendly. Without these innovations, societal change, as we know it today, would not exist. On the other hand, competition issues that have arisen from technological change alone have resulted in people's privacy being significantly reduced, and in many cases infringed. A good example occurred in 2017–2018 when Cambridge Analytica obtained and used large amounts of personal data and information of over 50 million people by accessing personal data.[53] Despite the privacy infringements, the personal data obtained was used generally for political purposes and to gain a competitive edge. This example highlights how the mining and harvesting of personal data can be used in almost any area of the economy. The power of a single organization, such as Cambridge Analytica was able to collect from Facebook users had never been experienced before.[54] While most of that personal data is unlikely to be defined within the national or European data protection law, arguably there is likely to be elements that fall within these laws. The broader issue is whether the personal data involved was stolen, or from individuals being misinformed. That question also related to the level of consent, if any, that was obtained from data subjects that their data could be used by Data Analytica? The evidence suggests that Facebook was the collector of the personal data. The evidence also suggests that Cambridge Analytica's mined the data without the permission of Facebook, but more importantly, without the permission of the data subjects. Thus, not only was there the potential for large scale breach of privacy, but the example demonstrates the ease with which an organization like Cambridge Analytica can obtain a competitive position, no matter what that market might be. In other words, by freely obtaining large quantities of data, even though this issue did not center on anticompetitive behavior, it demonstrates how an entity can gain a market edge.

The theory of harm, which is a well-established principle in competition law, poses further challenges when applied to anti-competitive practices that involves the use of data. Largely, it is an area that has not been fully tested, even though there is jurisprudence that has emerged in some jurisdictions, such as the EU.[55] It is argued that, applying the harm test to competition matters involving data would complement those other known harms, such as the infringement to privacy. However, further work is required to better understand what and where the harm commences and concludes. Additionally, measuring the actual harm of an infringement of privacy will be challenging both economically and socially.

A potential way forward calls for more work to be undertaken to better understand the various approaches taken by different states in regulating competition,

[53] Unterhalter, D *Data privacy: why internet giants elicit antitrust critiques*, The Cambridge Analytica furore vindicates fear that groups such as Facebook are not benign monopolies, https://www.businesslive.co.za/bd/opinion/2018-04-18-data-privacy-why-internet-giants-elicit-antitrust-critiques, accessed 22 June 2018.

[54] Walters, R., Zeller, B., Trakman, L, *Personal Data Law and Competition Law – where is it heading?* European Competition Law Review (2018).

[55] Ibid.

data protection and privacy.[56] A starting point is the definition of personal data and personal information. There needs to be a comprehensive study undertaken to better understand whether this definition is adequate in determining the extent to which personal data and information can be collected by Internet platforms and systems.[57]

Moreover, the important role that consent has in individual's allowing entities to harvest and mine their personal data.[58] As highlighted earlier in this book, as Internet platforms continue to provide wide-ranging levels of consent, the boundaries of consent become less clear. Furthermore, the consent provided by data subjects for their data to be provided to and used by a third, fourth or even fifth parties and so on, appears to be ever less informed the further removed the data subject is from the data users. Therefore, the question arises as to the legal implications arising from the use of data beyond the relationship between the data subject and the immediate data collector.

As highlighted by Walters, Zeller and Trakman, Entity AA has collected personal data from a data subject ZZ who provides consent for AA to use that personal data. Entity AA then enters into a contract which provides Entity BB with consent to use that personal data. Entity BB then sells that personal data to Entity CC. However, unless Entity AA had clearly stated that upon consent from the data subject, Entity AA can use the personal data in whichever manner they choose – consent from ZZ has not been granted for Entity AA to transfer or pass on ZZ's personal data to Entity BBB and beyond. In addition, Entity BB who has had no contact with ZZ, and therefore having no direct consent, is likely to be unaware of any consent provided or otherwise to the use of that personal data.[59] This is the unknown factor in most, if not all third, fourth, fifth, sixth party (and so on) transactions of personal data. This area in which the nature and extent of consent is unclear needs to be remedied, both through the law and by means of practical processes that better inform the data subject.[60] This includes personal data that is not defined by the law, particularly because, in most cases. Personal data that is on-sold may not be defined by national data protection laws. This is an important but abstract area of personal data in which data subjects are often oblivious to their data being collected and sold. In addition to the above, combining regulatory and non-regulatory tools in the transmission of data, such as through data portability, is slowly being accepted to strengthen personal data protection in competition law.

---

[56] Ibid.

[57] Ibid.

[58] Ibid.

[59] Ibid.

[60] Ibid.

13.4 Data Portability

Data portability is becoming increasingly more important as it allows for both terms of warranting control rights to data subjects and is found at the intersection between data protection and other fields of law of competition law.[61] It therefore constitutes a valuable case of development and diffusion of effective user-centric privacy enhancing technologies and a first tool to allow individuals to enjoy the immaterial wealth of their personal data in the data economy. It is outside of the scope of this book to examine every jurisdictions' approach to data portability in relation to competition law. This Chapter only briefly explores some of the issues arising out of the EU which has been chosen because of the leading role it has taken in all areas of data protection and privacy law.

DeHert *et al* argue that the impact of a right to data portability is very relevant both for businesses (in particular, e-businesses involved in the digital market, such as internet service providers) and for data subjects.[62] Therefore, from a business perspective, this impact is tangible in several fields. It is both a challenge to the traditional system of competition law[63] and a 'problematic opportunity' in terms of interoperability. From the user perspective, the impact of data portability is evident both in terms of control of personal data (and in general in the sense of empowerment of individuals to exercise their control rights), and in terms of a more user-centric interrelation between services. At the same time, it is a challenge to the rights of third party data subjects.[64]

The authors go onto to say that, in relation to the principle of interoperability of systems, Recital 68 of the GDPR states that data controllers should be encouraged to develop interoperable formats that enable data portability. Therefore, efforts imposed upon data controllers to promote fully interoperable digital systems are moderate: they should be encouraged and not obliged to develop these interoperable formats.[65] A further confirmation of this is the final part of Recital 68: data subjects should have the right to have the personal data transmitted directly from one controller to another, but only where technically feasible.[66] Thus, data controllers can prevent a full exercise of users' right to data portability if they prove that, in a given situation, the level of technological development of their organization makes not technically feasible the direct transmission of data to another controller, for example, because interoperable formats have not yet been developed.[67] The empower-

---

[61] De Hert, P., Papakonstantinou, V., Malgieri, G., Beslay, L., Sanchez, I *The right to data portability in the GDPR: Towards user-centric interoperability of digital services*, Computer Law & Security Review, Volume 34, Issue 2, (2018), pp. 193–203.

[62] Ibid.

[63] Ibid.

[64] Ibid.

[65] Ibid.

[66] Ibid.

[67] Ibid.

ment of data subjects has emerged as an important factor in providing individuals with a level of control and ownership over their personal data defined by the law. Recital 68 of GDPR reinforces this point, and states that the rationale of right to data portability is to further strengthen the control [of the data subject] over his or her own data.

In order to resolve the tension between the rights of data protection and competition DeHert *et al* argue that the GDPR allows two opposite options: a minimalist approach in which the object of data portability is only explicitly given to the controller, in a written form, and where right to data portability is inherently linked to the withdrawal of data from the controller.[68] Secondly, the extensive approach, where a wide interpretation of data provided (including data observed by the controllers) joined with the right to have data directly transferred from one controller to another (Article 20(2)) allows a fusing scenario, leading to user-centric platforms of interrelated services. The authors propose to adopt the extensive approach, considering that the rationale of this right (Recital 63) is to further strengthen control rights of the data subject on his or her own data and foster opportunities for innovation by means of sharing personal data between data controllers in a secure manner under the constant control of the data subject.[69]

However, there is criticism by some that the introduction of the GDPR has not assisted in achieving these objectives. Swire and Lagos argue that Article 20 of the GDPR[70] has a perverse anti-competitive effect, because it applies broadly to various organizations, including those that currently hold a dominant position in the market.[71] Even though the rule pertains to data portability, which has been designed to promote competition, Swire and Lagos argue that it will ultimately come down to how the courts interpret and apply Article 20. They further highlight that an additional factor that also needs to be considered is how Article 20 will be enforced. Currently, there is no jurisprudence or scholarly argument that can direct or substantiate how Article 20 operates.[72]

The introduction of data portability has begun to provide a solution to the complex interrelationship between personal data protection and competition law. However, it is only a recent addition to the legal framework, with the EU again taking the lead. It has been argued that, by providing data subjects with the option of portability from one controller to another, this will enhance competition between digital services and Internet platforms. However, due to its limited implementation

[68] Ibid.

[69] Ibid.

[70] Regulation (EU) 2016/679, Article 20 states, "The data subject shall have the right to receive the personal data concerning him or her, which he or she has provided to a controller, in a structured, commonly used and machine-readable format and have the right to transmit those data to another controller without hindrance (…)".

[71] Swire., P, Lagos., Y, "Why the Right to Data Portability Likely Reduces Consumer Welfare: Antitrust and Privacy Critique", Maryland Law Review, Vol. 72, No. 2, (2013), http://digitalcommons.law.umaryland.edu/cgi/viewcontent.cgi?article=3550&context=mlr, accessed 6 August 2018.

[72] Ibid.

and lack of jurisprudence on this concept, it remains to be seen how this will play out in the interrelationship between competition and rights in data protection and privacy law over the longer term.

The advancement in data protection law has created challenges and tensions in various areas of the law. On the other hand, data protection law is providing similar protections to that, for example, of competition law. Arguably, at the heart of competition law is consumer welfare, and more broadly the protection of consumer rights. Friedrich von Hayek makes the point that consumer welfare "*cannot be adequately expressed as a single end, but only as a hierarchy of ends, a comprehensive scale of values in which every need of every person is given its place.*"[73] The position taken by Hayek is important because there are many different areas of anti-competitive behavior that have emerged from the Internet; from the use of personal data. That includes personal data that is not necessarily defined by the law.

This Chapter will highlight some examples in which jurisdictions have recently had to grapple with (i) the abuse of power; (ii) web browsers; (iii) mergers and acquisitions; and (iv) predatory pricing, relating to the collection and use of personal data and data more generally. Due to the breadth and depth of the issues that arise under these areas of competition law, the following sections will only highlight examples from Australia, Europe, India and Singapore.

13.4.1 Abuse of Power and the Consumer

The abuse of power arising from the collection of personal data is vast and complex. It does not always mean that the personal data collected falls within the legal definition of personal data or personal information. Arguably, there are challenges with the way businesses operate, and the way they can over time obtain a greater market share from subversive behavior, about which consumers have no understanding.

Wolfie Christle argues that, when surfing the web, hidden pieces embedded in software transmit information about the websites visited, navigation patterns, and sometimes even keystrokes, scrolls and mouse movements to hundreds of third-party companies.[74] Furthermore, when carrying a smartphone, rich information about the user's everyday life, not only flows to Google, Apple, and a variety of app providers, but also to a significant number of third-party companies, again based on hidden software embedded by app providers.[75] The information obtained from these activities are vast and varied, and can include, but not limited to, a person's contacts, information about real-time app usage and movements, as well as data from all

[73] von Hayek, F *The Road To Serfdom: Text And Documents - The Definitive Edition,* 1st Edn, University Of Chicago Press (2007), p. 101.

[74] Christl, W *How Companies Use Personal Data Against People, Automated Disadvantage, Personalised Persuasion, and the Societal Ramifications of the Commercial Use of personal Information,* Working Paper of Crack Labs (2017).

[75] Ibid.

kinds of sensors recording motion, audio, video, and more. More importantly, companies can now find and target users with specific characteristics and behaviors in real-time, regardless of which service or device is used, which activity is pursued, or where the user is located at a given moment. Within milliseconds, digital profiles about consumers are auctioned and sold to the highest bidder.[76] Large consumer data brokers have started to partner with hundreds of advertising technology firms as well as with platforms such as Google and Facebook.[77] They combine data about offline purchases with online behaviors and provide services that allow other companies to recognize, link, and match people across different corporate databases. Businesses in all industries can use the services of data companies to seamlessly collect rich data about consumers, their personal information and data defined by law including the personal data and information that falls outside the current definition (s) and adds additional information to them, and utilizes the enriched digital profiles across a wide range of technology platforms.

Notwithstanding the above, below is an outline from the research undertaken by Crack Labs in relation to personal data being used to create and advantage in the market:

> Lists of people with names, addresses, or other contact information that group consumers by specific characteristics have historically been an important product sold by marketing data companies. Today these lists include people with low credit scores and with specific conditions such as cancer or depression. They originate from third-party companies which sell or rent information about their customers to data brokers. Lists have been used to sell products and services and to send direct mail to consumers, but also as a basis for other applications. In 2017, for instance, Amnesty International was offered a list of 1.8 million US Muslims; during an investigation on data companies, the organization also discovered offers for lists of "Ameri- cans with Bosnian Muslim Surnames" or "Unassimilated Hispanic Americans". The website dmdatabases.com offers email and mailing lists of wheelchair and insulin users, of people ad- dicted to alcohol, drugs, and gambling, as well as of people suffering from breast cancer, HIV, clinical depression, impotence, and vaginal infections. Nextmark offers consumer lists titled "Pay Day Loan Central – Hispanic", "Help Needed – I am 90 Days Behind With Bills", "One Hour Cash", "High Ranking Decision Makers in Europe", or "Identity Theft Protection Re- sponders".[78]

Apart from the quantities of personal data that is available today for the use of companies, the more problematic issue facing society generally, is that governments, industry and the community at large are unaware that these practices are being carried out. Thus, authorities, businesses and consumers will need to be even more vigilant because of the subtle way in which an internet provider can influence and manipulate the consumer is real and unlikely to go away.

---

[76] Ibid.

[77] Ibid.

[78] Christl, W *Corporate Surveillance In Everyday Life, How Companies Collect, Combine, Analyze, Trade, and Use Personal Data on billions*, Crack Labs (2017) p. 41.

13.4.2 Web Browser

The internet and the web browser have introduced an alternative route for competitors to deliver their services and applications to end users.[79] Instead of trying to develop a better browser and thus to compete on merits, large companies have countered the threat by leveraging the market power they enjoyed in markets, by limiting the number of web browsers available. The practice has not been limited to a single jurisdiction. However, with the development of the Internet taking hold in the United States, it is not surprising that large companies such as Microsoft had been called to account in the late 1990s for engaging in such anti-competitive practices.

In *Commission decision Microsoft (tying) (COMP/C-3/39.530)* [80] investigated the use of market power in the market for operating systems into the market for web browsers. The 2007 investigation followed a complaint by Opera Software ASA. Microsoft was accused of leveraging its market power in the market for operating systems into the market for web browsers. The complaint was similar to the US case involving Microsoft and Netscape. The Commission argued that:

> Microsoft's behaviour resulted in foreclosure of competition on the market for web browsers. Microsoft had a considerably larger market share than its competitors.[81]

The Commission further argued that:

> because of a certain degree of users' inertia, it required additional effort on behalf of distributors, vendors, and/or users to switch to using other browsers.[82]

At issue were attempts to foreclose future markets, as a result of large and powerful digital companies significantly limiting entrants into the market. Moreover, within digital markets, competition is predominantly based on innovation, and subsequently, there are continued concerns that such behavior and practices restrict, and in some cases totally stifle, the development of new products, services or business models. The resulting effect is that the consumer has significantly less choice, if not any choice at all.

In 2017, the United States passed laws to allow service providers to sell the browsing habits of customers.[83] Opponents to this, argue that is undermines consumer privacy, and that consumers will lose control of their personal data. They express concern that service providers can on sell personal data multiple times, to third, even fourth party recipients. Singapore, within its IP legislation, has established safeguards against anti-competitive behavior.[84] The Singapore Patents Act

---

[79] European Parliament, Directorate General For Internal Policies, Challenges for Competition Policy in a Digitalised Economy IP/A/ECON/2014–12, (2015), p 28–32.

[80] *Commission decision Microsoft (tying) (COMP/C-3/39.530)*.

[81] Ibid.

[82] Ibid.

[83] Solon, O *Here's how to protect your internet browsing data now that it's for sale*, The Guardian Australian Edition, (2017).

[84] Singapore Patents Act 1994, section 5(1).

enables licence conditions to be established, prohibiting the licensee from using a competitor's patented product.[85] Similar legislation exists in other countries, including the EU. It is outside the scope of this book to examine and compare the IP laws in all these countries.

A further issue has arisen from this practice of on selling personal data. In particular, access to digital platforms often seems to be available free of charge to the data subject. However, by providing the platform operators with personal data, the consumer as the data subject unwittingly pays a price. That price is in terms of switching costs, which results by individuals automatically giving up their personal data. Individuals do not even realize that their personal data may be at risk, while companies using that personal data can capitalize on it, notably by disclosing customer information for purposes that are not in consumers' best interests.[86] For instance, health apps from health insurance companies or online payment apps of credit card companies are used to gather data about a consumer's life style.[87] This information is used to set discriminatory prices or to deny a service. Another concern relates to multi-platform operators like Google and Apple who are developing online devices like bracelets, watches, and glasses that can be of great support in managing consumers' lifestyles.[88] At the same time these devices give platform operators all kinds of information about lifestyles, simply from accessing these apps and web browsers.

13.4.3 Mergers and Acquisitions

One of the biggest issues facing global markets is the potential for monopolies to form, and with particular ease in the digital economy. One of the most common ways larger organizations achieve this is by purchasing or merging with other organizations. Mergers and Acquisitions (M&A) have been common in the business world for decades. Businesses have either merged or acquired their competitors. The OECD reported that the number of mergers and acquisitions in the data sector had risen from 55 in 2008 to more than 160 by 2012.[89] However, there is an increasing concern in relation to privacy in M&A transaction. M&As have two major challenges when it comes to data. Firstly, is there is the concern that the merger or acquisition will be contrary to competition law. Secondly, is the concern about management and pre-contractual use of data, in signing contract and post contractual conduct.

---

[85] Ibid.

[86] European Parliament, Directorate General For Internal Policies, Challenges for Competition Policy in a Digitalised Economy IP/A/ECON/2014–12, (2015), p 28–32.

[87] Ibid.

[88] Ibid.

[89] Data-Driven Innovation: Big Data for Growth and Well-Being, OECD Publishing, Paris, 2015, https://doi.org/10.1787/9789264229358-en, accessed 17 December 2017.

One of the largest acquisition in the technology economy was WhatsApp by Facebook. The EU Commission learnt that WhatsApp had begun to link its data with the data of Facebook, which resulted in a privacy breach.[90] The Commission fined Facebook EUR 110 million for providing misleading information. In recognition of these consequences, companies and businesses considering merging with, or, acquiring another company now need not only to understand competition law, but also understand data protection law. In relying on the European Merger Regulation,[91] Article 14 and 6 allow fines of up to 1% of worldwide turnover.[92]

To mitigate against any breaches of data law, companies and businesses will also need to understand the cross-border transfer of data laws, outlined the preceding chapters. Nonetheless, this could an area where industry can regulate itself through effective contracts, data agreements, risk management systems and by ensuring the compatibility of data transfer from one system to another.

It is incorrect to assume that, once an M&A has concluded been signed, personal data can be easily shared. There are different rules across different countries, which need to be dealt with in the pre-contractual and final contractual arrangements. Certain jurisdictions, such as the EU, have very specific rules when transferring data to a third country that is outside the European Economic Area. Consent is one option in an M&A, however dealing with large quantities of data and information, an organization would find doing so costly and time consuming. The application of competition rules related to data protection need to be both commercial and private. They also need to span many other areas of law in respect of which further research will be needed on an ongoing basis.

In August 2017, Singapore undertook a comprehensive analysis of the data landscape in collaboration with Personal Data Protection Commission, Singapore ("PDPC").[93] The Intellectual Property Office of Singapore ("IPOS"), CCS also sought to explore the implications of the proliferation of data analytics and data sharing on competition policy.[94] The reported noted that:

> The benefits arising from the adoption of data analytics and data sharing may not be fully realized if businesses engage in anti-competitive conduct in the course of adopting data analytics and/or data sharing. It is thus crucial for competition policy and law to foster a level playing field for businesses.[95]

The report concluded that 'while there have been calls for competition law to be applied to promote data protection and privacy policy, this is not consistent with the roles and functions of Competition and Consumer Commission of Singapore (CCCS). In this regard, CCCS aims to ensure that markets are, and remain, competi-

---

[90] Official Journal of the European Union, L 24.

[91] Ibid.

[92] Ibid.

[93] Data: Engine for Growth – Implications for Competition Law, Personal Data Protection, and Intellectual Property Rights, Intellectual Property Office of Singapore, Personal Data Protection Commission, Singapore 16 August 2017.

[94] Ibid.

[95] Ibid.

tive by protecting the competitive process. Where data protection is a non-price competition factor, the treatment of personal data may affect how CCCS considers and assess the competitive dynamics of a market.[96] Nonetheless, in 2018 Singapore's competition watchdog fined Grab and Uber a total of SG$13 million over their merger, saying that the deal had led to the substantial eroding of competition in the ride-hailing market.[97] Uber was fined S$6.58 million, while Grab was fined SG$6.42 million.[98] The issue of competition in the market was not so much about the use of or acquisition of personal data, but that both companies had large quantities of personal data from consumers and not just in Singapore. For these reasons, it was successfully argued that, in acquiring and using such information, these companies had placed themselves at a competitive advantage.

Sakle and Chand highlight how in India, innovation and technology-driven markets such as ecommerce.[99] That is through the use of ride hailing apps and online wallets which have been growing swiftly and witnessing a progressive surge in M&A activity rendering such markets exposed to potential competition law concerns.[100] The authors go on to say that superior technology and 'internet of things' has permeated our social relationships, shopping habits and even societal norms to an extent that they have made Indian consumers more digital savvy but not sufficiently privacy skeptical as yet.[101] Thus, it may be an opportune time for the Competition Commission of India (CCI) to follow suit and as a preliminary step, conduct a study as to whether the topical issue of big data is really a significant problem that needs a legal response.[102]

In *Vinod Kumar Gupta v. WhatsApp Inc., Case No. 99 of 2016,*[103] two issues were raised. Apart from anti-competition behavior, the question of privacy was also considered. It was contended that WhatsApp was abusing its dominant position under section 4 of the Competition Act, by forcing users to share account and other information with WhatsApp's parent company Facebook, without admitting the exact nature of the disclosure to them, and further, by indulging in predatory pricing by not charging any fees for its services.[104] Section 4(1) of the Competition Act states

---

[96] Ibid.

[97] Singapore competition watchdog fines Grab, Uber S$13 million, https://www.channelnewsasia.com/news/singapore/grab-uber-fined-after-merger-deal-competition-watchdog-10751522, accessed 2 October 2018.

[98] Ibid.

[99] Sakle, A., Chand, ABig Data: Emerging Concerns under Competition Law *Practical Lawyer PL (Comp. L)* (2018).

[100] Ibid.

[101] Ibid.

[102] Ibid.

[103] *Vinod Kumar Gupta v. WhatsApp Inc., Case No. 99 of 2016.*

[104] Ibid.

no enterprise shall abuse its dominant position.[105] Further, the explanation to the section clarifies that the expression "dominant position" means a position of strength, enjoyed by an enterprise, in the relevant market, in India, which enables it to operate independently of competitive forces prevailing in the relevant market and affect its competitors or consumers in the relevant market in its favour.[106] The Commission found that, even though WhatsApp held a dominant position in the market, there had been no abuse of its position, since the disclosure had been made to the users regarding the sharing of information. There was also no evidence found that WhatsApp had contravened any of the provisions of section 4 of the Competition Act.[107]

The High Court of Delhi in W.P. (C) 7663/2016 dealt with this issue of privacy in *Karmanya Singh Sareen and Others Vs. Union of India and Others.* There, the Petitioners, who were the users of WhatsApp, and sharing personal data of subscribers with entities including 'Facebook' should be prohibited and protection. The court order dated 23 September 2016 observed that:

> "However, the contention of the petitioners is that the proposed change in the privacy policy of WhatsApp amounts to infringement of the Right to Privacy guaranteed under Article 21 of the Constitution of India. Even this cannot be a valid ground to grant the reliefs as prayed for since the legal position regarding the existence of the fundamental right to privacy is yet to be authoritatively decided *{Vide: K. S. Puttaswamy (Retired) and Anr. v. Union of India & Ors., (2015) 8 SCC 735}*. Having taken note of the inconsistency in the decisions on the issue as to whether there is any "right to privacy"[108] guaranteed under our Constitution, a three Judge Bench in K.S. Puttaswamy (supra) referred the matter to a larger Bench and the same is still pending. Be that as it may, since the terms of service of "WhatsApp" are not traceable to any statute or statutory provisions, it appears to us that the issue sought to be espoused in the present petition is not amenable to the writ jurisdiction under Article 226 of the Constitution of India.[109]

However, it is noted that the fact that under the Privacy Policy of "WhatsApp", the users are given an option to delete their "WhatsApp" account at any time, in which event, the information of the users would be deleted from the servers of "WhatsApp". The court was of the view that it is always open to the existing users of WhatsApp who do not want their information to be shared with Facebook to opt for deletion of their account.[110] This is an example of a case in which competition and privacy arise from the use of personal data, although they were treated as very different issues.

---

[105] Ibid.

[106] Ibid.

[107] Ibid.

[108] Ibid.

[109] Ibid.

[110] Ibid.

13.4.4 Predatory Pricing

Predatory pricing is commonly used by countries to benefit their markets, and by organization to squeeze out competitors. Richard Posner defines predatory pricing as a predation of pricing actions by a dominant undertaking aiming to remover the effective competitor from the market.[111] The digital economy is an example of an industry that is expected to generate large quantities of data that will be used commercially.

The European Court of Justice has developed four elements in order to determine predatory pricing. The first element is whether the price of the product covers all the costs. In *Tetra Pak International SA v. EU Commission,*[112] the court highlighted that this included the average total costs, average avoidable costs, average variable costs, and long run average incremental costs. However, in the case when an undertaking sets prices higher than average variable-avoidable costs, it is then necessary to prove the illegal intent of the undertaking in engaging in predatory pricing.[113] The court in Case T – 340/03 *France Télécom v. Commission* held:

> that when undertaking sets prices higher than average variable costs and smaller than average total costs the undertaking has an opportunity to eliminate even effective competitors without experiencing losses.[114]

Apart from these core principles, the Court of Justice has placed an emphasis on "intent" as a crucial test. In the earlier case of Case C – 62/86, *AKZO Chemie BV v. EU Commission,*[115] the court of Justice held that the intention to eliminate competitor was a decisive factor in the recognition of predation. Arguably, that earlier case provided the bases for the courts to consider what the intention of the parties would be when determining the average variable costs. In Case T – 340/03 *France Télécom v. Commission,*[116] the Commission reinforced its earlier 1991 judgement and argued that it is possible to indicate an 'intention' based on the facts of the case even without any direct evidence on the competitor.

In Australia, intention is also viewed as a critical principle and in *Boral Besser Masonry Limited (now Boral Masonry Ltd) v Australian Competition and Consumer Commission (Boral),*[117] the court held that:

> 'intent' is at the heart of the offence in relation to section 46 of the *Trade Practices Act 1974 (Cth)* (TPA).[118]

---

[111] Prosner R (1976) *Antrust Law: An economic Perspective,* University of Chicago Press, p. 189

[112] Case C – 333/94, *Tetra Pak International SA v. EU Commission*, [1996], para. 4, *Deutsche Post AG, Decision of the European Commission of March 20, 2001,* OJ L 125/27, and 40. Post Danmark A/S v Konkurrencerådet, Case C-209/10 [2012], para. 17.

[113] Case T – 340/03 *France Télécom v. Commission*, 2003, para. 197.

[114] Ibid.

[115] Case C – 62/86, *AKZO Chemie BV v. EU Commission*, [1991], para. 71–77.

[116] Case T – 340/03 *France Télécom v. Commission*, 2003.

[117] *Boral Besser Masonry Limited (now Boral Mason,ry Ltd) v Australian Competition and Consumer Commission (Boral),* [2003] HCA 5.

[118] Ibid.

McHugh J argued that, when a corporation with substantial market power cuts prices below cost, in order to recoup the costs and losses at a later stage, abuses its market power to charge supra-competitive prices, such action is, by definition, obtaining a level of market power and dominance.[119]

The Australian Competition and Consumer Commission (ACCC) alleged that Boral Besser Masonry (BBM) and its parent company, Boral, had contravened s 46 of the TPA (misuse of market power) by pricing below avoidable cost in order to drive out a competitor (C&M Brick).[120] The ACCC alleged BBM had a substantial degree of power in the market for concrete masonry products in metropolitan Melbourne.

Subsequently, in 2017, Australia introduced the *Competition and Consumer Amendment (Misuse of Market Power) Act 2017,*[121] which repealed the original section 46 in the TPA, and replaced with a new provision that introduces the "effects test".[122] That is, the test is applied to determine the level, if any, of anti-competitive behavior. Nonetheless, the new provision continues to be concerned with the Misuse of Market Power and prohibits an organization that has market power from taking advantage of that power. Specifically, the provision, while not specifying intent as a key principle, implies an intention to eliminate a competitor, prevent entry into a market, deter or prevent engagement within the market. Moreover, the new provision removes the 'take advantage' element. It replaces it with a 'purpose or likely effect test', prohibiting an organization having market power from engaging in conduct with the 'purpose effect or likely effect' of restricting competition. Importantly, the new provision requires the Court to consider the conduct of the purpose or effect of increasing competition, and purpose or effect of lessening competition. The changes, only coming into effect in November 2017, have yet to be tested by the courts. The 'effects' test is likely to be similar to the 'intent' test; however, the effects test will determine the impact based on the practices employed, rather than the intent of the party or parties.

Singapore law is comparable to the EU in determining what constitutes predatory behavior. Singapore will assess the facts in deciding whether the pricing is below cost, the 'intention' is to eliminate a competitor, and the feasibility of recouping losses.[123] However, there are a number of exclusions in section 47. These exemptions would affect how the intention of the parties will be assessed by the courts. For instance, intention will not apply where there is an undertaking entrusted with the operation of services of general economic interest or relates to a revenue producing monopoly. In addition, this is unlikely to apply where the conduct is in conflict with an international obligation. The exemption applies to a compelling reason of public

[119] Ibid.

[120] Ibid.

[121] Competition and Consumer Amendment (Misuse of Market Power) Act 2017, No. 87, 2017.

[122] Australian Competition Law, https://www.australiancompetitionlaw.org/legislation/provisions/2010cca46.html, accessed 20 December 2018.

[123] Competition Act 2004, section 47.

policy is very broad terms and allows the Minister to determine whether there will be a significant economic or social impact.

The Competition Act 2004 of Singapore regulates, amongst others, anti-competitive agreements, abuse of a position of dominance, and anti-competitive mergers, similar to most other jurisdictions.[124] Section 34 prohibits agreements being established to restrict competition. Any agreement that fixes a purchase or selling price and other trading conditions also does not meet the requirements of section 34. Even though section 47 assesses the intention of the parties, any intentions may not be considered as a factor to determine whether the agreement is restrictive.[125]

In 1999, Indonesia established the Law Number 5 on the Prohibition of Monopolistic Practices and Unfair Business Competition (ICL) was enacted.[126] The ICL reflects a similar approach to its neighbour Singapore. An organization cannot take a dominant position directly or indirectly to prevent consumers from obtaining goods and services competitively. The test is based on the need to determine the 'intent' of the parties.[127]

Predatory pricing is unlikely to go away. Government, industry and the consumer will need to be vigilant in responding to the introduction of new and ever more powerful technology. The digital economy could see even greater monopolies and oligopolies develop, particularly as automation is adopted by industry sectors.

13.5 Conclusion

This Chapter has provided a brief introduction to some of the issues that are, have and likely to continue to arise between personal data and competition law. To date, only the EU, Australia, Malaysia and Singapore have begun to recognize the link between personal data and competition law. However, these developments are far from comprehensive. There is a long way to go before they fully understand and respond to the continually changing use of personal data and implement a robust legal framework to ensure that competition is not restricted to a few. However, it is well understood that the horse may already have bolted with only a few entities claiming absolute dominance in the market – such as Google, Facebook Twitter and others. The current legal framework does not address the issues raised in this Chapter.

The ongoing need for regulator (s) to balance economic needs along with innovation, as well as protect personal data and privacy, will continue to face challenges. That balance may never be resolved. In other words, the balance between stifling innovation and protecting people's personal data and privacy is likely to remains

[124] Competition Act 2004, sections 34, 47 and 54.

[125] Wong., B, Yong Quan., Y, *Object Restrictions in Singapore Competition Law*, Singapore Journal of Legal Studies, 2017, PP. 169–191.

[126] State Gazette of the Republic of Indonesia, 33 of 1999.

[127] Ibid, Article 25.

tenuous as innovation becomes increasingly dependent on large scale dissemination, transfer and trade in all forms of data, including personal data. This trade will not only heighten the potential for privacy breaches, but is also likely to be used to obtain a market advantage.

The concept of consent and the definition of personal data and personal information are key to strengthening the interrelationship between competition, data protection and privacy law. As this book demonstrates, the concept of consent is fast becoming important to personal data across many areas of the law. Both the concept of consent and the definition of personal data should not be seen as barriers to innovation or consumer protection. Friedrich von Hayek makes the point that consumer welfare:

> cannot be adequately expressed as a single end, but only as a hierarchy of ends, a comprehensive scale of values in which every need of every person is given its place.[128]

The position taken by Hayek is important because there are many different areas of anti-competitive behavior that has developed over the Internet and from the use of personal data.[129] That includes personal data that is not necessarily defined by the law. However, more work is needed to better prepare the community for the digital economy and potential competition issues that may arise. More work is also required to better project and understand whether consent in its current form is adequate, along with understanding whether the definition of personal data or personal information (depending on the jurisdiction) meets current and future needs – particularly in relation to competition and data protection.

Furthermore, work is also needed to better develop the theory and application of harm in relation to data protection and unfair competition. What is certain is the fact that personal data and data is being used more generally to create anti-competitive environments. Unfortunately, however, the broader community is mostly unaware that this is occurring. This continued technological evolution and changes in technology are likely to make it even more challenging for data subjects to understand and measure the harm to them.

The introduction of data portability has begun to provide a solution to the complex interrelationship between personal data protection and competition law. However, it is only a recent addition to the legal framework, with the EU again taking the lead. It has been argued that, by providing data subjects with the free portability of their personal data, as defined in law, from one controller to another, [provides a tool for the] data subjects are better able to contribute to competition between digital services and the interoperability of platforms. While this is appearing to be a step in the right direction, due to the limited implementation and jurisprudence regarding this concept, it remains to be seen how this will influence the interrelationship between competition and rights in data protection and privacy law.

---

[128] von Hayek, F *The Road To Serfdom: Text And Documents - The Definitive Edition,* 1st edn, University Of Chicago Press (2007), p. 101.

[129] Ibid.

In the vision of Wolfie Christl, today more than at any other time in history, the ubiquitous streams of behavioral and personal data collected at the individual level by a plethora of services across many fields, even across international borders, of life. They are linked and utilized to monitor and analyze every interaction of a consumer (data subject) that might be relevant to a company's customer management and acquisition efforts.[130] Thus, enhancing the opportunity for companies to obtain a dominant position in the market. Wolfie Christl highlights a critical milestone that occurred in 2012, when Facebook started to link its profile data with information about offline purchases in stores. Oracle's Datalogix system has allowed companies, for the first time, to measure how Facebook ads affect store visits and purchases. Today, companies try to capture as many "touchpoints" across the whole "customer journey", from online and mobile to in-store purchases, direct mail, TV ads, and call center calls, as possible.[131]

One way to respond to these issues is to conduct more in-depth analysis and research on data protection and competition in the immediate future. Different data protection and competition laws can achieve similar results, namely to protect the individual consumer, albeit in different ways and from different perspectives.[132] Although there appears to be significant overlap between the objectives of these laws, they can arguably converge and be harmonized in the future to provide greater accountability to businesses, no matter in what jurisdiction they are located.[133] A starting point is for countries to set aside their different economic and social policies and objectives and work towards establishing a balanced policy approach that would pave the way for legal harmonization. This approach can ensure that there is adequate competition in the new digital economy.

However, in those jurisdictions where data protection and privacy laws are underdeveloped, there is a long way to go. Therefore, the jury is out regarding the direction personal data and competition will take, and where the balance between them will be settled. There are many unanswered questions because the area continues to evolve and change. The issues raised in this Chapter are not confined to a single nation state. The issues between competition and data protection law may also require an international response.

---

[130] Christl, W *Corporate Surveillance In Everyday Life, How Companies Collect, Combine, Analyze, Trade*, and Use Personal Data ON billions, Crack Labs (2017) p. 73.

[131] Ibid.

[132] Walters, R., Zeller, B., Trakman, L, *Personal Data Law and Competition Law – where is it heading?* European Competition Law Review, (2018).

[133] Ibid.

References

Alessandro, A., Varian, H *Conditioning prices on purchase history,* Marketing Science (2005) 24(3): pp. 367–381

Avi, A., Tucker, C *"Online Advertising."* In The Internet and Mobile Technology Advances in Computing, (2011) 81, pp. 290–337

Bernasek, A., Mongan, D *Our Massive New Monopolies: Amazon, Google and Facebook Have the Power to Move Entire Economies,* Salon, (2015) https://www.salon.com/2015/06/07/our_massive_new_monopolies_amazon_google_and_facebook_have_the_power_to_move_entire_economies

Burnside, A *'No Such Thing As A Free Search: Antitrust And The Pursuit Of Privacy Goals'* (2015) Competition Policy International, https://www.competitionpolicyinternational.com/assets/Uploads/BurnsideCPI-May-15.pdf

Christl, W *How Companies Use Personal Data Against People, Automated Disadvantage, Personalised Persuasion, and the Societal Ramifications of the Commercial Use of personal Information,* Working Paper of Crack Labs (2017)

Christl, W *Corporate Surveillance In Everyday Life, How Companies Collect, Combine, Analyze, Trade, and Use Personal Data on billions*, Crack Labs (2017) p. 41

Cooper, J *Privacy and Antitrust: Underpants Gnomes, the First Amendment, and Subjectivity,* George Mason University School of Law, (2015)

Colangelo, G., Maggiolino, M Data Accumulation and the Privacy- Antitrust Interface: Insights from the *Facebook* case for the EU and the U.S. Transatlantic Technology Law Forum, Stanford Law School and the University of Vienna School of Law, (2018)

De Hert, P., Papakonstantinou, V., Malgieri, G., Beslay, L,. Sanchez, I *The right to data portability in the GDPR: Towards user-centric interoperability of digital services*, Computer Law & Security Review, Volume 34, Issue 2, (2018), pp. 193–203

Ishiara, T *The Privacy, Data Protection and Cybersecurity Law Review*, (2017) https://thelawreviews.co.uk/edition/the-privacy-data-protection-and-cybersecurity-law-review-edition-4/1151289/japan

Lande, R *The Microsoft- Yahoo Merger: Yes, Privacy is an Antitrust Concern*, FTC: WATCH, (2008) p. 1

Ohlhausen, M., Okuliar, A'Competition, Consumer Protection, and The Right [Approach] to Privacy', 80 *Antitrust Law Journal* 121 (2015)

Prosner R (1976) *Antrust Law: An economic Perspective,* University of Chicago Press, p. 189

Sakle, A., Chand, ABig Data: Emerging Concerns under Competition Law *Practical Lawyer PL (Comp. L)* (2018)

Solon, O *Here's how to protect your internet browsing data now that it's for sale*, The Guardian Australian Edition, (2017)

Stucke, M., Grunes, A *Big Data and Competition Policy* New York: Oxford University Press, (2016)

Swire, P *Submitted Testimony to the Federal Trade Commission Behavioral Advertising Town Hall*, (2007), http://ftc.gov/os/comments/behavioraladvertising/071018peterswire.pdf

Swire, P., Lagos, Y "Why the Right to Data Portability Likely Reduces Consumer Welfare: Antitrust and Privacy Critique", Maryland Law Review, Vol. 72, No. 2, (2013), http://digitalcommons.law.umaryland.edu/cgi/viewcontent.cgi?article=3550&context=mlr, accessed 6 August 2018

Unterhalter, D. *Data privacy: why internet giants elicit antitrust critiques*, The Cambridge Analytica furore vindicates fear that groups such as Facebook are not benign monopolies, https://www.businesslive.co.za/bd/opinion/2018-04-18-data-privacy-why-internet-giants-elicit-antitrust-critiques

von Hayek, F *The Road To Serfdom: Text And Documents – The Definitive Edition,* 1st Edn, University Of Chicago Press (2007), p. 101

Walters, R., Zeller, B., Trakman, L, *Personal Data Law and Competition Law – where is it heading?* European Competition Law Review (2018)

Wong, B, Quan, Y, *Object Restrictions in Singapore Competition Law*, Singapore Journal of Legal Studies, 2017, pp. 169–191

Chapter 14
Conflict of Laws, Transnational Contracts in Personal Data

Abstract This Chapter explores whether the law of contracts is an adequate mechanism for personal data protection pertaining to transnational commercial trade. The question arises whether Data Protection has become a new frontier in transnational contract law. It is well understood that the transition into the new digital economy has begun and is moving at a rapid rate. Subsequently, data protection is having to be considered by organizations when establishing contracts, both domestically and internationally. In this regard it is understood that national contract laws are far from consistent or uniform. Following recent case law, this Chapter examines whether the Convention on the International Sale of Goods (CISG) can be a mechanism to strengthen the protection of data in transnational contracts.
This Chapter does not deal directly with protecting an individual's privacy. However, where it is proven that the law of contract can apply, the resulting affect will be strengthening compliance of personal privacy. The Chapter puts the reader in a position of a practitioner who may wish to explore using the CISG and the UPICC to include specific contractual clauses for the transnational commercial trade in personal data. Additionally, this Chapter addresses the issue of applicable law in personal data protection in relation to contracts. It expressly excludes privacy from the conflict of laws argument. Finally, this Chapter only uses Australia, Indonesia, European Union and Singapore as working examples.

14.1 Introduction

As noted in previous chapters, personal data is increasingly being used by companies to enhance their profits.[1] To assist in this process, contract law is being considered as a way to strengthen the governance and compliance of personal data. The problem is that domestic legislation protecting data is not uniform and, in some

---

[1]Zech H (2017) *Data as a Tradeable Commodity – Implications for Contract Law*, Josef Drexl (ed.), Proceedings of the 18th EIPIN Congress: The New Data Economy between Data Ownership, Privacy and Safeguarding Competition, Edward Elgar Publishing, pp. 1–15.

© Springer Nature Singapore Pte Ltd. 2019

R. Walters et al., *Data Protection Law*,
https://doi.org/10.1007/978-981-13-8110-2_14

jurisdictions it is non-existent. Chapter 15 considers two areas of contract law that are of concern in relation to personal data and contracts.

The first examines the issue of the applicable law in personal data protection in relation to online contracts between the individual purchasing a good or service in a country other than in their country of residence. In other words, do the long standing rules of conflict of laws apply? Chapter 1 has argued that the protection of privacy is a by-product of the introduction of data protection law. Therefore, when examining the conflict of laws, this Chapter distinguishes between data protection and privacy. It argues that privacy is not relevant to the conflict of laws debate because, the choice of is about entering into an online contract that transcends national borders, whereby a data subject enters into a contract by agreeing to the host's terms and conditions. The second considers whether a way has been forged to include personal data in international (transnational) contracts and to subject those data contracts to transnational laws such as the Convention on the (CISG) or the UNIDROIT Principles of International Commercial Contracts (UPICC).

Personal data is becoming a commodity that can be bought or sold.[2] The resulting effect is that, to some extent, privacy is also becoming a commodity, which is being traded. However, the transaction in privacy is not the same as the traditional sale of goods or services. Joseph Jerome argues that browser add-ons such as Privacy Fix try to show users their value to companies,[3] and a recent study has suggested that free Internet services offer $2600 in value to users in exchange for their data. Interestingly, this figure tracks closely with a claim by Chief Judge Alex Kozinski that he would be willing to pay up to $2400 per year to protect his family's online privacy.[4] On the other hand, Federico Zannier decided to mine his own data to see how much he was worth. Zannier recorded all of his online activity, including the position of his mouse pointer and a webcam image of where he was looking, along with his GPS location data for $2 a day and raised over $2700.[5] Thus, privacy does have a financial value. The costs and value of protecting personal data and privacy arguably comes down to consumer demand. It is well understood that people will forego their privacy, when a provider charges a minimal cost which the consumer believes to be of value.

Personal data as a tradeable commodity can be summarized as having two key components. The first is that personal data from a legal point of view is meant to be commercialized by the original right holder.[6] Secondly, personal data is protected

[2] Jerome J (2013) *Buying and Selling Privacy,* Big Data's Different Burdens and Benefits, 66 Stan. L. Rev.

[3] Ibid.

[4] Kozinski, A Federal Judge, *Would Pay $2400 A Year, Max, For Privacy*, https://www.huffingtonpost.com/2013/03/04/alex-kozinski-privacy_n_2807608.html, accessed 2 November 2018.

[5] Zannier, F *A Bite of Me*, Kickstarter, http://www.kickstarter.com/projects/1461902402/a-bit-e-of-me, accessed 11 November 2018.

[6] Schwartz P, *Managing Global Data Privacy* A Report from Privacy Projects (2009) p. 18, https://www.brookings.edu/wp-content/uploads/2016/06/internet-data-and-trade-meltzer.pdf, accessed 4 May 2018.

by legal rules and therefore is tradeable when it is allocated to a certain person. This tradability of personal data, directly or indirectly also includes the trade in privacy. However, it is contended that because personal data is protected, the by-product of that protection, is that a data subject's privacy is also protected over the Internet, at least to some extent. But this does not mean that the data subjects privacy is totally protected, and we are not espousing that privacy itself has a commercial value. That commercial value resides in the personal data.

Considering the development and advances in technology, the issue of tradable data including privacy data will not diminish but increase. To that end Paul Schwartz highlights that it is estimated that the trade in cloud computing services (data) alone was US $1.5 billion in 2010 and he predicts this to climb by 600 percent by 2020.[7] The importance is that cloud computing is a cutting edge computer service that is used to collect, use, process and transact data. If this figure is realized then, arguably, the commercial trade in data (commercial and personal) could easily be worth more than $900 billion dollars over the next decade.[8] This would be a significant contribution to global economic activity.

Quantum technology appears to be the next frontier in technological development, post blockchain. Quantum technology which is fundamentally different from traditional computer technology because they leverage quantum mechanics to do calculations, could easily decrypt the advanced encryption that are widely used today.[9] Bauer argues that, where encrypted data, which may include personal data and contracts, are safe from today's hackers, that data is potentially vulnerable to hackers in the future, through their use of quantum technology. Securing data will require protection against quantum algorithms, or a system of public and private keys that erase themselves over time.[10] A further unresolved question is in deciding the nature and extent of the impact that quantum technology will have to the general rules related to conflict of laws.

This poses significant challenges for personal data that forms part of transnational and local contracts. Marc Van Allen and Umer Chaudhry[11] maintain that companies offering quantum based solutions for contract and procurement management in the future will need to become familiar with the government's data rights legal framework. This advanced technology will likely impact on personal data that forms part of contracts, but also extends to competition and intellectual property law. However, it is out of the scope of this Chapter to explore this issue further.

Nonetheless, at this current time, blockchain is being used increasingly by entities to manage contracts, otherwise referred to as smart contracts. Smart contracts,

[7] Ibid.

[8] Ibid.

[9] Bauer, M *Quantum Computing is Coming for Your Data (2017)* https://www.wired.com/story/quantum-computing-is-coming-for-your-data, accessed 26 October 2018.

[10] Ibid.

[11] Van Allen, M., Chaudhry, U *Quantum computing is about to disrupt the government contracts market*, Bloomberg Government, https://about.bgov.com/blog/quantum-computing-emerging-technology-bound-disrupt-government-contracts-market/, accessed 26 October 2018.

are not new, and can be used to automate such a series of transactions.[12] However, like many technological systems, platforms and infrastructure, this may or may not be applicable within a decade. Jean Bacon *et al* highlight how smart contracts date back to a computerized transaction protocol that executes the terms of a contract. The term 'smart contract' may confuse lawyers, because lawyers traditionally refer to contracts that are developed offline and not on or within a system or platform. In other words, the lawyer has for decades come to know and learn that contracts are negotiated in person and prepared in paper format, not within or over the Internet. Smart contracts can be used to automate agreements between parties according to the set of instructions written into their codes. In many ways, smart contracts resemble the stored procedures and/or triggers, event-condition- action rules, which are common in relational databases. Bacon *et al* point out that smart contracts aim to capture in software the semantics of potentially complex interactions. However, the challenges involved in correctly capturing these semantics as smart contracts include validation and verification.[13] They believe that a smart contract is not a legally enforceable *promise*, but an automated mechanical process. The authors further point out that while this may be true at the level of the computer-readable code, it is unlikely to reflect smart contract use in practice. Furthermore, in practice, the creator of a smart contract will ordinarily need to explain his offer to human counterparties in human- intelligible language.[14]

Importantly, a further issue highlighted by *Bacon et al* is that even machine-to-machine smart contracts' terms may not be legally binding in all cases. This is because many jurisdictions limit parties' contractual freedom by determining that certain contractual terms are not enforceable, for instance in order to address power asymmetries between the contracting parties—such as between producers/retailers and consumers; landlords and tenants; or employers and employees—or because the terms are otherwise unconscionable.[15] The smart contract will always do exactly what it says in its code. However, the legal contract between the parties is likely to include obligations beyond the code itself, based on other communications.[16] If that is the case, not all of the obligations can be captured fully and correctly by the underlying smart contract. There may be a mismatch between what the parties have agreed and what the smart contract's code executes, resulting in non-performance.[17] Smart contracts are by their nature limited to those contractual terms that can be specified in computer-readable code, and further limited by any constraints imposed by the blockchain system in which the contract operates.[18]

[12] Jean Bacon., J Michels., D, Millard., C, Singh, J *Blockchain Demystified: A Technical and Legal Introduction to Distributed and Centralised Ledgers*, 25 RICH. J.L. & TECH., no. 1, (2018).

[13] Ibid.

[14] Ibid.

[15] Ibid. For example, a smart contract that does not give a consumer a right of withdrawal or refund may fall foul of consumer protection law.

[16] Ibid.

[17] Ibid.

[18] Ibid.

It is understood that these forms of contracts will generally self-execute, which essentially facilitate, verify, execute and enforce the terms of a contract.[19] The authors are of the view that this removes the need for human intervention as far as monitoring compliance and enforcement of the contract are concerned. Arguably, this is no difference to the standard form of contract that are still used today. However, what remains unclear is where contracts automatically self-execute when certain conditions have been met. Therefore, similar to many other areas of the law where data protection and personal data law intersect, further work is needed. What can be confirmed though, is the fact the data protection law will likely apply to smart contracts. In other words, as bacon and other highlight, many blockchain operators will fall within the territorial scope of those countries that have established specific data protection laws. They go onto say that because anybody can use an open a platform, operators of such platforms may be deemed to offer services to data subjects. Thus, it could be argued that the nodes and miners who collectively support the Bitcoin network offer a payment service to data subjects. But, the question arises, could data protection laws be circumvented to alleviate this issue? Anecdotally, and provided that data protection laws exist, and platforms are located in countries such as the EU, Australia and Singapore, operators (controller-processors) located in one of these countries, could attempt to prevent data subjects located outside of these countries from using their platforms. Thus, this is an area in which national laws may need to be reconciled through an international convention or code to which countries accede. However, it is outside the scope of this chapter to explore this prospective development.

A further issue arises where data subjects make online purchases from one country such as Australia to Singapore. Maja Brkan argues that, as data subjects and consumers, we often receive a standardized set of rules to which they consent by agreeing to their general terms and conditions, mostly because there is no *real* choice of law available which would give data subject-consumers a say in determining the applicable law.[20] Put another way, if a company selling products or services online from India or Indonesia to customers in Australia and Europe the question is whether the laws of Australia, where the purchaser is located, would apply. The next section provides a working example pertaining to terms and conditions that form part of online contracts which it applies to and applies to the EU, Australia, Indonesia and Singapore.

14.1.1 Conflict of Laws

Data subjects are not only businesses; they are also consumers. The trend in purchasing practice indicates that they are more likely than not to make their purchases online. These online purchases can be within a single country or across single or

---

[19] York, H., McMillan, M., Wong, K *Blockchain and Smart Contracts: The dawn of the Internet of Finance?* Communications Law Bulletin, Vol 35.3 (2016).

[20] Brkan, M *Data Protection and Conflict-of-laws: A Challenging Relationship* EDPL (2016).

multiple international borders and a data subject might agree on an actual or implied agreement or approval to rules and laws which are not applicable domestically. They can be undertaken over a standard computer, laptop, IPad or IPhone. Arguably, the terms and conditions constitute a contract of purchase, unless they are not brought to the attention of the consumer in a clear fashion as prescribed by the relevant consumer protection laws. It is out of scope to resolve whether and how the courts or law will apply contractual terms and conditions, when those terms and conditions have not been clearly stated.

However, in *Apple Corps Ltd -v- Apple Computer Inc* Justice Mann noted that:

> The evidence before me showed that each of the parties were overtly adamant that it did not wish to accept the other's jurisdiction or governing law, and could reach no agreement on any other jurisdiction or governing law. As a result, [the relevant agreement] contains no governing law clause and no jurisdiction clause. In addition, neither party wanted to give the other an advantage in terms of where the agreement was finalized. If their intention in doing so was to create obscurity and difficulty for lawyers to debate in future years, they have succeeded handsomely.[21]

Creating obscurity serves no party well in cross border issues except as a means to avoid a contractual obligation. Otherwise, all obscurity accomplishes is tie up the courts and other, and can become quite a costly exercise to resolve which the adoption of a governing law clause by contract, could readily resolve. This section will confirm whether general conflict of law rules apply to online purchases that transcend national borders. This section will only look at the jurisdictions of the EU, Australia, Singapore and Indonesia to evaluate the working example.

(i) *European Union*

Firstly, on the issue of conflict of laws, Directive 95/46/EC, predating the GDPR, specifically referred to the 'applicable law'. Article 4 of Directive 95/46/EC stated that national law applicable in each member state shall apply the national provisions it adopts pursuant to this Directive to the processing of personal data where:

(a) the processing is carried out in the context of the activities of an establishment of the controller on the territory of the Member State; when the same controller is established on the territory of several Member States, he must take the necessary measures to ensure that each of these establishments complies with the obligations laid down by the national law applicable;
(b) the controller is not established on the Member State's territory, but in a place where its national law applies by virtue of international public law; and
(c) the controller is not established on Community territory and, for purposes of processing personal data makes use of equipment, automated or otherwise, situated on the territory of the said Member State, unless such equipment is used only for purposes of transit through the territory of the Community.[22]

[21] *Corps Ltd -v- Apple Computer Inc [2004] EWHC 768.*

[22] Directive 95/46/EC of the European Parliament and of the Council of 24 October 1995 on the protection of individuals with regard to the processing of personal data and on the free movement of such data, OJL 281.

The GDPR, on the other hand, has no corresponding provision specifying the applicable law of a particular member state for the processing of personal data.[23] inferring that it intended to unify the law governing the processing of data under its own rules.[24] Brkan argues that the conflict of laws issue (s) have largely been addressed by the GDPR in relation to third countries. However, she notes that within the EU there still remains a potential issue when conflict of law issues to arise. Brkan is of the opinion that:

> different Member States can have more or less favorable civil law rules on causal link or quantification of damage. Since the regulation does not specify which law is applicable in case of absence of specific unified rules, such cases might lead to forum shopping in favour of regimes of certain Member States. The conundrum on applicable law would, in such cases, be solved on the basis of general conflict-of-law rules. However, the problem with this approach is that only Rome I Regulation could be applicable and not Rome II Regulation, since the latter excludes from its scope issues related to privacy, as explained above. In tortious claims, it therefore seems that the national conflict-of-laws of the court deciding on the issue would be pertinent to determine the applicable law.[25]

However, the conflict of laws and applicable rules, as Brkan points out, become very important because the GDPR provides member states with the ability to supplement its rules. For instance, member states may add specific requirements with regard to the lawfulness of data processing, lower the age of child's consent, or further limit processing of genetic, biometric or health data.[26] Thus, the GDPR does

---

[23] Brkan, M *Data Protection and Conflict-of-laws: A Challenging Relationship* EDPL (2016).

[24] Ibid. Brkan highlights that if the controller has an establishment in the EU, the regulation applies, according to its Article 3(1), 'to the processing of personal data in the context of the activities of an establishment of a controller or a processor in the Union, regardless of whether the processing takes place in the Union or not. Article 3 contains two important elements. On the one hand, it can be seen that the first part of the rule for territorial application of EU data protection legislation partially remained the same as in Data Protection Directive: processing of data in the context of the activities of a controller or processor, established in the Union. Therefore, with regard to this issue, the legal questions concerning the interpretation of this provision also remain the same, in particular the meaning of the phrase 'in the context of the activities' and 'establishment'. It should be noted that – just as in Data Protection Directive the notion of establishment is defined in Recital 19 GDPR as 'the effective and real exercise of activity through stable arrangements'.

[25] Ibid. Brkan adds that Article 3(1) – regardless of whether the processing takes place in the Union' broadens the applicability of GDPR much further than the current regime. The place of processing or an activity closely related to processing is currently an important factor for determination of applicable law. Within the GDPR, the place of processing becomes an unimportant criterion for such determination. Rightly so, given the fact that data can be processed anywhere in the world, in particular from a technical perspective (servers being located in a different country than the headquarters of a company). It can be established that the criterion of processing is still important in that it still has to be done in the context of the activities of an establishment of a controller or a processor. However, that processing can be done in a third country, not within the EU, as long as the establishment of the controller is within the EU.

[26] Ibid. Recital 20 of GDPR it becomes clear that it should be 'apparent that the controller is *envisaging* the offering' of goods/services to European data subjects. Through this recital, it is also clarified that, whereas mere access to a website or e-mail address are not sufficient for the GDPR to apply, other criteria, such as the mention of Member State's currency or offering of goods/services in a language of this Member State could point to controller's intention to offer goods/services to European data subjects.

not make clear in what circumstances the national laws of member states will apply.[27] The lack of clarification and potential gap between the applicable law of the member state and its general rules, poses another challenge for data protection law, particularly within the EU, even though the applicable law with third countries appears to be somewhat settled.[28]

The question in relation to third countries is somewhat different. According to Article 3(2)(a) of GDPR, a controller has to comply with the rules established in the regulation if his activities relate to the offering of goods or services to data subjects within the Union. Brkan[29] further argues that the third-country controller only has to offer goods or services within the Union in order for the GDPR to apply. Therefore, this provision seems likely to bring all providers of Internet services under the scope of the EU Regulation as soon as they interact with data subjects in the European Union'.[30] In accordance with the general conflict-of-law rules of Rome I Regulation, the parties, in principle, have the freedom to choose which law will govern their contract. However, under Article 3(2)(a) GDPR, it is this regulation that is applicable, since the services are offered in the Union. Nonetheless, in relation to issues arising from a breach of contract, the applicable law will be determined based on Rome I Regulation. Brkan argues that the GDPR will apply regardless of whether an agreement has been established or otherwise and that the Rome II Regulation is not applicable to data privacy issues. It is clear from Article 1(2)(g) of the that Regulation. The applicability of the Rome I Regulation in data privacy issues has not been explored by the Court prior to this case. [31]

Alex Mills highlights some potential issues related to conflict of laws with regard to social media, within the EU. Mills focused on analyzing Facebook's and Twitter's general terms and conditions.[32] He noted that, on the issue of privacy in data protection and conflict of laws, the Court of Justice of the European Union, these terms and conditions raised private international law issues stemming from "vertical contractual relationships" between the social media platform and final users.

Mills highlights *Case322/14, Jaouad El Majdoub v CarsOnTheWeb. Deutschland GmbH.* [33] The court asked the CJEU whether 'click-wrapping', by which a purchaser agreed to the general terms and conditions of sale on a website by clicking on a hyperlink which opened a window, met the requirements of Article 23(2) of Brussels I.[34] The court was allegedly chosen over the courts in Leuven, Belgium, in

---

[27] Ibid.

[28] Ibid.

[29] Ibid.

[30] Svantesson, D *Extraterritoriality in Data Privacy Law* (Ex Tuto 2013), p. 107, in Brkan, M *Data Protection and Conflict-of-laws: A Challenging Relationship* EDPL (2016).

[31] *Case 191/15, Verein für Konsumenteninformation v. Amazon [2016]* par. 73–80.

[32] Rquejo, M *Jurisdiction, Conflict of Laws and Data Protection in Cyberspace,* (2017), http://conflictoflaws.net/2017/jurisdiction-conflict-of-laws-and-data-protection-in-cyberspace/, accessed 2 November 2018.

[33] *Case322/14, Jaouad El Majdoub v CarsOnTheWeb.Deutschland GmbH* 2015.

[34] Ibid.

the vicinity of which the seller's parent company has its head office. The buyer sued in Germany, the domicile of the German subsidiary (as well as the domicile of the buyer, a car dealer). The buyer claimed that the contract was with the subsiduary, not the parent company, and that choice of court had not been validly made.[35] The court stated:

> Article 23(2) of Council Regulation (EC) No 44/2001 of 22 December 2000 on jurisdiction and the recognition and enforcement of judgments in civil and commercial matters must be interpreted as meaning that the method of accepting the general terms and conditions of a contract for sale by 'click-wrapping', such as that at issue in the main proceedings, concluded by electronic means, which contains an agreement conferring jurisdiction, constitutes a communication by electronic means which provides a durable record of the agreement, within the meaning of that provision, where that method makes it possible to print and save the text of those terms and conditions before the conclusion of the contract.[36]

Notwithstanding this decision, jurisdictional agreement can be obtained by clicking 'I agree', which is considered as being sufficient to satisfy the requirements of Article 25 of Brussels. Professor Mills highlights, in particular, the difficult position of social media users within the current legal framework.

Contrary to case law on dual purpose contracts social media users are conceived as "consumers" under the Brussels I and the Rome I Regulations. As a result, social media users are left at the mercy of choice of court and choice of law clauses unilaterally drafted by social media providers.[37] This situation is further complicated by the location where the social media companies are located and legally registered.

In using Facebook as an example, the companies are registered as Facebook in the United States (US) and also in Ireland. For instance, a data subject who is an Australian, located in Australia takes action against Facebook Australia. Facebook Australia may have little responsibility because its principal operation involves promotional activities only. In addition, one must also look at the terms and conditions (clauses) to better understand the choice of jurisdiction governing proceedings and the choice of law determining the applicable law. Currently for data subjects located in the US and Canada are required to deal with Facebook's registered office in the US. Therefore, they will be subject to the jurisdiction of the US. Whereas, for every other data subject, no matter what country they are located, they are required to deal with Facebook, who is registered in Ireland. That being the case, they are likely to be subject to the jurisdiction of Ireland, and arguably, EU law (provided Ireland are not impacted from Brexit) that the commercial terms and conditions specify that the any dispute will be resolved by the court in California, US. The question arises as to whether the jurisdiction is outside the EU, and also whether the EU rules apply in

---

[35] Ibid.

[36] Ibid.

[37] Rquejo, M *Jurisdiction, Conflict of Laws and Data Protection in Cyberspace,* (2017), http://conflictoflaws.net/2017/jurisdiction-conflict-of-laws-and-data-protection-in-cyberspace/, accessed 2 November 2018.

third country jurisdiction, here California or Ireland.[38] In effect, does Article 25 of the Brussels regulations apply solely to disputes brought before member states, of which the U.S. is not, or do those regulations apply even if the case is subject to jurisdiction in a Californian court? As Mills points out, this area of the law remains inconclusive.[39] Also uncertain is whether a court in California would apply EU rules including those contained in the GDPR, or opt for Californian law, With these issues coming to light in the EU, the question is whether, and if so, when conflict of law issues will also arise in Australia, Singapore and Indonesia.

(ii) *Australia*

The common law approach to contracts in Australia has similarities to the United Kingdom. It is generally accepted that under private international law, contracting parties may choose the system of law to govern their contract.[40] A question arises over whether a court in Australia would imply a choice of law where the term and conditions of online purchases do not include an express choice law. In *Akai Pty Ltd v People's Insurance Co Ltd ('Akai')* [41] the decision delivered by plurality stated:

> It is not a question of implying a term as to choice of law. Rather it is one of whether, upon the construction of the contract and by the permissible means of construction, the court properly may infer that the parties intended their contract to be governed by reference to a particular system of law.[42]

Nonetheless, this is not a new issue in Australia. In *Oceanic Sun Line Special Shipping Co Inc v Fay*[43] the High Court had to consider whether conditions on a ticket were incorporated into a contract. Importantly, those 'conditions' included an exclusive choice of court clause in favour of Greek courts. Brennan J argued:

> The question whether a contract has been made depends on whether there has been a consensus ad idem and the terms of the contract, if made, are the subject of that consensus.... In deciding whether a contract has been made, the court has regard to all the circumstances of the case including any foreign system of law which the parties have incorporated into their communications, but it refers to the municipal law to determine whether, in those circumstances, the parties reached a consensus ad idem and what the consensus was... There is no system other than the municipal law to which reference can be made for the purposes of answering the preliminary questions whether a contract has been made and its terms.[44]

---

[38] Facebook Commercial Terms and conditions, https://www.facebook.com/legal/commercial_terms, accessed 2 November 2018.

[39] Mills, A *Jurisdiction, Conflict of Laws and Data Protection in Cyberspace* (Part 2) https://www.youtube.com/watch?v=NYt6SFUkeYU, accessed 2 November 2018.

[40] *Akai Pty Ltd v People's Insurance Co Ltd* (1996) 188 CLR 418.

[41] Ibid, at 441.

[42] Ibid, at 441.

[43] *Oceanic Sun* (1988) 165 CLR 197, 225 (Brennan J), 261 (Gaudron J).

[44] Ibid.

The test and decision in Ocean Sun Line has been reinforced in *Hargood v OHTL Public Company Ltd*[45] where the plaintiff suffered an injury whilst on holiday in Bangkok, Thailand in June 2013. The plaintiff was a guest at the Mandarin Oriental Hotel, and was participating in a Thai cookery class conducted by the Hotel when the floorboards on which she was standing gave way, causing her to injure her shoulder. The plaintiff booked the holiday through a travel agent, and received an email confirming the reservation from the Hotel.[46] The Court found that:

> in this reservation the parties had agreed to all the necessary terms of the contract, including the cost and dates of the booking. The contract was made at the time the reservation was made, not at the time of check-in. The Guest Registration Form (GRF) did not amount to a variation of the original contract or a collateral contract because there was no fresh consideration given for either position.[47]

The court went further stating that:

> 'even if "the charges" referred to the room rate, the fact that the hotel requires an incoming guest to sign that the rate is agreed cannot alter the fact that the agreement, here, had been reached in the reservation made at an earlier time. When these charges were paid casts no light on when the contract was made. The submission that there was a collateral contract or a variation of the contract originally made at the time of reservation must be rejected. There is no evidence of any consideration for either position, and the Defendant does not assert any such consideration.
>
> The result is that the exclusive jurisdiction notation on the GRF is not a term of the contract between the parties. The contract was made at the time the reservation was made and not at the time of check-in at the hotel. The parties had agreed in the reservation to all the necessary terms of the contract including the dates on which the rooms were reserved and the cost of the rooms. Significantly, the reservation of the rooms was guaranteed by the American Express card.'[48]

More recently in 2017, the Federal court of Australia reinforced the above. The case of *Gonzalez v Agoda Company Pty Ltd*[49] provides an example for understanding the terms and conditions set out on a website before entering into a contract with a foreign party. The facts of the case highlight how Gonzalez booked a hotel in Paris through Agoda, using an online Singaporean booking intermediary.[50] In making the online booking, Gonzalez accessed Agoda website from her home computer, located in Sydney, Australia. Agoda's terms and conditions specify that when booking using their service, it is stated:

> The Terms and the provision of our services shall be governed by and construed in accordance with the laws of Singapore without reference to Singapore conflict of laws rules, and any dispute arising out of the Terms and our services shall exclusively be submitted to the

[45] *Hargood v OHTL Public Company Ltd* [2015] NSWSC 446, 23–30.

[46] Ibid.

[47] Ibid.

[48] Ibid.

[49] *Gonzalez v Agoda Company Pty Ltd* [2017] NSWSC 1133.

[50] Ibid.

> competent courts in Singapore. The Contracts (Rights of Third Parties) Act (Cap. 53B) is expressly excluded and shall not apply to the Terms.[51]

Therefore, the only law that would apply to the contract was the law of Singapore. Furthermore, the terms also required that any disputes were to be submitted to the courts of Singapore. Gonzalez was provided with a link to these terms, and clicked the 'book now' button which was positioned below the words: I agree with the booking conditions and general terms by booking this room.[52]

However, while staying at the hotel, Gonzalez slipped in the bathroom and fractured several bones in her leg. She claimed it was due to the shower screen causing soapy water to leak into the bathroom.[53] The resulting effect, was that the bathroom floor became wet and very slippery. Subsequently, Gonzalez sued in the New South Wales Supreme Court, claiming damages under Australian consumer and contract laws She argued that:

> the exclusive jurisdiction clause was not incorporated into the contract properly. She ran a range of arguments relating to the website, including the lack of a button displaying the words 'I agree', and absence of a statement regarding the exclusive application of Singaporean law.[54]

The court noted that Gonzalez was bound by the terms of the contract because she had 'signed' the contract, regardless of whether she had read it.[55] Arguably, it is unlikely that the issues that have emerged in the EU are likely to be replicated in Australia, because the supranational law that EU member states are obliged to follow diverge from Australian common law.

Australia does not have the same contractual requirements as the EU, and neither do Singapore or Indonesia. As highlighted above, with the adoption of the *Hague Principles*, Douglas and Loadsman argued that with the introduction of the International Civil Law Act would make Australian private international law somewhat less parochial and would bring Australia closer to the international community. However, at the time of writing this Chapter, the legislation had not been passed by the Australian Parliament. In summary, it appears that the common law in Australia is somewhat settled in relation to the determinative jurisdiction and the conflict of laws rule. Regarding the issue of jurisdiction, the 1988 Oceanic dispute is cited as a precedent in which an Australia court adopted the *lex fori* in determining whether the terms and conditions issue formed part of the contract. Regarding the conflict of laws, the parties are entitled to choose the law of the country specified in the contract, provided that they are reasonable informed or otherwise aware of the terms and conditions there, that those conditions are certain in nature, and that good consideration (a bargained-for-exchange) was exchanged between the parties. It is outside the scope of this chapter to consider legal issues when the parties do not

[51] Ibid.

[52] Ibid.

[53] Ibid.

[54] Ibid.

[55] Ibid.

adopt an express choice of law, or the recipient of the contract terms and conditions did not reasonable agree to them.

(iii) *Singapore*

Similar to Australia, the Singapore legal system is based on the English common law. By Singapore following the English common law of contract, it enjoys two benefits: the benefit of a system of principles that have proven extremely stable yet flexible in light of technological change and, the benefit of hindsight and the resulting ability to better appreciate the legal problems involved in online transactions.[56] However, there are differences in contract law. This extends to conflict of law because the United Kingdom (UK) currently rely on the European law of Rome I Regulation and Rome II Regulation. It is out of scope to analyze whether there will be any impact to the UK as a result of Brexit, including the likely future inapplicability of Rome I and II in the U.K. Nonetheless, Singapore has established the International Arbitration Act 1994 (IAA). In 2016, Singapore adopted *The Hague Convention Principles on Choice of Law in International Commercial Contracts.* Chapter 143A of the IAA shall decide the dispute in accordance with such rules of law as are chosen by the parties as applicable to the substance of the dispute. Failing such a choice, the arbitral tribunal shall apply the law determined by the conflict of laws rules which it considers applicable.

In *Chwee Kin Keong v Digilandmall.com Pte Ltd,* V K Rajah JC raised the question as to whether the traditional common law principles apply in cyberspace, but there was uncertainty over how they apply?[57] The court noted that:

> The individual also visited the Digilandmall website to familiarise himself with their standard terms and conditions. He acknowledged having had conversations with the other plaintiffs about "how much money we can sell the printer and how much we can make and about storage space" as well as "how many units we intend to buy." Prior to placing his order, he was again contacted by the second plaintiff. The second plaintiff made an enquiry as to the terms and conditions governing purchases through the HP website while the fifth plaintiff was perusing the conditions of the Digilandmall website. After the second plaintiff read out some of the terms and conditions he had found, the fifth plaintiff told him that the contract was binding upon a successful purchase order being received.[58]

Even though there were discussions regarding the website terms and conditions governing the purchases, the individuals denied that there was any discussion between them on even the possibility of an error having taken place. In referring to Internet contracts, the court held that:

> There is no real conundrum as to whether contractual principles apply to Internet contracts. Basic principles of contract law continue to prevail in contracts made on the Internet.

[56] Mik, E *Terms of Use: Reflections on a Theme* (2014). Asian Law Institute 11th Conference, 28–30 May 2014, Kuala Lumpur. Research Collection School of Law.

[57] *Chwee Kin Keong v Digilandmall.com Pte Ltd [2004] SGHC 71 at 91.*

[58] Ibid – at 38 to 60.

> However, not all principles will or can apply in the same manner that they apply to traditional paper-based and oral contracts.[59]

More importantly, the in referring to the relevant legislation established by Singapore that governs Internet contracts, highlighted that:

> The Electronics Transaction Act (Cap 88, 1999 Rev. Ed) ("ETA") places Internet contractual dealings on a firmer footing. Section 11 expressly provides that offers and acceptances may be made electronically. Section 13 of the ETA deems that a message by a party's automated computer system originates from the party itself. The law of agency and that pertaining to the formation of contracts are expressly recognised in s 13(8) of the ETA as continuing to apply to electronic transactions. This provision acknowledges that the essential framework of an electronic contract needs to be considered in the usual manner; in other words, principles of contract formation, consideration, terms and conditions, choice of law and jurisdictional issues need to be examined.[60]

However, and while there was no an exhaustive discussion in relation to consent, the court took the position that, as the now stands, mistakes that are not fundamental or which do not relate to an essential term do not vitiate consent. Mistakes that negative consent do not inexorably result in contracts being declared void. In some unusual circumstances where a unilateral mistake exists, the law can find a contract on terms intended by the mistaken party.[61] Arguably this early case has set the scene and parameters of how online contractual terms and conditions will be applied in Singapore. Thus, it is well understood in Singapore that, where there is an express choice of law that forms part of a contract (terms and conditions), the laws to which have been specified will be determined and applied by Singapore courts. Moreover, where there is no express terms in the contract determining the jurisdiction and/or applicable law, the closest and most real connection with the transaction and the parties is likely to apply.[62]

Notwithstanding the above, generally, the Singapore courts respect the law and jurisdiction that is specified in in a contract. In *John Reginald Stott Kirkham v Trane US Inc*[63] the appellants and respondents entered into a dispute regarding a distributorship agreement governing the distribution of American air conditioning systems and services in Indonesia.[64] The court adopted a cautious approach to granting an injunction, even though the matter did not involve an online contract.

Nonetheless, under common law, a choice of law and/or jurisdiction is treated as a contractual agreement. This was affirmed in *AbdulRashid bin AbdulManaf vHii Yii Ann,*[65] in which the Singapore High Court considered the effectiveness of such clauses in determining the appropriate forum, as well as the effect which prior negotiation may have on the interpretation of that clause. The facts were that the agree-

---

[59] Ibid.

[60] Ibid, at 91 to 93.

[61] Ibid, at 107.

[62] Unfair Contract Terms Act 1994, section 27.

[63] *John Reginald Stott Kirkham v Trane US Inc [2009] 4 SLR 428.*

[64] Ibid.

[65] *AbdulRashid bin AbdulManaf vHii Yii Ann [2014] SGHC 194.*

ment contained terms relating to jurisdiction. Clause 6.1 stated that the agreement was to be governed by and construed in accordance with the laws of England. Clause 6.2 stated that the stipulation that "the Parties hereby irrevocably submit to the 'non-exclusive' jurisdiction of the courts of the State of Queensland, Australia. Thus, the Singapore court applied the common law approach as the applicable law that governed the choice of jurisdiction and law provided for in main contract.

However, and while the court indicated that it would evaluate the intention of the parties, it stated that, in the absence of an express clause within the terms or conditions, non-exclusive jurisdiction clause posed some challenges in resolving the dispute. The court highlighted the need to understand the intention of the parties. This is part of the standard examination in contract law. However, the court also highlighted that there is also a need to examine whether the terms and conditions were clearly able to be understood, by the parties, amongst other clauses – no matter whether the contract is under the tradition model or online. Thus, the law in Singapore is comparable to the law in Australia, including some uncertainty as the materiality of factors beyond the consent of the parties to contract.

(iv) *Indonesia*

Contract law in Indonesian[66] is not substantially omitted to law in Australia, Singapore and the EU and is part of its private law system. Indonesian Private Law was inherited from the Dutch colonial Government, through two codifications – the Civil Code (*Indonesian: Burgerlijk Wetboek*) and the Commercial Code (*Indonesian: Wetboek van Koophandel*). Nevertheless, the principal laws governing online contracts are the Law No. 11 of 2008 on Electronic Information and Transactions, as amended by Law No. 19 of 2016 on the Amendment to Law No. 11 of 2008 on Electronic Information and Transactions. Also applicable is Government Regulation No. 82 of 2012 on the Application of Electronic Systems and Transactions.

Article 47 of the Government Regulation No. 82 of 2012 on the Application of Electronic Systems and Transactions (GR No. 82/2012) requires a valid electronic contract to have been subject of mutual consent, amongst other requirements. Click-wrap, browse-wrap and shrink-wrap contracts are not specifically regulated under Indonesian laws and regulations in its provisions regulating the terms and conditions for online contracts, Article 50 (3) of GR No. 82/2012. However, the terms and conditions in a contact are considered to be enforceable under Indonesian Law if the parties intend to bind themselves to those terms. In accordance with Article 50, consent is considered to have been satisfied by ticking/clicking on the "I Agree" box.

Therefore, similar to the EU and common law counterparts, Suharnoko argues that the choice law will apply in determining when, where and how the parties have specified the applicable law.[67] Suharnoko goes on to explain that the civil law tends

---

[66] Sinta Dewl Rosadi, LLB (Unpad), LLM (Washington College of Law, American University), Ph.D (Unpad), Associate Professor in Law at Faculty of Law University of Padjadjaran, Bandung, Indonesia, provided input and verified the information in this section.

[67] Suharnoko *Contract Law in A Comparative Perspective*, Vol. 2, Indonesia Law Review (2012).

to take subjective approach to the formation and interpretation of contracts.[68] The contract is based primarily on that which the parties intended, rather than the literal interpretation of the words they actually used Article 1343 the Civil Code states that 'if the wording of an agreement is open to several interpretations, one shall ascertain the intent of the parties involved rather than be bound by the literal words'.[69]

Nevertheless, Article 16, Article 17, and Article 18 of Algemeene Bepalingen van Wetgeving voor Nederlands Indie (AB) Staatsblad 1847 No 23 of 1847, do not expressly imply that the contractual law between nationals is the law that had been chosen and agreed between the parties by contract.[70] With the introduction of the private International Law Bill (PIL), Article18 AB is treated as equal to Article 12 of PIL, in stipulating that a legal act is formally valid if it satisfied the requirements of the substantive law which governs it and the law of the state where it is performed. A legal transaction that is performed by persons who are in different states is formally valid if it satisfies the formal requirements of the law which governs the legal act itself, or the law of either state, or the law of the states where either of the parties has his habitual residence.

Notwithstanding the above, another layer of complexity that has recently arisen in Indonesia, is the need for contracts to be drafted in the Indonesian language. In the case of *PT Bangun Karya Pratama Lestari v Nine AM Ltd*[71], the Indonesian Supreme Court handed down a decision that a contract not drafted in the Indonesian language was null and void. The court, complied with Article 3, of Law Number 24 of 2009 on National Flag, Language, Emblem and Anthem, known as "the Language Law," It requires that memoranda of understanding, contracts or agreements which involve Indonesian government institutions, Indonesian private entities or Indonesian citizens shall be in Bahasa Indonesia, for example, the Indonesian language ("Bahasa").[72] The court said that:

> Article 31 (2) of Law 24/2009 explicitly allows execution of an agreement in more than one language. Whilst this law seeks to regulate the use of Bahasa, in practice it means that any contract with any governing law, as long as it involves an Indonesian party, must be drafted in Bahasa, in addition to the foreign language. Law 24/2009 further provides that the implementation of the law will be further stipulated by an implementing regulation, which will be issued within two years after the release of Law 24/2009. We understand that, to date, this implementing regulation has not been released.[73]

The court went further by highlighting how on 28 December 2009, the Minister of Law and Human Rights issued a Clarification to Law Firms, in which the Minister

---

[68] Ibid.

[69] Ibid.

[70] Allagan, T *Indonesian Private International Law: The Development after More than a Century*, Indonesian Journal of International Law, Vol. 14 No. 3, (2017) pp. 381–416.

[71] *451/Pdt.G/2012/PN.Jkt.Bar.*

[72] Ibid, Jones Day, https://www.jonesday.com/files/Publication/202d219d-d9e4-4656-b25d-3071c32a870d/Presentation/PublicationAttachment/e5a54e8d-9240-4111-a178-374d0be20912/Indonesia_High_Court_Upholds.pdf, accessed 5 November 2018.

[73] Ibid.

opined that Article 31(1) of Law 24/2009 did not apply to private commercial agreements and that accordingly, these could continue to be drafted in English in accordance with parties' intentions.[74] The Minister was also of the view that the actual implementation of Article 31(1) would have to await the issuance of a Presidential Regulation, as mandated by Article 40.[75] However, the Presidential Regulation was issued, it provided that the language requirement under Article 31(1) would essentially be unenforceable.

The West Jakarta District Court held a loan agreement between an Indonesian borrower and a foreign lender was unenforceable for failure to comply with the Language Law. The loan agreement concerned was drafted in English only. The Court determined that:

> Article 31 of Law 24/2009 required every contract involving an Indonesian party, whether public or private, to be made in Bahasa. In the absence of a Bahasa translation, the loan agreement violated Article 31 of Law 24/2009, which resulted in the contract having an illicit cause. Although Law 24/2009 does not expressly set out the consequences if it is not complied with, the Court relied on Article 1335 read with Article 1337 of the Indonesian Civil Code to find that the loan agreement was null and void. Article 1335 provides, amongst other things, that a contract concluded pursuant to an illicit cause is invalid. Article 1337 further states that a cause is illicit if it is prohibited by law or if it violates morality or public order.[76]

The position held by the West Jakarta District Court's was reaffirmed by the Jakarta High Court and subsequently by the Indonesian Supreme Court in Judgment No. 48/Pdt/2014/PT.DKI dated seventh May 2014 and Judgement No. 601 K/Pdt/2015 dated 31st August 2015 respectively. The provisions of Law 24/2009 cannot be ignored in contracting with an Indonesian counterparty. The provisions of Law 24/2009 may arise at the enforcement stage where the correct language (s) has not been used.

Arguably, the above decision would likely ensure that even online terms and conditions for the sale of products and services from Indonesia to third countries need to be in the language of Indonesian and possibly English. However, this has not been confirmed, as such, further work in this area is required.

In summary, Indonesia contracts generally applies international standard for conflict laws that is based on consent and choice of law. That is, as long as both parties agree to the choice of law then those choice of law will be apply offline as well as online. The online choice of law is only regulated generally and not specifically. If both parties have not conclusively agreed on the choice of law when a dispute arises then the court will decide on the applicable law. Until now there are no cases yet relating to this issue and the *Bangun Karya Pratama* case also has not yet followed by online case. In conclusion, this is an area of law that has had little to no attention to date in Indonesia, and further research is required.

---

[74] Ibid.

[75] Ibid.

[76] Ibid.

(v) *Growing use of Online Contracts*

The argument regarding for the use of personal data in contract terms and condition of contracts online is only going to expand, as technology expands. The conflict of laws appears to be settled within the EU, however, they are not when contract (terms and condition) are developed and located in third countries. In common law countries of Australia and Singapore, the rules regarding conflict of laws is well settled. However, few issues have arisen in which the court have had to decide on these matters. The message from Indonesia, is somewhat different. Even though it has adopted international norms and rules pertaining to contracts, Indonesian law does not demand that contract terms and conditions to be in the language of Indonesian. Where the terms and conditions are not clear, there is the possibility that cases are likely to be set aside, particularly, when the matter is decided by a court located in Indonesia, or where a court outside Indonesia adopts Indonesian law. Therefore, it is our view that even though much of the law is settled in Australia and Singapore, as this area of the law evolves in other jurisdiction, additional research may be required in common law countries. Arguably, further work is required, in Indonesia to better understand the impact of contractual relationships online when organizations apply terms and conditions, that oblige people to comply with, not matter where they are located.

The use and application of personal data does not stop with online contracts. Organizations are using contracts to manage the trade in and exchange of personal data, although little attention to personal data has been applied or used within transnational contract law. As personal data begins to gain an intellectual property rights and begins to gain momentum, individuals and entities are likely to identify legal mechanisms to protect the trade and use of that data. Therefore, it is likely that domestic and transnational contract law will play a role in transnational trade in personal data.

As part of the transnational trade in personal data continues to grow, another legal mechanism that is available to individuals and entities to manage the cross border trade in personal data, could be the CISG. The CISG has grown in popularity, and is now used by many organizations, whether located in the West, East, Middle East or Asia. The CSIG is also well placed to be used for transnational contracts, albeit in a limited form, as the CISG only applies to the sale of goods. Coupled with the UPICC, the CSIG can become an effective tool to strengthen the trade in personal data.

14.1.2 CISG – UPICC

The trade in personal data is, as stated throughout the book, becoming a reality. It is expected to generate enormous economic activity. Thus, the trade in digital data, especially in the commercial use of personal data requires alternative legal mechanisms to manage that data. This is particularly the case in the transnational sale of this type of data. Corley argues that with:

the newfound ease of collecting and transferring personal information, businesses have been able to collect, analyse, and package this sensitive data to sell to advertisers and other entities as a commodity.[77]

This is not new and in 1999 it has been noted that the Clinton Administration has worked very hard to persuade Internet economy firms to adopt privacy policies and practices to make users more comfortable about engaging in ecommerce transactions in cyberspace, these efforts have done little to overcome the inertia of the current technical and economic environment that is generally hostile to privacy interests.[78]

A transnational law – built on the example of the Cape Town Convenient – would provide a proper pathway for the personal data to be registered. It can then be purchased and sold both within the nation state and across international borders.[79] Ciani argues that "what is crucial in order to realize this economic value is to ensure a possibility to make data available to third parties on the basis of transfer or licence agreement."[80] Arguably a transfer or licence agreement presupposes the granting of copyright to the original data owner. Steps in that direction have already commenced. The EU Commission, *Communication Building a European Data Economy*, [81] "considered the possibility of a legislation on a data producer's right as a possible way to incentive sharing data initiatives, enhance new business models for the exploitation of the data and unlock their economic value".[82] Ciani argues it is generally accepted that freedom of contract should be "king" in this area and this idea has been strengthened after the CJEU's 2015 decision in *Ryanair*, according to which if a database is not protected by the database right, freedom of contract applies, subject to any restrictions imposed by competition laws or national laws.[83]

Moreover, Samuelson correctly points out that one of the virtues of a contractual approach to protecting information privacy is that it can accommodate the multiple interests people have in personal information. The contextual nature of determinations about the appropriateness of collection or use of personal data, the significance of consent as a factor in determining appropriate uses, and the evolutionary nature of social understanding about information privacy, as it evolves and applies within

[77] Corley, M *The Need for an International Convention on Data Privacy: Taking a Cue from the CISG,*

41 Brook. J. Int'l L. 721 (2016), 722.

[78] Samuelson, P *Privacy As Intellectual Property? Privacy as Intellectual Property*, 52 Stan. L. Rev. 1125 (1999), 1126.

[79] Chesterman S (2012) *After Privacy: The Rise of Facebook, the Fall of WikiLeaks, and Singapore's Personal Data Protection Act 2012*, Singapore Journal of Legal Studies,

[80] Ciani abobve n at 288.

[81] COM (2017) 9 final.

[82] Ciani, J Governing Data Trade in Intelligent Environments: Taxonomy of Possible Regulatory Regimes between Property and Access Rights, *Intelligent Environments 2018., 285, 286.*

[83] Ibid.

contracts. It is a flexible, adaptable, market-oriented way to allow individuals to control uses of personal data.[84]

It follows that any business in transnational trade can utilize the CISG to manage contracts of sale of goods. It is also well known and understood that the CISG is not a code and has gaps because the drafters did not find consensus on all issues governing a contract. It is important to understand how much or how far the CISG satisfies the contractual expectations of parties in relation to the protection of data. Arguably it will serve as a starting point in the discussion of whether the CISG, or for that matter, transnational law in general can resolve at least some of the problems in the protection of data. The ensuing discussion seeks to determine whether the CISG can be used to strengthen the governance of data protection within transnational contracts. In proposing the CISG to regulate the sale of personal data, it will be argued that data is a good and not a service, and hence that copyright can be attached to personal information which is traded.

(i) *GDPR*

A preliminary issue is that the CISG obliges a seller to deliver goods free from any third-party claims. The importance of a discussion of the CISG in data protection is twofold. First, can the CISG assist in protecting data or does it have which must be filled by domestic law? Second, how far is the GDPR of assistance when the CISG is the governing law?

The GDPR[85] can apply extraterritorially in relation to dealing in "goods or services" with individuals in the EU. These regulations are new and hence untested (Chap. 3). From the little guidance that is available, it is likely that, having EU-based customers or contacts, or a general website that is accessible from the EU, would automatically mean that the GDPR applies. However, if one proactively markets to individuals in the EU or take steps to position one's website to attract individuals in the EU, one is likely to be affected by the regulation.[86] In other words, a company can be based and operate in and from Singapore, and deal with organisations located in Switzerland or Slovenia, and as such the Singaporean organisation will be affected by the GDPR. Arguably this is of importance when considering that gaps within the CISG that need to be filled with the otherwise applicable laws.[87]

(ii) *Intellectual Property*

Articles 41 to 43 of the CISG directly address the issues of "industrial property or any other intellectual property".[88] However, no definition can be found within the CISG as to intellectual property. It is left to domestic law to do so. The CISG

[84] Samuelson, P *Privacy As Intellectual Property? Privacy as Intellectual Property*, 52 Stan. L. Rev. 1125 (1999), 1126.

[85] Council Regulation 2016/679.

[86] Debevoise & Plimton, https://www.debevoise.com/insights/publications/2018/05/gdpr-should-i-care, accessed 20 June 2018.

[87] Ibid.

[88] Article 41 Convention on the International Sale of Goods.

addresses the seller, requiring that he must sell goods which are free from "any claim of a third party based on industrial property or other intellectual property".[89] The effect is that the seller must, at least. Take care that his goods are not infringing any data protection laws specifically "under the law of the State where the goods will be resold or otherwise used."[90] Arguably, therefore, any sale into the EU would be subject to the definition of intellectual property as noted in the GPDR.

Article 43 needs to be read in conjunction with Article 41 as it demonstrates that the function of Article 42 is to limit the seller's strict liability. The conclusion is that the "seller's lack of knowledge of a defect, which is part of a third-party claim is irrelevant".[91] It automatically triggers a potential claim by the buyer.

Article 42 notes that the seller is not only responsible for claims in relation to breaches of property rights by third parties, but he is also responsible to make sure that a third party does not possess an intellectual property rights over the goods. In effect the seller must indemnify the buyer, should a third party decide to enforce his/her property rights. This point is an obvious one, as Article 42 clearly states that the goods must be free from third party rights or claims. If not, the seller is in breach of his obligations, that is, he/she is in breach of contract. The purpose of Article 42 is to protect the normal expectations of a buyer that he is not purchasing a lawsuit.[92] This observation is still valid today, as it was in 1999.

An Austrian case[93] noted that "the general burden of proof pursuant to the CISG was on the party that wanted to rely on a provision in its favour, unless reasons of equity would demand otherwise".[94] Therefore, Article 42 became important, leading the French Court of Cassation to state that "the trial judges found that the buyer could not, as a professional, have been unaware of the counterfeit; therefore, the buyer acted with knowledge of the property right invoked.[95] Article 42(2)(a) states that the obligation of the seller does not in all circumstances extend to delivering goods free from any intellectual property right.

Intellectual property rights are territorial in nature. The reason was that "it would constitute a disproportionate and unnecessary obligation upon the seller"[96] to warrant his obligation on a worldwide scale. Even so, this view has now changed, as many data protection laws such as the EU GDPR has been applied extraterritorially.

---

[89] Article 42 Convention on the International Sale of Goods.

[90] Article 42(1)(a) Convention on the International Sale of Goods.

[91] Honnold, O *Uniform Law for International Sales under the 1980 United Nations Convention*, Kluwer, (1999) p. 295.

[92] Ibid, 265.

[93] Austrian Supreme Court (*Oberster Gerichtshof*) 12 September 2006 [10 Ob 122/05x].

[94] Pace https://iicl.law.pace.edu/cisg/case/austria-ogh-oberster-gerichtshof-supreme-court-austrian-case-citations-do-not-generally-12, accessed 2 June 2018.

[95] France 17 December 1996 Supreme Court (*Ceramique Culinaire v. Musgrave*), http://cisgw3.law.pace.edu/cases/961217f1.html, accessed 20 June 2018.

[96] Janal, R., *The Seller's Responsibility for Third Party Intellectual Property Rights under the Vienna Sales Convention*, in Andersen, C and Schroeter U., (eds) Sharing International Commercial Law across National Boundaries. Hill Publishing (2008), pp. 203–206.

Put simply, it is becoming increasingly understood that a seller of personal data must take note of data protection laws and depending on where the buyer resides – the effect of that person's liability might be wider than under the CISG.

Nonetheless, intellectual property rights are also different to any claims for non-conforming goods. Article 35 of the CSIG provides that intellectual property rights are a result of domestic public law, applying the law of the state where the goods are eventually destined. The Austrian Supreme Court supports this point, where the court noted:

> The seller merely has to guarantee a corresponding conformity in certain countries, but not on a worldwide level. It is primarily liable for any conflict with property rights under the law of the State in which (not: "into which"!) it is being resold or in which it is supposed to be used, provided that the parties took this State into consideration at the time of the conclusion of the sales contract.[97]

The burden of proof in this respect is on the buyer. Arguably, when the parties take a particular state into consideration at the time of the conclusion of the contract Article 43 of the CSIG is automatically invoked. That is, the buyer has then lost his right to rely on the provisions of Article 41 or 42 of the CSIG. It is our view that the time is right for personal data to be afforded an intellectual property right. The respective data protection laws discussed in this book do go some way to providing for personal data to be afforded an intellectual property right. Doing so will, in part, reinforce the position that data (personal) protection and subsequently privacy can by supported by the CISG.

However, what constitutes an intellectual property right (s)? Intellectual property rights are based on public law, with the definition being contained in the WIPO Rules.[98] Intellectual property systems vary considerably from state to state. However, this statement needs to be moderated as a result of the multiple international treaties in the domain that play an important role in the light of Article 7(1) of the CISG.[99] The WIPO's definition is the most extensive definition and covers all the rights which owe their existence to an activity of human mind in the fields of industry, science, literature and art (Article 2(viii)) of the WIPO Convention). Article 2(vii) WIPO states that "intellectual property" shall include the rights relating to "literacy,

---

[97] Austrian Supreme Court (*Oberster Gerichtshof*) http://cisgw3.law.pace.edu/cases/060912a3.html

[98] Rauda C, Etier G, *Warranty for Intellectual Property Rights in the International Sale of Goods*, Vindobona Journal of International Commercial Law and Arbitration, Issue 1 (2000) pp. 30–61.

[99] The Paris Convention for the Protection of Industrial Property (1967), the Universal Convention of Copyright (1971) and the Berne Convention for the Protection of Literary and Artistic Works (1971). The Secretariat's Commentary refers to Article 2(viii) of the Convention of the World Intellectual Property Organization of 14 July 1967 (WIPO). This rule is very important for finding the definition of intellectual property law in the sense of the CISG for the words industrial or other intellectual property. These terms were introduced by the Finnish deputy during the deliberations on Article 42 CISG, and are rooted in a proposition of the WIPO on the project of Article 42 CISG (at the time being Article 40 of the New York project).

artistic and scientific works; performance of performing artists, phonograms and broadcasts, inventions in all fields of human endeavor; industrial designs; trademarks, service marks, and commercial names and designations; protection against unfair competition, and all other rights resulting from intellectual activity in the industrial, scientific, literary or artistic fields.[100] This is very broad description and is a useful guide when the issue of definition under the CISG might arise. However, as noted above, the definitions and scope of the GPPR is equally relevant and arguably superior in cases of contracts.

Furthermore, the English and Wales Court of Appeal in *Coogan v News Group Newspapers Ltd & Anor*[101] ruled that confidential personal information is intellectual property under section 72 of the *Senior Courts Act 1981*.[102] The information had to be 'information which have a confidential quality, relates to commerce,[103] and of commercial value.[104] Therefore, personal data that is confidential and of commercial value could form part of contractual agreements. Furthermore, in applying the rules set out in this case, personal data that is intellectual property, commercial in nature and confidential, can arguable come under the CISG. Nonetheless, it must be noted that there is no other case law, at the time of writing this book, that discusses the same issues of whether personal information (data) constitutes intellectual property.[105] That being so, it may leave the courts to decide that the above case, at some time in the future was incorrectly decided, and therefore, that the CISG may not apply.[106] Intellectual property in personal data[107] is a highly contestable area of law. However, it is argued that the current legal framework supports this position.

(iii) *Party rights of claims*

Article 42 notes that the seller is not only responsible for claims in relation to breaches of property rights by third parties, but is also responsible to make sure that a third party does not possess an intellectual property rights over the goods. In effect, the seller must indemnify the buyer should a third party decide to enforce his/her property rights. This point is an obvious one, as Article 42 clearly states that the

[100] Convention Establishing the World Intellectual Property Organization http://www.wipo.int/treaties/en/text.jsp?file_id=283854, accessed 20 June 2018.

[101] [2012] EWCA Civ 48.

[102] Ibid, para 22.

[103] Ibid, para 23.

[104] Cornish W, Llewelyn D, Aplin T, *Intellectual property: patents, copyright, trademarks and allied rights*, Sweet & Maxwell, 2010, in *Coogan v News Group Newspapers Ltd & Anor* [2012] EWCA Civ 48, paras 36–38.

[105] Zeller B., Walters R., Trakman, L *Data Protection – a new frontier for transnational contract law?* Journal Law and Commerce, University of Pittsburgh School of *Law (2018)* – under review.

[106] Ibid.

[107] Trakman L, Walters R, Zeller B, *Is Privacy and Personal Data set to become the new Intellectual Property?* International Review of Intellectual Property and Competition Law (2018).

goods must be free from third party rights or claims.[108] If not, the seller is in breach of his/her obligations, namely, is in breach of contract. The conclusion is that the CISG is well placed to protect the sale of goods online which are subject to intellectual property as defined under the relevant domestic law.[109] However, can the CISG cover the sale of data in general? Could personal data or privacy become a good for the purposes of the CISG?

(iv) *Personal data a good (CISG)?*

A complicating factor is whether personal data can be constituted as a good. It is widely understood that the drafters of the CISG did not define goods.[110] Hall J. stated in *South Central Bell Telephone Co v Sidney J Barthelemy,*[111] which are equally applicable when considering software under the CISG. He noted:

> The software itself, i.e. the physical copy, is not merely a right or an idea to be comprehended by the understanding. The purchaser of computer software neither desires nor receives mere knowledge, but rather receives a certain arrangement of matter that will make his or her computer perform a desired function. This arrangement of matter, physically recorded on some tangible medium, constitutes a corporeal body.[112]

Joeseph Lookofsky argues that the CISG could be applicable to computer software.[113] Lookofsky goes on to say that the CISG applies to diverse forms of software licensing, but that goods, like software, frequently involve a mix of sales (goods) and services. He takes the positon that:

> [t]hough we cannot see or touch it, a computer program is not really all that different from a tractor or a micro-wave oven, in that a program—designed and built to process words, bill customers or play games—is also a kind of "machine". In other words, a computer program is a real and very functional thing; it is neither "virtual reality" nor simply a bundle of (copyrighted) "information." Once we recognize the functional nature of a program, we begin to see that the CISG rules (on contract formation, obligations, remedies for breach etc.) are well-suited to regulate international sales of these particular "things".[114]

Therefore, the CISG must only ever apply to goods that are tangible. Lookofsky further highlights, the refusal by the German Court to characterize, a "scholarly market analysis" as CISG goods lends logical support to such a broad generalization. A market analysis and a computer program are two very different things.[115]

---

[108] Ibid.

[109] Ibid.

[110] Ibid.

[111] *South Central Bell Telephone Co v Sidney J Barthelemy,* 643 So. 2d 1240.

[112] Ibid at 1246.

[113] Lookosfky points out that the CISG is an elastic document and it ought not be stretched beyond its essential design.

[114] Ibid, 276.

[115] Lookosfky J *In Dubio Pro Conventione? Some Thoughts about Opt-Outs*, Computer Software and Preemption Under the CISG, 13 Duke J. Int & Comp. L. 258 (2003), pp. 274–277 Decision of OLG Köln, 26 August 1994, RIW 1994, 970, CLOUT Case 122, available at http://cisgw3.law.pace.edu/cisg/wais/db/cases2/940826g1.html – holding that a contract calling for a "scholarly analysis of a certain segment of the German market for express delivery services" did not consti-

Goods of merchandise include, but not limited to, something that can be physically seen, touched and used, such as a television or fridge. Teija Poikela argues that a possible dispute over whether electricity is tangible (a quantum) or intangible (a wave) was avoided by the exclusion of electricity. However, she goes onto say that the sale of gas is within the CISG.[116] Thus, gas constitutes a good. In our view, an individual can rarely see, touch or feel gas, other than in its liquid form. Like gas not in liquid form, data cannot be touched or felt but it can be seen, once it is printed onto paper or is visible on a computer screen. The feeling of personal data is when the person has been impacted or incurred a harm, from their data being used or disclosed unlawfully.[117] On the other side, there is an argument that that data can constitute a good.

Article 2 of the CISG is of assistance as it provides clarity that goods must be tangible, corporeal things, and not intangible rights. Article 2(d) excludes stocks, shares, investment securities and instruments evidencing debts, obligations or rights to payment. It must be noted that a transaction that is transmitted through or in a computer or by or in computer software, is not specifically excluded by the CISG. Were the CISG to address these issues, it would provide greater breadth in its scope of application and wider coverage. Either way, data, whether the data is commercial or personal would not totally be excluded from the CISG. Moreover, custom software, Internet downloads, and standard mass-market licenses can all be brought within the confines of the CISG's, including networks. Therefore, if one's view is to base one's argument using Lookofsky, our view is that data falls within the confines of property.

Personal data, by definition under EU and national law is gaining ground as intellectual property rights.[118] Most jurisdictions including Australia, the European Union, Malaysia, Japan and Singapore have implemented data protection and privacy laws. These laws define and provide the basis for individuals to have a level of control over their personal data. That control arguably provides a level of ownership over that data. Thus, in our view the CISG deal adequately with intellectual property, and hence data protection (personal and commercial) can fall within the sphere of this important international trade Convention. In conclusion, we are of the view that the CISG can be used to protect personal data in relation to transnational contracts, provided that the data is connected to or is part of the goods.

---

tute a contract for the "sale of goods". In this connection, the court noted that a sale of goods is characterized by the transfer of property in an "object"; though the analysis results were embodied in a written report, the main concern of the parties was the right to use the ideas therein.

[116] Poikela T *Conformity of Goods in the 1980 United Nations Convention of Contracts for the International Sale of Goods* Nordic Journal of Commercial Law (2003).

[117] Ibid.

[118] Trakman L, Walters R, Zeller B, *Is Privacy and Personal Data set to become the new Intellectual Property?* International Review of Intellectual Property and Competition Law (2018).

14.2 Conclusion

The emerging law in regard to contracts and data protection is complex and can be best described as evolving over time. It is unlike any other area of law, because some of the technology, such as quantum technology, is in its development stage, and once this enters the market, it could radically change the landscape for transnational contract and data protection law.

The issue related to personal data and contracts, is whether the conflict of laws rules is adequate in resolving issues pertaining to individual data subjects making online purchases whilst being located in different countries. Increasingly, an ever-wider range of economic, political and social activities are moving online, encompassing various technologies that are transforming the way business is conducted. This is also impacting on the way people interact and transact, including in the trading goods and services that involves government, enterprises and other stakeholders including business.

There is the opportunity for further exploration of how the CISG could be used to strengthen the governance of personal data in transnational contracts. This position is supported by the fact that earlier in this book, it has been argued that personal data csn, and should be, afforded an intellectual property right. However, it remains to be seen whether Indonesia, for example, require these types of contracts to be in both Indonesian or English, in light of the recent 2015 case law there. Arguably, the CISG is capable of dealing with enforcement of one aspect of data namely intellectual property. The CISG is very clear that a breach of Article 42 is a breach of contract and that damages will flow from such a breach. Importantly the CISG takes a "business like" approach by protecting the innocent third party. The protection of personal data once harvested and traded is in flux. The issue is that he law has not caught up with the new reality. Van Erp put it succinctly when he noted 'Next to the "real" world, we now have the "virtual" world, which is just as realistic as the physical world around us. This virtual world demands a rethinking of classical property law, particularly the numerus clauses of legal objects'.[119]

How much and how far the CISG complements the GPPR needs to be seen and is far from being tested. The benefit of doing so, comes from the fact that a large number of countries accept the CISG and UPICC as part of their framework for managing cross border contracts.[120] Therefore, as personal data continues to be traded and forms part of contracts, transnational contracts that apply the CISG and UPICC could be used to strengthen the governance of personal data. Finally, it must be noted that as most jurisdictions continue to apply the agreed international standards for transnational contract (terms and conditions), even though they are online.

---

[119] van Erp, S Ownership of Data: The Numerus Clausus of Legal Objects, *Brigham-Kanner Property*
Rights Conference Journal 6 (2017), PP. 235, 235–236.

[120] Trakman L, Walters R, Zeller B, *Is Privacy and Personal Data set to become the new Intellectual Property?* International Review of Intellectual Property and Competition Law (2018).

References

Allagan, T (2017) *Indonesian Private International Law: The Development after More than a Century*, Indonesian Journal of International Law, Vol. 14 No. 3, pp. 381–416

Brkan, M (2016) Data Protection and Conflict-of-laws: A Challenging Relationship EDPL

Bauer, M (2017) *Quantum Computing is Coming for Your Data* https://www.wired.com/story/quantum-computing-is-coming-for-your-data, accessed 26 October 2018.

Ciani, J Governing Data Trade in Intelligent Environments: Taxonomy of Possible Regulatory Regimes between Property and Access Rights, *Intelligent Environments 2018., 285, 286.*

Cornish W, Llewelyn D, Aplin T, (2010) *Intellectual property: patents, copyright, trademarks and allied rights*, Sweet & Maxwell

Corley, M *The Need for an International Convention on Data Privacy: Taking a Cue from the CISG,* 41 Brook. J. Int'l L. 721 (2016), 722.

Honnold, O (1999) *Uniform Law for International Sales under the 1980 United Nations Convention*, Kluwer, p. 295

Janal, R., (2008) *The Seller's Responsibility for Third Party Intellectual Property Rights under the Vienna Sales Convention*, in Andersen, C and Schroeter U., (eds) Sharing International Commercial Law across National Boundaries. Hill Publishing, pp. 203–206.

Jean Bacon., J Michels., D, Millard., C, Singh, J (2018) *Blockchain Demystified: A Technical and Legal Introduction to Distributed and Centralised Ledgers*, 25 RICH. J.L. & TECH., no. 1

Jerome J (2013) *Buying and Selling Privacy,* Big Data's Different Burdens and Benefits, 66 Stan. L. Rev.

Lookosfky J (2003) *In Dubio Pro Conventione? Some Thoughts about Opt-Outs*, Computer Software and Preemption Under the CISG, 13 Duke J. Int & Comp. L. 258, pp. 274–277

Mik, E (2014) *Terms of Use: Reflections on a Theme*, Asian Law Institute 11th Conference, 28-30 May 2014, Kuala Lumpur. Research Collection School of Law.

Mills, A Jurisdiction, *Conflict of Laws and Data Protection in Cyberspace* (Part 2) https://www.youtube.com/watch?v=NYt6SFUkeYU, accessed 2 November 2018.

Poikela T (2003) *Conformity of Goods in the 1980 United Nations Convention of Contracts for the International Sale of Goods* Nordic Journal of Commercial Law

Rquejo, M *Jurisdiction, Conflict of Laws and Data Protection in Cyberspace,* (2017), http://conflictoflaws.net/2017/jurisdiction-conflict-of-laws-and-data-protection-in-cyberspace/, accessed 2 November 2018.

Samuelson, P *Privacy As Intellectual Property? Privacy as Intellectual Property*, 52 Stan. L. Rev. 1125 (1999), 1126

Suharnoko *Contract Law in A Comparative Perspective,* Vol. 2, Indonesia Law Review (2012)

Schwartz P, (2009) *Managing Global Data Privacy* A Report from Privacy Projects p. 18, https://www.brookings.edu/wp-content/uploads/2016/06/internet-data-and-trade-meltzer.pdf, accessed 4 November 2018.

Svantesson, D *Extraterritoriality in Data Privacy Law* (Ex Tuto 2013), p. 107, in Brkan, M Data Protection and Conflict-of-laws: A Challenging Relationship EDPL (2016)

Van Allen, M., Chaudhry, U *Quantum computing is about to disrupt the government contracts market*, Bloomberg Government, https://about.bgov.com/blog/quantum-computing-emerging-technology-bound-disrupt-government-contracts-market/, accessed 26 October 2018.

van Erp, S Ownership of Data: The Numerus Clausus of Legal Objects, *Brigham-Kanner Property Rights Conference Journal* 6 (2017), PP. 235, 235–236

York, H., McMillan, M., Wong, K (2016) *Blockchain and Smart Contracts: The dawn of the Internet of Finance?* Communications Law Bulletin, Vol 35.3

Zannier, F *A Bite of Me*, Kickstarter, http://www.kickstarter.com/projects/1461902402/a-bit-e-of-me, accessed 11 November 2018.

Zech H (2017*) Data as a Tradeable Commodity – Implications for Contract Law*, Josef Drexl (ed.), Proceedings of the 18th EIPIN Congress: The New Data Economy between Data Ownership, Privacy and Safeguarding Competition, Edward Elgar Publishing, pp. 1–15

Chapter 15
Personal Data and Cybersecurity [Crime]

Abstract This Chapter highlights how personal data has become an important tool in cyber-crime. This Chapter also discusses the issues associated with the collection and use of personal data by law enforcement agencies investigating criminal offences. The Chapter brings together the discussions already highlighted in Chaps. 13 and 14 that relate to personal data being stolen to enhance the ability for organisation to increase their market position and obtain intellectual property. The issues surrounding personal data and criminal law are vast and varied. It is outside the scope of this Chapter to explore all these variables. Even so, to date, there has been little scholarly work on the relationship between personal data and criminal law.

It has become apparent while writing this book that data protection law, either directly or indirectly, collides with many other areas of the law, such as competition and intellectual property, as well as indirectly criminal law. Similar issues arise between data protection and cyber-security law. This is because, data protection laws do one thing, namely to protect and facilitate the collection and use of personal data, while cyber-security law addresses the criminal activity undertaken through computer systems and infrastructure. At issue is the tension between cyber- security law, strategies and initiatives established by jurisdictions with the need to protect the right of an individual from criminal activity, along with privacy over the Internet. The developing cybersecurity laws are beginning to consider personal data. These laws are also likely to enhance the interrelationship between data protection, competition and intellectual property law. For instance, big data is being used by businesses to collect and analysis large quantities of data that contain personal data, which have a commercial value. Having to consider personal data in cybersecurity allows enforcement agencies to have potentially greater access to information in which criminal activity was undertaken. However, there is a likely cost to online privacy, and the potential for individuals to have their privacy infringed. What is not fully understood, and upon which further work is needed, is the level of privacy infringement on all levels, and the need to deal with them affectively. For example, this is needed when Internet systems and infrastructure are poor and personal data stolen, relates to a criminal enterprise. In other words, patterns of criminal behavior undertaken by

© Springer Nature Singapore Pte Ltd. 2019
R. Walters et al., *Data Protection Law*,
https://doi.org/10.1007/978-981-13-8110-2_15

individuals and entities to enhance their economic or social position – whether politically or commercially. The multilayered approach to regulating personal data that is obtained and used illegally in criminal activity, only further complicates this area of the law. This Chapter will provide a working example of the EU, Australia, India, Indonesia, Malaysia, Japan, Singapore and Thailand on this issue.

15.1 Introduction

The transformative effect of digital and communications technologies, in particular social media, has been a well-documented focus of interdisciplinary study.[1] The introduction of personal computer workstations in the early 1980s, and following the launch of the 'world wide web' in 1991, the criminal law study of computer and cybercrimes has likewise rapidly expanded.[2] However, there has been very few comparatively studies that look at how personal data is playing a role in cybercrime. This has been reinforced by Holt and Bossler who point out that the preceding 20 years of criminal law research has predominantly focused on the study of the 'impact of technology on the practices of offenders, factors affecting the risk of victimization, and the applicability of traditional theories of crime to virtual offences.[3] Stratton *et al* argue that criminological engagement with computer and cybercrime, to date, has been largely insular; and lacking in a critical and interdisciplinary engagement with disciplines such as sociology, computer science, politics, journalism, and media and cultural studies. They suggest, it is particularly detrimental to advancing a new generation of scholarship concerning technology, crime, deviance and justice in our digital age.[4] Thus, there is a need for more work to comparatively understand the issues surrounding personal data and the intersection with cybercrime-security within the criminology framework. The comparative study of personal data and criminology can be best described as multilayered. There appears to be no single specific legislative instrument that deals with personal data that has been obtained and used in criminal activity, whether that be for identity theft, fraud, or to gain a commercial advantage in the market.

One of the main reasons why there is little scholarly work analyzing personal data in cybercrime is because data protection laws are doing one thing, protecting and facilitating the collection and use of personal data, while cybercrime law is addressing the criminal activity undertaken through computer systems and infrastructure. As highlighted in Chaps. 13 and 14, personal data can be stolen or misused for the purpose of obtaining data and information that may have an intellectual

[1] Stratton G., Powell A., Cameron., R *Crime and Justice in Digital Society: Towards a 'Digital Criminology'?* International Journal for Crime, Justice and Social Democracy 6(2): (2017) pp. 17–33.

[2] Ibid.

[3] Holt, T., Bossler, A *An assessment of the current state of cybercrime scholarship. Deviant Behavior* 35(1) (2014) pp. 20–40 DOI:https://doi.org/10.1080/01639625.2013.822209

[4] Ibid.

property. Therefore, data protection law(s) either directly or indirectly collide with many other areas of the law. However, it is argued that the influence of the development of data protection law in the EU has played a role in redirecting the how privacy over the Internet is protected. In other words, the traditional notion of protecting personal data for privacy purposes alone, is no longer viable, and laws have developed to assist in other areas in which personal data is being used. With the introduction of the GDPR in 2018, the EU has sought to extend data protection law to provide, not only greater control over personal data, but also to protect that data from criminal activity. For instance, the introduction of the right to data portability, while not new, accomplishes two things. It regulates the transfer of personal data amongst controllers and provides a mechanism whereby organizations need to collaborate on the interoperability of personal data.

Jonathan Clough highlights how central to the power of digital technology now enables the storage of enormous amounts of data in a small space, and to replicate that data with no appreciable diminution of quality. Storage and processing power which would once have occupied rooms, will now fit into a pocket. He goes onto say that copies of images or sound may be transmitted simply and at negligible cost to potentially millions of recipients. Furthermore, he notes that the convergence of computing and communication technologies has made this process seamless, with the ability to take a digital image with a mobile phone and then upload it to a website within seconds.[5] This issues raised by Clough are not new, but he highlights how the collection of vast amounts of data leave open the potential for cyber-security and criminal activity to be undertaken using this data.

Clough points out that there are two other important principles related to cybersecurity and cybercrime, namely, principles relating to anonymity and to global reach. Anonymity is an obvious advantage for an offender, and digital technology facilitates this in a number of ways. Offenders may deliberately conceal their identities online by the use of proxy servers, spoofed email or IP addresses of anonymous emailers.[6] The internationalization of data protection and privacy law has taken place in a relatively short period of time. Thus, the globalization of cybercrime over the Internet has grown significantly in the past couple of decades. Modern computer networks have challenged that paradigm. As individuals now communicate across international borders with ease, offenders may be present, and cause harm, anywhere there is an Internet connection.[7] The result is that cybercrime is vast in its scope of application. It may include a fraudulent scheme, or stealing personal data and data in general to obtain intellectual property secrets or to gain a competitive edge in the market. Thus, like data protection law, cybercrime and security law also presents enormous challenges to law enforcement, government and industry (and legal harmonization).

---

[5] Clough, J Principles of Cybercrime, *Faculty of Law, Monash University* Cambridge University Press, New York (2010).

[6] Ibid.

[7] Ibid.

Therefore, a further issue is how personal data laws have created a tension in which individuals demand that their personal privacy over the Internet be protected from criminal activity. In particular, data protection laws simplify the process to facilitate the transfer of personal data internationally, including for criminal purposes. Most jurisdictions discussed in this book provide for the transfer of personal data outside the nation state. This can be undertaken in various ways, through contracts, and organization to organization distribution, amongst others. The transfer of personal data between the public and private sectors, is also at issue, because in many countries public and private sector agencies do not have the infrastructure to fully secure this data. This leads to a potential failure within the system in which large amounts of personal data are stolen, including peoples' personal identities and used in criminal activities. On the other side, personal data is also being used by law enforcement agencies around the world to combat criminal activities. These agencies view such personal data as very beneficial to fighting crime both online and offline, not unlike the benefit that cyber-criminals associate with that data. However, the risk is a collision in the law arising from these two contraindicated beneficiaries of data transmission. That occurs when both the victim and the perpetrator demand the protection of "their" personal data (information), as that data is defined in law. Both are also likely to do so in during the criminal investigative stage and also when a court hands down its decision.

In 2013, it was found that 'personal data' theft constituted more than 65% of all fraud cases in the United Kingdom.[8] The UK's Fraud Prevention Service (CIFAS) has reported that:

> more than 50% rise in the number of cases where fraudsters unlawfully hijacked individuals' accounts and operated them for their own gain. CIFAS believe that there were 38,428 cases of 'facility takeover fraud' recorded by its 260 members in 2012, up from 25,070 in 2011. CIFAS highlight examples of facility takeover fraud could be where criminals steal individuals' security details through computer hacking, intercepting their physical mail or from online "social engineering", which is where individuals are coerced into divulging confidential information. "The fraudulent use of identity details (either those of an innocent victim or completely fictitious ones) is the biggest and most perturbing fraud threat," CIFAS said in a statement. "50% of all frauds identified during 2012 relate to the impersonation of an innocent victim or the use of completely false identities."[9]

While the theft of data often entails confidential information about a person, such as identity theft, that theft may also involve data that is not defined as personal data in law and therefore not legally protected.

This Chapter will, therefore, highlights areas of data protection law that were identified in prior jurisdictional Chapters that go some way to assist in preventing and addressing cybercrime and cybersecurity threats through the application of concepts and principles regulating data portability, assessments, consent, definition of personal data, and notification of breach (see country Chaps).

[8] Personal data theft behind 65% of all fraud cases, says United Kingdom Fraud Prevention Service https://www.out-law.com/en/articles/2013/january/personal-data-theft-behind-65-of-all-fraud-cases-says-uk-fraud-prevention-service/, accessed 9 November 2018.

[9] Ibid.

However, a particular tension in the laws of these different jurisdictions is attributable to the lack of legal harmonization and policy convergence in data protection law. This is highlighted, earlier in the book, in recognizing that nation states have different economic and social needs. Countries with established data protection laws have sought to ensure that their laws are compatible with the law in other jurisdictions. The purpose, or simply the result, is to facilitate sharing of information and data in conducting cross-border investigations into the misuse of personal data.

15.1.1 Technology

Stratton *et al* highlight how digital technologies offer opportunities for a range of actors to explore and investigate criminal behaviour in both online and offline settings.[10] They go onto to say that data which is stored or transmitted on digital devices are increasingly being used to assist in identifying offences, such as in challenging an alibi or in proving intent to commit a crime. The authors argue that digital evidence extends well beyond personal computers, to mobile, personal and wearable devices that expand the repertoire of investigators and traditional law enforcement agencies. For instance, wearable fitness technologies have been introduced as evidence in criminal trials to identify the location of key figures at the time of the crime.[11]

Importantly, digital evidence (including personal data) is collected and used in different ways that require a greater understanding of the data investigation process. Such investigations also raise important new questions about how evidence is collected, retained and regulated in relation to privacy and the liberty of individuals whose personal data is the subject of criminal investigation.[12] Stratton, Powell and Cameron argue that, where online platforms such as Facebook, provide government agencies with new opportunities for investigation, the ensuing monitoring and policing constitutes a form of surveillance creep directed at investigating complex and interconnected breaches of the law. This includes surveillance of social media data by adopting a data tools in order to collect and analyze texts, photos, videos, and other materials shared via social media systems, such as Facebook and Twitter.[13]

[10] Stratton G, Powell A and Cameron R *Crime and Justice in Digital Society: Towards a 'Digital Criminology'?* International Journal for Crime, Justice and Social Democracy 6(2): (2017) pp. 17–33.

[11] Ibid.

[12] Ibid.

[13] Hollywood., J Michael., J. Vermeer., M, Woods., D, Goodison., S, Jackson, B Using Social Media and Social Network Analysis in Law Enforcement, Creating a Research Agenda, Including Business Cases, Protections, and Technology Needs, https://www.rand.org/content/dam/rand/pubs/research_reports/RR2300/RR2301/RAND_RR2301.pdf, accessed 8 November 2018.

Hollywood *et al* argue that social network surveillance is a type of data analysis that investigates social relationships and structures that are represented by networks (which can also be called graphs). Given that social media reflects personal relationships, it is a key source of personal data that can be easily analysed. Conversely, social network analysis is one key type of social media analysis. These data sources inevitably identify individuals, by their personal data that has been defined in law. Hollywood *et al* highlight the need for law enforcement agencies to have the correct legal resources by which to conduct covert and undercover operations using social media analysis. More importantly, there is need for that data to be protected from both external and insider threats. These law enforcement needs are also confirmed through global developments that demonstrate the continued erosion of personal security through the exposure of personal data to technological vulnerabilities, instability and uncertainty. These threats are increasingly evidence in the world in which cyber-piracy threatens to undermine personal security and privacy *en masse*.[14]

Raul believes that the coming years are likely to bring increased attention to connected devices, autonomous vehicles, artificial intelligence, machine learning, big-data analytics and predictive algorithms. These novel areas hold serious future implications for security (such as in hacking cars and medical devices), having uncertain, abstract and ethereal impacts on personal autonomy, privacy and profiling.[15] He elaborates upon these concerns in relation to data transfer disputes, data localization trends (which have developed in data protection laws), government demands for decryption and access to underlying software code and algorithms, election hacking and fake news. He contends that these developments will lead to challenges in digital trade and arguably even give rise to political stability. For Raul, therefore, the intersection of cybersecurity, counter-terrorism, privacy and human rights remains fraught and subject to abuse, hypocrisy, as well as deficient checks and balances in different jurisdictions.[16]

Arguably, issues related to online privacy includes personal data protection. Coupled with cybersecurity, and the continued development of laws in both areas, are continued tensions highlighted above that are unlikely to subside anytime in the immediate future. Ever changing technology will exacerbate, rather than redress, the ongoing clashes arising in attempting to resolve the tensions underlying them.

Accordingly, this Chapter will conduct a brief examination of these laws to identify the convergence and harmonization among them, as well as the ad hoc and fragmented approach to data protection. It will use, as its working examples, the law in the jurisdictions discussed in this book, the EU, Australia, India, Indonesia, Malaysia, Japan and Singapore. It will also demonstrate the extent to which Cybercrime and security is global, in transcending beyond a single jurisdiction.

---

[14] Raul, AC *Privacy, Data Protection and cybersecurity Law Review*, 4th Edit, Law Business Research Ltd. (2017), pp. 2–5.

[15] Ibid.

[16] Ibid.

15.1.2 Data Protection & Cybersecurity

(i) *European Union*

Cybercrime and cybersecurity is not new to Europe's legal framework. It can be traced taking back to 1981, when the Council of Europe's Convention on Cybercrime 2001[17] was introduced. This Convention forms an important part of the overall framework and requires that personal data be protected. Article 2 and 3 have become important as the basis for protecting protect computer systems from being illegally accessed. In addition, legislative measures were adopted to regulate the illegal intercepting non-public transmissions of computer data to, from or within a computer system, including electromagnetic emissions from computer systems carrying such data. It is arguable that such computer data would constitute personal data today, even though personal data did not have the same level of importance to that of other commercial data in 1981. Since then, cybercrime law, followed by the 1995 Directive on data protection and the 2018 GDPR, have affirmed the principles established in 1981. Apart from the member state of the EU all ratifying the 1981 Convention, only Australia[18] and Japan[19] have signed and ratified the Convention of those jurisdictions that are discussed in this book.

The introduction of the EU GDPR in 2018, has arguably strengthened the landscape for personal data protection, both within the EU and outside. The GDPR now applies to companies that are located within and outside the EU, particularly where that organization is processing data in relation to European citizens. The GDPR has also enhanced the concept of consent. It requires that consent to collect and use personal data be acquired from the data subject. Furthermore, consent for children under the age of 16 is required from their parents, unless individual EU Member States lower the age of consent (but not younger than 13). The resulting effect of the GDPR is that controllers and processors now automatically assume greater responsibility, and need to better justify the collection, protection and processing of data to the extent that they do not have or cannot obtain consent.[20] Importantly, in exercis-

[17] Council of Europe's Convention on Cybercrime 2001 *European Treaty Series - No. 185,* 1981 Council of Europe Convention for the Protection of Individuals with regard to Automatic Processing of Personal Data. The preamble goes onto state Recalling Committee of Ministers Recommendations No. R (85) 10 concerning the practical application of the European Convention on Mutual Assistance in Criminal Matters in respect of letters for the interception of telecommunications, No. R (88) 2 on piracy in the field of copyright and neighbouring rights, No. R (87) 15 regulating the use of personal data in the police sector, No. R (95) 4 on the protection of personal data in the area of telecommunication services, with particular reference to telephone services, as well as No. R (89) 9 on computer-related crime providing guidelines for national legislatures concerning the definition of certain computer crimes and No. R (95) 13 concerning problems of criminal procedural law connected with information technology.

[18] Australia signed and ratified the Council of Europe's Convention on Cybercrime 2001, in 2013.

[19] Japan signed and ratified the Council of Europe's Convention on Cybercrime 2001, in 2012.

[20] Note for further discussion regarding the GDPR and the concept of consent, see Chaps. 10 and 11.

ing their data portability rights, companies will increasingly need to collaborate to protect the privacy of data subjects.[21] Their data portability right will also not exclude data inferred or derived by the controller, nor be restricted to data communicated by the data subject directly to the company.[22]

Raul notes that companies will be obliged to undertake privacy impact assessments where they process high-risk data; for example, profiling based on sensitive data such as health information. He adds that, where the risk of processing cannot be mitigated by privacy-enhancing measures, the company may need to consult with the relevant data protection authority (DPA). The requirements of the GDPR now place greater responsibility on organizations to consider the need for data protection officers (DPOs) to be established. Moreover, the mandatory requirement for notification of breach of the GDPR, within 72h, which places even greater accountability on organizations to interact with and inform data subjects whether their personal data has been potentially breached. This new requirement, arguably, provides a greater level of accountability and to a lesser extent, transparency for the control over personal data on data controllers. In other words, the controller has responsibility, to some extent, for ensuring the personal data is safe and secure, and not easily available for use in criminal activity.

As part of the overall cybercrime framework for the EU, in 2014, the European Parliament adopted a proposal for the (security of network and information systems) SNIS Directive.[23] The SNIS Directive is part of the European Union's Cybersecurity Strategy, which was developed to provide a framework for tackling network and information security incidents and risks across the EU and member states. Key elements of the SNIS Directive include:

- new requirements for 'operators of essential service' and 'digital service providers';
- a new national strategy;
- designation of a national competent authority; and
- designation of computer security incident response teams (CSIRTs) and a cooperation network.[24]

The SNIS Directive also provides the basis for greater consideration of personal data in cybersecurity breaches. Preamble (63) states that personal data in many cases is compromised as a result of security incidents. In this context, competent authorities and data protection authorities should cooperate and exchange information on all relevant matters to tackle any personal data breaches resulting from such incidents. Furthermore, at Preamble (72) the processing of information might require the sharing of information on risks and incidents within the Cooperation

---

[21] Japan signed and ratified the Council of Europe's Convention on Cybercrime 2001, in 2012.

[22] Ibid.

[23] Directive (EU) 2016/1148 of the European Parliament and the Council of 6 July 2016 concerning measures for a high common level of security of network and information systems across the Union. Official journal of the European Union, L 194, 19.7.2016, p. 1–30.

[24] Ibid.

Group and the CSIRTs network and require that the national competent authorities or the CSIRTs be notified. In addition, the data processing should comply with Directive 95/46/EC of the European Parliament and the Council and Regulation (EC) No 45/2001 of the European Parliament and of the Council. In complying this Directive, Regulation (EC) No 1049/2001 of the European Parliament and of the Council should also apply. Preamble (75) states further, that this Directive respects the fundamental rights, and observes the principles, recognized by the Charter of Fundamental Rights of the European Union, in particular the right to respect for private life and communications, the protection of personal data, the freedom to conduct a business, the right to property, the right to an effective remedy before a court and the right to be heard.[25] It is also required that this Directive be implemented in accordance with the rights and principles identified above.[26] It is arguable the tension lies in complying with these directives. On the one hand, there is a need to protect online privacy. On the other hand, there is the need to make available personal data to a point at which a crime can be detected and redressed.

Article 2 of the SNIS Directive states that the processing of personal data pursuant to this Directive shall be carried out in accordance with Directive 95/46/EC. Even though it refers to the GDPR processor, Article 2 remains relevant today. Processing personal data by EU institutions and bodies pursuant to this Directive shall be carried out in accordance with Regulation (EC) No 45/2001.[27] Article 15 requires member states to ensure that the competent authorities have the necessary powers and means to assess the compliance of operators of essential services with their obligations under Article 14 and the effects thereof on the security of network and information systems.[28] Member States shall ensure that the competent authorities have the powers and means to require operators of essential services to provide:

(a) the information necessary to assess the security of their network and information systems, including documented security policies; and
(b) evidence of the effective implementation of security policies, such as the results of a security audit carried out by the competent authority or a qualified auditor and, in the latter case, to make the results thereof, including the underlying evidence, available to the competent authority.[29]

When requesting such information or evidence, the competent authority shall state the purpose of the request and specify what information is required. Following the assessment of information or results of security audits referred to in paragraph 2, the competent authority may issue binding instructions to the operators of essential services to remedy the deficiencies identified. The competent authority shall work in close cooperation with data protection authorities when addressing inci-

[25] Ibid.

[26] Ibid.

[27] Ibid.

[28] Ibid.

[29] Ibid.

dents resulting in personal data breaches.[30] This approach adopted by the EU is likely to enhance the interrelationship between data protection, competition and intellectual property law. Having to consider personal data in cybersecurity allows enforcement agencies to have potentially greater access to information where criminal activity is being undertaken. It can be argued that this approach is likely to be at the cost of the level of privacy protection online that can be provided. However, that level of infringement is not readily understood.

(ii) *Australia*

The *Privacy Act 1988* contains broad extraterritorial application and applies to the overseas activities of Australian organizations and foreign organizations that are linked to Australia.[31] APP 11 requires an organization to take such steps as are reasonable in the circumstances to protect information from misuse, interference and loss; and from unauthorized access, modification or disclosure. In addition, APP 11 extends to taking reasonable steps to protect information that an organization holds against cyberattacks. Details of the APP 11 requirements are provided in Section III.[32]

Section 26WA states that an eligible data breach occurs when there is unauthorized access to, unauthorized disclosure of, or a loss of, personal information held by an entity.[33] Section 26WC of the Privacy Act notes that, if an APP entity has disclosed personal information about one or more individuals to an overseas recipient; and Australian Privacy Principle 8.1 applied to the disclosure of the personal information; and the overseas recipient holds the personal information; this Part has effect as if the personal information were held by the APP entity. In addition, where the APP entity is required under section 15 not to do an act, or engage in a practice, would constitute a breach of the Australian Privacy Principle 11.1 regarding personal information.

APP 6 sets out when an APP entity may use or disclose personal information. An APP entity can only use or disclose personal information for a purpose for which it was collected (the "primary purpose") or for a secondary purpose, if an exception applies. Thus, the reference to disclosure in APP 6 does not extend to unauthorized access. This only highlights the enormous vacuum and the fragmented approach

---

[30] Ibid, Article 2 and 15.

[31] An organization is considered to have links to Australia link when the organization is a company incorporated in Australia, or if the organization carries on business in Australia and collects or holds personal information in Australia.

[32] Australian Privacy Principle 11, An APP entity must take reasonable steps to protect personal information it holds from misuse, interference and loss, as well as unauthorized access, modification or disclosure. Where an APP entity no longer needs personal information for any purpose for which the information may be used or disclosed under the APPs, the entity must take reasonable steps to destroy the information or ensure that it is de-identified. This requirement applies except where: the personal information is part of a Commonwealth record, or the APP entity is required by law or a court/tribunal order to retain the personal information.

Many of the issues discussed in this Chapter are discussed in more detail in the Office of the Australian Information Commissioner's (OAIC) *Guide to securing personal information.*

[33] Privacy Act section 26WA.

that is applied, not only to data protection, but also to protecting personal information (data) in conjunction with cybersecurity.

In addition, the Australian Government, over the past few years have released a number of initiatives to assist in the management of personal information and cybersecurity. For instance, the Protective Security Policy Framework (PSPF) has been developed to assist Australian Government entities to protect people, information and assets, at home and overseas. The PSPF articulates the government's protective security policy. It also provides guidance to entities to support the effective implementation of the policy across the areas of security governance, personnel security, physical security and information security.[34]

In 2018, the Attorney-General reissued the *Directive on the Security of Government Business* (DSGB)[35] to reflect the new PSPF. The directive articulates the government's requirements for protective security to be a business enabler that supports entities to work together securely, in an environment of trust and confidence. The directive establishes the PSPF as a policy of the government, which non-corporate Commonwealth entities are required to apply as it relates to their risk environment.[36] More importantly, the DSGB identifies a number of areas where personal information is to be managed according to the privacy laws, such as ensuring consent has been obtained.[37] With the introduction of the requirement for notification of breaches, there is now an express obligation on entities to notify the Office of the Australian Information Commissioner, of the affected data subjects in the event of an eligible data breach.[38]

(iii) *Malaysia*

The protection of personal data in Malaysia from criminal activity, is multilayered, and similar to other nation states. The Personal Data Protection Act 2010 (PDPA) establishes a cross-sectoral framework for the protection of personal data in relation to commercial transactions which can be expanded to cyber-attacks. The PDPA imposes strict requirements in collecting or processing personal data (see Chap. 7).[39] Following the EU, Malaysia arguably has developed laws to consider its diverse religious and cultural heritage, along with its economic imperatives.[40]

The Commissioner of the Department of Personal Data Protection (the Commissioner), is responsible for the implementation of the Act. It also provides a level of privacy protection to data subjects. Unlike other jurisdiction, Malaysia appears to have, to some extent, embodied the principles of cybersecurity into its

[34] Australian Government, *Protective Security Policy Framework*, https://www.protectivesecurity.gov.au/Pages/default.aspx, accessed 8 November 2018.

[35] Australian government Attorney General Department, https://www.protectivesecurity.gov.au/directive/Pages/directive-security-government-business.aspx, accessed 10 November 2018.

[36] Ibid.

[37] Ibid.

[38] Ibid.

[39] Personal Data Protection Act 2010.

[40] Ibid.

PDPA. In other words, the security principle for data protection, requires that an organization ensure both technical and organizational security measures are established to safeguard the personally identifiable information. The PDP Commissioner has also issued the Personal Data Protection Regulations 2013 and the Personal Data Protection Standard 2015, which together require that data users comply with specific security standards.

Malaysia adopts a co-regulatory approach, similar to Australia. ISO/IEC 27001 Information Security Management System (ISMS), an international standard is used to address information technology systems risks, such as hacker attacks, viruses, malware and data theft. As a result, he ISMS is considered to be the leading standard for cyber risk management in Malaysia.[41] Sectorial legislation for the finance and banking sectors, have also been established to provide further protection for the collection and use of personal data. Furthermore, as Raul points out, the intersection between privacy (data protection) and cybersecurity is also manifest in the extent of tolerance for government surveillance activity, the PDPA does not constrain government access to personal data. For example in law enforcement and in combating terrorism.[42] However, this issue is out of scope of cybersecurity and personal data that is evaluated in this Chapter.

Cybersecurity Malaysia, MyCERT Incident Statistics, estimate that in 2016 alone there were over 8000 reports on cyber-related incidents.[43] It was unconfirmed whether they all pertain to personal data. Nevertheless, the National Cybersecurity Policy is Malaysia's integrated cybersecurity implementation strategy to ensure the critical national information infrastructure is protected to a level that is commensurate with the risks faced.

BNM (Banking Nagara Malaysia) has also issued a circular on 'Managing Cybersecurity Risks', under which financial institutions are required to adhere to the 'Minimum Measures to Mitigate Cyber threats' to:

- assess the implementation of multilayered security architecture;
- ensure security controls for server-to-server external network connections;
- ensure the effectiveness of the monitoring undertaken by Security Operation Centre to view security events, including incidents of all security devices and critical servers on a 24/7 basis; and
- subscribe to reputable threat intelligence services to identify emerging cyber threats, uncover new cyber-attack techniques and provide counter measures.[44]

[41] Ibid.

[42] Ibid.

[43] Mycert Incidents Statistics www.mycert.org.my/statistics/2016.php, accessed 8 November 2016.

[44] Communications and Multi Media Act 1998, sections 231, 233, 234, 235. Personal Data Protection Act 2010. Sectoral regulators such as Securities Commission Malaysia have been actively tackling issues relating to cybersecurity in relation to their relevant sectors by issuing guidelines and setting standards for compliance.

The PDPA does not constrain government access to personal data, as discussed in Section VI. The reasons given to justify broad government access and use include national security, law enforcement and the combating of terrorism.

The legal landscape for cybercrime-security in Malaysia is currently fragmented. CyberSecurity Malaysia was established in 2005, and frequents APEC meetings in relation to data protection and privacy. CyberSecurity Malaysia plays an active role in the economic development in the Asia Pacific region, particularly in the development of data privacy in electronic commerce (electronic commerce development data privacy).[45]

Malaysian courts have already decided cases involving fraud in the use of personal data. In *Basheer Ahmad Maula Sahul Hameed v PP*[46] the two accused persons, were husband and wife in which the wife worked in a bank. The co-accused's were convicted under section 4(1) of the CCA for using a debit card belonging to an airplane accident victim in order to withdraw cash from an ATM machine and to transfer funds from several other victims' online banking accounts without their authorization.[47]

Notwithstanding the above, the issue of identity theft is provided for by law in the jurisdictions studied, including in Malaysia. Section 416 of the Malaysian Penal Code applies to identity theft, and creates an offence to 'cheat by personation'. Identity theft occurs when a person cheats by pretending to be some other person, or by knowingly substituting one person for another, or representing that he or any other person is a person other than he or such person really is. To date, there have been no formal cases reported in Malaysia regarding the theft of identity, even though there have been anecdotal reports that this activity is frequent.

Similar to other jurisdictions, the Malaysian Government has developed a personal data framework that allows its underlying concepts and principles to apply across different areas of the law. As legal developments in Malaysia demonstrate, the complexity in protecting personal data within a broader framework of cybersecurity will continue to challenge nation states. What can be seen in Asian countries, is the establishment of separate Cybersecurity Agencies to provide policy and legal oversight for government.

(iv) *Japan*

In Japan, the Act on the Protection of Personal Information (APPI) primarily handles the protection of data privacy issues. The APP imposes obligations on business operators handling personal information to make and keep accurate records for a certain period when they provide third parties with personal information. In addition, the Act requires business operators handling personal information to verify third parties' names and how they obtained personal information upon receipt of personal information from those third parties.[48] The Act also imposes criminal liability for providing or stealing personal information with a view to making illegal profits. Articles 15, 16, 18, 19 to 25, 27 to 36, 41, 42 (1), 43 and 76 of the Act apply

---

[45] Cybersecurity Malaysia, http://www.cybersecurity.my/data/content_files/46/1634.pdf, accessed 12 November 2018.

[46] *Basheer Ahmad Maula Sahul Hameed v PP* [2016] 6 CLJ 422.

[47] Ibid.

[48] Act on the Protection of Personal Information 2016, Article 25, 26, 83.

to the provision of a good or service to a person in Japan, where the business operator handling personal information has acquired that information, in or from, a foreign country. Furthermore, consent is required for the transfer of personal information to a third party. However, there was no specific provision regarding international data transfers in the APPI. To deal with the globalization of data transfers, the APPI requires consent for the international transfer of personal information.

The business operator handling personal information must also take necessary and proper measures to prevent the leakage, loss or damage of personal data, arguably including theft and cyber-attacks for the purpose of retrieving personal data.[49] The measures of control applied to redress such activities may be systemic, human, physical or technical.[50] Unlike other jurisdiction that require notification of a breach, Japan does not. However, the APPI Guidelines do emphasize the need for action to be taken in response to data breaches, etc. and that they should be described separately from the guidelines. More specifically, the Basic Act on Cybersecurity 2014 provides a framework for cybersecurity throughout Japan. That Act does not mention personal data or personal information. Article 1 provides that, with the intensification of threats against cybersecurity on a worldwide scale, and with the progression of the Internet and other advanced information and telecommunications networks, the use of personal and general information must be managed to ensure that it is not subject to criminal activity.[51] This reference to information being managed is arguably consistent with the definition of personal information as described by the APPI.[52]

Article 7 of the UCAL prohibits phishing, while Article 4 of the UCAL prohibits obtaining any identification code through phishing.[53] These actions are punishable in accordance with Article 12 by imprisonment of up to 1 year, or a fine of up to JPY 500,000.[54] In addition, any person who gains illegal benefits by using identification codes obtained by phishing is subject to imprisonment of up to 10 years under Article 246–2 of the Penal Code. Identity theft is treated in the same way as phishing.

The Cybersecurity Management Guidelines recommend: knowing who should be notified if a cyber-attack has caused any damage; gathering information to be

---

[49] Ibid, Article 20.

[50] Guidelines on Protection of Personal Information in the Employment Management (Announcement No. 357 of 14 May 2012 by the Ministry of Health, Labour and Welfare). Guidelines Targeting Financial Sector Pertaining to the Act on the Protection of Personal Information (Announcement No. 63 of 20 November 2009 by the Financial Services Agency). Guidelines Targeting Medical and Nursing-Care Sectors Pertaining to the Act on the Protection of Personal Information (Announcement in April 2017 by the PCC and the Ministry of Health, Labour and Welfare). General Guidelines regarding the Act on the Protection of Personal Information dated November 2017 (partially amended March 2017).

[51] Basic Act on Cybersecurity 2014, Article 1.

[52] Act on the Protection of Personal Information 2016.

[53] Ibid.

[54] Ibid.

disclosed; and promptly publishing the Incident, taking into account its impact on stakeholders. If the Incident involves any disclosure, loss, or damage of Personal Information handled by a business operator, then, according to the guidelines issued by the Personal Information Protection Committee regarding the APPI, the operator is expected to promptly submit to the PPC a summary of such disclosure, loss or damage, and planned measures to prevent future occurrences. Arguably, this loss of information would also constitute personal data.

(v) *Singapore*

Chapter 8 discussed the Personal Data Protection Act 2012 which provides the framework for protecting personal data in Singapore. Throughout 2017, Singapore had been working, not only to strengthen the governance around personal data, but also, to develop a Cyber Security Act. This new legislation came into effect in 2018 and provides a much needed framework for the governance and oversight of cyber-security. The *Cyber Security Act 2018* establishes four key objectives that include:

- Strengthening the protection of Critical Information Infrastructure (CII) against cyber-attacks. CII are computer systems directly involved in the provision of essential services. Cyber-attacks on CII can have a debilitating impact on the economy and society. The CII sectors are: Energy, Water, Banking and Finance, Healthcare, Transport (which includes Land, Maritime, and Aviation), Infocomm, Media, Security and Emergency Services, and Government; and
- Empowering the Commissioner of Cybersecurity to investigate cybersecurity threats and incidents to determine their impact and prevent further harm or cybersecurity incidents from arising; and
- Establishing a framework for sharing cybersecurity information, and facilitates information sharing, which is critical as timely information helps the government and owners of computer systems identify vulnerabilities and prevent cyber incidents more effectively; and
- Establishes a licensing framework for cybersecurity service providers for penetration testing and managed security operations center monitoring.[55]

Arguably, the Energy, Water, Banking and Finance, Healthcare, Transport (which includes Land, Maritime, and Aviation), Infocomm, Media, Security and Emergency Services, and Government sectors, all deal with personal data and are subject to the PDPA. The legislation does provide a solid foundation for regulating the technology sector, which will complement the PDPA. However, the government is exempt from the provisions of the PDPA.

Moreover, unlike other jurisdictions such as Australia, Indonesia and the EU, Singapore does not require organizations to report breaches in relation to the PDPA. Nevertheless, the Ministry Authority Singapore established a series of guidance notes for the financial sector. The finance sector is very important to the Singapore economy, and deals with large quantities of personal data on a daily

[55] Cybersecurity Act 2018 (No. 9 of 2018), Cybersecurity Agency Singapore.

basis. The *Computer Misuse Act 1993* (CMA).[56] The CMA also strengthened offences for personal information. Section 8 provides criminal penalties where a person uses personal information that is obtained illegally, such as through hacks, to commit or facilitate crimes, such as identity fraud. An example is *Lim Siong Khee v Public Prosecutor*[57], where the accused hacked the victim's email account by answering correctly the hint question to successfully retrieve passwords and to gain unauthorized access. He was sentenced to 12 months' imprisonment.

Under section 4 of the CMA[58], it is an offence to secure unauthorized access to any computer program or data with the intent to commit an offence involving property, fraud or dishonesty. This offence is punishable on conviction by a fine not exceeding SG$50,000, or imprisonment for a term not exceeding 10 years or to both. This offence was tested in *Public Prosecutor v S Kalai Magal Naidu* 226[59], an individual was convicted under section 4 for conducting searches on her bank employer's computer systems to make cash withdrawals from a victim's bank account.[60]

Furthermore, in the case of *Public Prosecutor v Tan Hock Keong Benjamin*[61] it was revealed that an individual found and used another person's debit card to make a number of purchases. The court held that:

> he knew that by doing so, he would cause unauthorized modification to the contents of a computer, namely the data stored in the bank's servers, such that the online purchase would be approved.[62]

Apart from further provisions within the Penal Code, the theft of personal data can also constitute an offence under the PDPA. Section 51 of the PDPA provides that it is an offence for an organization or individual to dispose of, alter, falsify, conceal or destroy personal data. The punishment could result in a fine up to SG$5000 in the case of an individual, and up to SG$50,000 in any other case.

The CMCA, PDPA and Penal Code have extraterritorial application. Section 11 of the CMCA[63] specifies that the CMCA provisions have effect against any person, irrespective of nationality or citizenship, and even if the person is outside or within Singapore, if:

(a) the accused was in Singapore at the material time of the offence;
(b) the computer, program or data was in Singapore at the material time of the offence; or
(c) the offence causes or creates significant risk of serious harm in Singapore.[64]

---

[56] Act 9 of 2018 wef 31/08/2018.

[57] [2001] 1 SLR(R) 631.

[58] Act 9 of 2018 wef 31/08/2018.

[59] *Public Prosecutor v S Kalai Magal Naidu* [2006] SGDC 226.

[60] Ibid.

[61] *Public Prosecutor v Tan Hock Keong Benjamin* [2014] SGDC 16.

[62] Ibid.

[63] Act 9 of 2018 wef 31/08/2018.

[64] Ibid.

Additionally, the extraterritorial application of CMCA offences that cause "serious harm" to Singapore, also forms an important element of its cybercrime-security. This broad provision provides that anyone who targets a computer, program or data located in Singapore commits an offence under the CMCA.[65] The PDPA requires organizations to protect personal data in their possession or under their control, by making reasonable security arrangements to prevent unauthorized access, collection, use, disclosure, copying, modification, disposal or similar risks. If the organization does not comply with this requirement, the Personal Data Protection Commission can give that organization directions to ensure compliance; for example, directing it to pay a penalty of up to SG$1 million.

(vi) *India*

The sectorial approach taken by both India and Indonesia in relation to personal data and privacy has resulted in a lack of specific cybersecurity legislation. Chapters 4 and 5 both states that both India and Indonesia have no specific data protection laws. Therefore, their current laws have not fully adopted the concepts of data portability, assessments, consent and notification of breach, in the same way as other jurisdictions. In India, the IT Act imposes a limited number of regulatory requirements in relation to personal data and cybersecurity.[66] However, the Act does provide options for the data subject and data processor to determine a standard for the protection of personal data. Furthermore, even though India promotes the effective use of co-regulation such as ISO 2700, there has been little attention given to this issue. This is likely to change under the Draft Bill India (see Chap. 5) because it is likely to adopt many of the concepts and principles that the EU GDPR has established. Nevertheless, the Reserve Bank of India[67] has released guidelines in relation to the security of information over electronic banking and cyber fraud. While not comprehensive, they require the use of encryption technology.

Arguably, India does have laws that provide explicit penalties for hacking, identity theft, cyber terrorism, privacy violations and impersonation or publication of obscene material. The Ministry of Communication and Information Technology has established the Computer Emergency Response Team (ERT), which has responsibility for cybersecurity breaches and other malicious activity. The ERT collects, analyses and disseminates information in relation to cyber activity, enabling it to forecast and provide alerts of cybersecurity incidents and their actual or potential impact. However, in the absence of specific data protection laws in India, it is increasingly vulnerable to the loss and theft of its citizen's personal data.

Notwithstanding the above, the Indian IT Act deals extensively with several types of offences that directly or indirectly relate to cybercrimes. Specifically, in

---

[65] Ibid.

[66] Information Technology Act 2000.

[67] Guidelines on Information security, Electronic Banking, Technology risk management and cyber frauds, https://rbidocs.rbi.org.in/rdocs/content/PDFs/GBS300411F.pdf, accessed 10 November 2018.

relation to data protection, sections 72[68] and 72A of the IT Act provide for a level of criminal redress. Section 72 is limited in scope as it prescribes a penalty only against those persons who have been provided delegated power under the IT Act. Section 72A[69] of the IT Act is broader in its scope as it imposes a penalty on any person, whether a private or public entity, for the disclosure of personal information without the consent of the person concerned. Section 72A comes into effect only when a person has secured access to such personal information while providing services under the terms of a lawful contract.[70]

A case concerning the theft of a person's personal identity and information arose *State of Odisha v. Jayanta Das* G.R. There, the court considered sections 66 and 67 of the ITA.[71] It highlighted that sect. 66(C) of Information Technology Act imposes criminal liability on anyone who fraudulently and dishonestly makes use of the electronic signature, password or any other unique identification feature of any person. The court argued that:

> identity theft means the phenomenon of filing another person identity, and is one of the fastest growing sector of crime in the world. Furthermore, in considering section 67 of I.T. Act which provides the Commission of a person who publishes or transmit or possessed to the public pornographic or obscene materials in electronic form – sending offensive e-mail postings containing defamatory messages.[72]

The court said that:

> section 67(A) of Information Technology Act provides the punishment for a person who publishes or transmitted in the electronic form any material which contains sexually explicit act or conduct.

The court issued a penalty of 6 years' imprisonment and a fine on charges of forgery, identity theft and cyber pornography for creating a fake profile on a pornographic website in the name of the complainant's wife.

---

[68] Information Technology Act 2000, section 72, IT Act states that save as otherwise provided in this Act or any other law for the time being in force, if any person who, in pursuance of any of the powers conferred under this Act, rules or regulations made thereunder, has secured access to any electronic record, book, register, correspondence, information, document or other material without the consent of the person concerned discloses such electronic record, book, register, correspondence, information, document or other material to any other person shall be punished with imprisonment for a term which may extend to2 years, or with fine which may extend to one lakh rupees, or with both.

[69] Ibid, provides that save as otherwise provided in this Act or any other law for the time being in force, any person including an intermediary who, while providing services under the terms of lawful contract, has secured access to any material containing personal information about another person, with the intent to cause or knowing that he is likely to cause wrongful loss or wrongful gain discloses, without the consent of the person concerned, or in breach of a lawful contract, such material to any other person, shall be punished with imprisonment for a term which may extend to 3 years, or with fine which may extend to five lakh rupees, or with both.

[70] Ibid.

[71] *State of Odisha v. Jayanta Das* G.R. Case No. 1739/2012 T.R. No. 21/2013.

[72] Ibid.

Finally, under section of the ITA, a violation of privacy by intentionally or knowingly publishing/transmitting a private image of a person without his/her consent is punishable by imprisonment of up to 3 years, or a fine of up to INR 200,000, or both penalties. Also, under section 72A of the ITA, the disclosure of personal information obtained while providing contractual services, with the intent/knowledge that wrongful loss/gain will result, is punishable with imprisonment of up to 3 years, or with a fine of up to INR 500,000, or both.

(vii) *Indonesia*

Across the Indonesian archipelago, the sectorial approach to data protection is very similar to that of India. The cybersecurity laws there are sectorial and dispersed according to the industry sector, such as, telecommunication, banking and finance. However, and even though Indonesia is developing specific data protection laws, its Electronic Information and Transactions Law No. 11 of 2008 (EIT) has some provisions that deal with cybercrime. These include, for example, identify theft, hacking, denial of server attacks, phishing and breach of copyright amongst others.[73] Article 2 provides for the extraterritorial scope of the EIT, whereby a person can face criminal proceedings where their actions outside Indonesia pose a threat to the interests of Indonesia. However, it is unclear whether this extends to personal data, and whether Indonesia would consider stolen personal data from the someone in Indonesia as being in its interests. That may well be the case where that data is combined with other national interest data and information.

Chapter 5 highlights how the EIT is supported by the Ministry of Communications and Information [MOCI] Regulation. Regulation 82 and MOCI Regulation, which amongst other provisions, guarantees the confidentiality of the source code of the software to ensure agreement on minimum service level and information security. Additionally, Regulation 82 further provides confidentiality of the information technology services being used, as well as security and facility of internal communication security. There is a requirement for privacy and personal data protection of users to ensure that the appropriate lawful use and disclosure of the personal data is undertaken. Another element in this process is ensuring that the data subject has provided a level of consent, as long as it is related to the purpose of obtaining and collecting personal data.

Notification of a breach is also part of the current legal framework, although different to other countries. Article 15 (2) provides that the provider of an Electronic System must give written notification to the owner of personal data, upon its failure to protect the personal data. In addition, Article 20 (3) provides that the provider of an Electronic System must make the utmost effort to protect personal data and to immediately report any failure/serious system interference-disturbance to a law enforcement official or the Supervising and Regulatory Authority of the relevant sector. Furthermore, Article 28 (c) of the MOCI Regulation provides that a written notice to the Personal Data Owner is required if there is a failure in protecting the secrecy of the personal data. Additionally, the Electronic System and must protect

[73] Electronic Transactions Law No. 19 of 2016, Articles 2, 4, 5, 30, 35.

the secrecy of personal data and it can be conducted electronically, provide the data subject has provided consent. This arguably includes cyber-attacks.

To that end, in 2018 the Indonesian Government established a Cybersecurity Agency, and will focus on tracking cybercrimes and identify perpetrators.[74] However the extent of responsibility of that agency remains unclear, largely because of the general focus given to it in the lead up to the future elections. Nevertheless, the establishment of that agency, is consistent with other national governments recognizing the need to tighten controls and regulations in order to respond to t current and future threats posed by cyber technology to personal data and data generally.

(viii) *Thailand*

Although, the current framework is somewhat diluted because of the lack of personal data laws. Since 2007, the Computer Crime Act 2007 ("CCA") has been in operation. The Thailand Penal Code, B.E. 2499, 1956 also plays a critical role in providing penalties and controls over cybersecurity and crime.[75]

Section 5 of the CCA deals with hacking. It provides that, whoever illegally accesses a computer system that has specific security measures and such security measures are not intended for that person's use, is liable to imprisonment not exceeding 6 months, or to a fine not exceeding THB 10,000, or both. Furthermore, section 7 of the CCA provides that, whoever illegally accesses computer data that has specific security measures which are not intended for that person's use, is liable to imprisonment not exceeding 2 years, or to a fine not exceeding THB 40,000, or both.

Section 14 of the CCA is one of the most important provisions in the Act, because it places some controls over the selling of forged information. It provides that a person must not engage in selling forged electronic cards, or dishonestly, or deceitfully inputting into a computer system computer data which is distorted or forged, either in whole or in part, or computer data which is false, in such a manner that is likely to cause injury to the public, but which does not constitute a crime of defamation under the Criminal Code. Furthermore, it prohibits the inputting into a computer system of data which is false, in such a manner likely to cause damage to the maintenance of national security, public safety, national economic security, or public infrastructure serving national public interest, or to cause panic amongst the public. It also prohibits information and data that is considered to be vulgar from being inputted into systems and published.

In relation to identity theft or identity fraud, including personal data and information, the current framework provides no specific offence. However, identity theft/fraud is considered as the act causing damage to the computer data of another person under section 9 of the CCA. In addition, section 16 of the CCA provides that, whoever inputs into a publicly accessible computer system data that will appear as an image of another person and that the image has been created, edited, appended

---

[74] Chisholm, J Indonesia launches cyber agency to combat country's extremism and fake news, http://sea-globe.com/indonesia-cybersecurity/, accessed 8 November 2018.

[75] Thailand Penal Code, B.E. 2499, 1956.

or adapted by electronic means or whatsoever means, and in doing so is likely to impair the reputation of such other person or exposes such other person to hatred or contempt, would be liable to imprisonment not exceeding 3 years and a fine not exceeding THB 200,000, or both. Section 14(1) of the CCA provides that, whoever dishonestly or deceitfully inputs into a computer system computer data which is distorted or forged, either in whole or in part, or computer data which is false, in such a manner that is likely to cause injury to the public (not defamation) under the Penal Code, would be liable to imprisonment not exceeding 5 years and not exceeding THB 100,000, or both (phishing).

The fragmented approach taken towards personal data and data protection in Thailand, makes it difficult to determine whether the above-mentioned provisions of key Thai law will provide a level of security and safeguards for personal data. It is our view that the current approach goes some to including personal data. The law also consistently refer to computer system-computer data. Computer data, arguably, includes any and all forms of data that computer systems collect and use, including personal data.

In 2018, Thailand became the destination of choice to establish the ASEAN -Japan Cybersecurity Capacity Building Centre (AJCCBC) which is scheduled to open in Bangkok.[76] This center will play a vital role in mitigating the regulation of cybercrime across the region. The center is projected to develop a cybersecurity workforce, particularly in ASEAN governmental agencies and in the Computer Emergency Response Team (CERT) in each of the ASEAN countries in order to enhance cybersecurity awareness, strengthen information security and data protection, as well as promote information sharing.[77] While taking a broad approach to the deep issues associated with cybercrime-security, personal data, will likely form a component of its work.[78] This Center will complement the current work and support ThaiCERT in handling computer security incidents across Thailand.

15.2 Conclusion

Personal data protection and privacy, while not often studied as part of cybercrime-security, is increasingly becoming an important part of government policy. The rise in computer hacking and incursions into Internet technology, is leading to more and more cases in which people's personal data being compromised. In some cases, this has resulted in identity fraud and theft, and the abuse of personal information defined by the law to establish market dominance.

[76] Toomgum, S *Cybersecurity Centre on way,* The Nation, http://www.nationmultimedia.com/detail/Startup_and_IT/30342035, accessed 10 November 2016.

[77] Ibid.

[78] Ibid.

As personal data heads towards gaining property rights and intellectual property rights, that data can also be stolen and misused in cybercrime activities. Moreover, the ease with which digital media may be shared has led to an explosion in, for example, copyright infringements. In addition, as personal data becomes more tradeable, and with the high profits already made from this data, there are incentives for individuals and entities to steel that data. Today, most people depend on the use of a computer and the Internet to go about their daily lives. This technology is creating opportunities for them to gain access to information, whether general or personal. A risk in securing such access is that the use of personal data can cause significant disruption and damage (socially and economically) to the state, business or an individual data subject.

Governments around the world and in the jurisdictions discussed in this book are developing strategies to combat cybercrime and increase security online. The evolution of data protection law is beginning to reflect growing concerns surrounding the misuse of computers and its infrastructure to unauthorized access to private information. These have expanded well beyond traditional concern that computers could be used to commit economic crimes.

However, as technology advances at such a rapid rate, governments are lagging behind in addressing current and future cybercrimes in this growing area. There is need to better understand where the gaps and slippage is for data breaches through what infrastructure and systems that support the Internet. This includes, but is not limited to, machine learning, analytics, algorithms, devises that are used to connect systems (routers and others), and autonomous systems (motor vehicles, medical equipment, credit/debit cards and house hold appliances). These areas pose a significant risk to the loss of personal data.

The consequences of a cyber-attack are severe and cannot be underestimated. They can cause significant economic and social disruptions that are likely to result in the loss of, or, damage to data (both commercial and personal) and damage to the reputation to individuals and entities. Therefore, the value of having a robust integrated cybersecurity and data protection policy and legal framework is likely to become ever more important in the year ahead, should the international community agree that protecting personal data is both important economically and socially.

The intersection of personal data, privacy, human rights and cybercrime-security poses many challenges to business and government. These areas are subject to large scale abuse. The lack of comprehensive checks and balances only heightens the need for greater certainty in markets, and more pervasive enforcement policy by governments. As with many other areas of data protection and privacy law, global law and policy on regulating cybercrime is also fragmented. The result is that individuals and entities are able to breach Internet infrastructure and systems for the purpose of obtaining and using personal data criminally. This fragmented approach over criminal liability for violations of data protection and privacy law is unlikely to be resolved anytime soon. The unfortunate result is the likelihood that the abuse of personal data will increase, not decrease. The call is for greater legal convergence and harmonization to redress the serious deficiencies in the regulation of cybercrime.

References

Clough, J Principles of Cybercrime, *Faculty of Law, Monash University* Cambridge University Press, New York (2010)

Holt, T., Bossler, A (2014) *An assessment of the current state of cybercrime scholarship. Deviant Behavior* 35(1) pp. 20–40 DOI:https://doi.org/10.1080/01639625.2013.822209.

Hollywood., J Michael., J. Vermeer., M, Woods., D, Goodison., S, Jackson, B Using Social Media and Social Network Analysis in Law Enforcement, Creating a Research Agenda, Including Business Cases, Protections, and Technology Needs, https://www.rand.org/content/dam/rand/pubs/research_reports/RR2300/RR2301/RAND_RR2301.pdf.

Stratton G., Powell A., Cameron., R (2017) *Crime and Justice in Digital Society: Towards a 'Digital Criminology'?* International Journal for Crime, Justice and Social Democracy 6(2): pp. 17–33

Part VI

Chapter 16
International & Regional Institutions

Abstract This Chapter briefly discusses the international, regional frameworks and institutions that currently deal with data protection and privacy. Arguably, the popularity of not only the Internet but also the portable devices such as the Iphone, laptop computers, Ipads, home security and camera systems, even televisions and other household appliances have now become even smarter. In other words, these modern day devices can all access the Internet. Many of these devices travel with the data subject across international borders daily. Moreover, they have created a new living environment or life structure in which personal data become highly valuable in the market economy and critical for personal development, that is no longer national or regional, but international. The internationalization of these devices has forced the internationalization of privacy as it pertains to personal data that, is collected and used over the Internet.

However, the international response is fragmented, adhoc, incoherent and lags along way behind the developments in technology. This Chapter does not compare how the international and regional frameworks have or have not been applied by each country. Rather, beginning with the United Nations, this Chapter looks briefly at those regional institutions that are developing data protection policy, such as APEC, ASEAN, OECD, ICDPPC, WEF and the Commonwealth of nations. This Chapter also examines the gaps that exist at the international level, in this area of law and policy. This Chapter explores what the United Nations has been doing in the area of data protection and privacy. These international and regional frameworks and institutions have been and continue to be influential in guiding and directing how nation states develop their data protection and privacy laws. Finally, this Chapter will briefly highlight how there is an opportunity for the use of Trade Agreements to assist in managing personal data, protecting that data along with privacy.

© Springer Nature Singapore Pte Ltd. 2019

R. Walters et al., *Data Protection Law*,
https://doi.org/10.1007/978-981-13-8110-2_16

16.1 Introduction

Globalization has enabled the movement of people, goods and services across international borders with ease. Combine globalization with technology and there are no international borders. Technology (the Internet) does not recognize the Australian border or likewise the borders of India, Indonesia, Japan, Malaysia, Singapore, Thailand or the European Union. Data flows (both personal and commercial) in large quantities to and from these countries daily. The exploitation of personal and commercial data by the private and some public sectors has caused widespread concern regionally and internationally. Thus, there is a growing need for the legal rules to protect the processing of personal and commercial data to form part of international and regional laws. The international and regional legal frameworks are not consistent. As stated in Chap. 1, the EU is effectively dragging other countries and regions to adopt their principles and legal framework, if countries want to participate with and across the EU.

There are many regulatory conflicts throughout the Asia, Pacific and European Union regions, nationally and regionally regarding personal data and privacy protection. At all three levels data protection and privacy laws remains very fragmented. This will, if not addressed, over time, create uncertainty to governments, industry, investors because individuals may perceive the current regulatory being not secure enough to facilitate safe and effective transfer and trade of commercial and personal data. It also places minors who are the largest users of the Internet in potentially compromising positions. This Chapter will explore the current international and regional approach to data protection, and subsequently privacy. It begins by discussing and highlighting how data protection in the international sphere, arguably, begins with the united Nations.

16.2 International Law and Regional Programs

The international community understands the extent of the issues that have currently surfaced and will continue to pose challenges across the world in regards to data protection. However, to date there is no single Treaty or Convention that deals with personal data, in the same way as many other international issues. Moreover, there is no clear Model Law that has been prepared that can guide not only nation states but also the private sector in the management and use of personal data (see Chap. 17). Arguably, this is because of the varied nature in which data is collected, collated, used, stored and transacted across the public and private sectors. Rather, the international organizations that have been involved in data protection have focused their efforts on developing, concepts, principles and guidance for countries to consider whether to include them into their national laws. However, the question arises are these adequate enough? Beginning with the United Nations, it is well understood that privacy has, for decades been part of the international legal framework. Thus, data protection, being a more recent phenomenon, has only been a relatively new consideration.

16.3 United Nations

The United Nations or the United Nations Conference on Trade and Development (UNICAD), is responsible for policy-oriented analytical work on the development implications of information and communication technologies (ICTs). Secondly, it is responsible for the preparation of thematic reports on ICT for development and promotes international dialogue on issues related to ICTs for development, to measure the information economy and to design and implement relevant policies and legal frameworks. However, the UNICAD has to balance the need of development, with the broader social policy needs of human rights in all areas of policy, including economic development.

The protection of privacy under the international legal framework dates back to just after world war two (WWII). Article 12 of the Universal Declaration of Human Rights 1948 states that 'no one shall be subjected to arbitrary interference with his privacy, family, home or correspondence, nor to attacks upon his honour and reputation. Everyone has the right to the protection of the law against such interference or attacks'. Article 17 of the International Covenant on Civil and Political Rights 1966, and to date, is ratified by 167 States, provides that no one shall be subjected to arbitrary or unlawful interference with his or her privacy, family, home or correspondence, nor to unlawful attacks on his or her honour and reputation. It further states that everyone has the right to the protection of the law against such interference or attacks.

In 1975, Declaration on the Use of Scientific and Technological Progress in the Interests of Peace and for the Benefit of Mankind was established.[1] Article 6 of the Declaration states that all States shall take measures to extend the benefits of science and technology to all strata of the population and to protect them, both socially and materially, from possible harmful effects of the misuse of scientific and technological developments, including their misuse to infringe upon the rights of the individual or of the group, particularly with regard to respect for privacy and the protection of the human personality and its physical and intellectual integrity.[2] Kinfe Micheal Yilma believes that the Declaration sought to absolve states and their

---

[1] Micheal Yilma K *The United Nations data privacy system and its limits*, International Review of Law, Computers & Technology, (2018).

[2] Proclaimed by General Assembly Resolution 3384 (XXX) of 10 November 1975.

agents from any responsibility for violation of digital privacy. The emphasis on sovereignty and international peace and security is also indicative of the political tone of the Declaration.

Between 2013 and 2015, the UN strengthened its role in privacy protection through the publication of a statement on Digital Rights, and the establishment of a Special Rapporteur for the right to privacy. In December 2013, Resolution 68/167, was adopted by the United Nations General Assembly, expressing concern for the negative impact that surveillance and interception of communications may have on human rights.[3]

By July 2015, the Human Rights Council appointed the first- ever Special Rapporteur on the right to privacy. This has been one of the most important appointments regarding privacy and data protection under the UN. The Special Rapporteur is mandated to:

(a) gather relevant information, including on international and national frameworks, national practices and experience, to study trends, developments and challenges in relation to the right to privacy and to make recommendations to ensure its promotion and protection, including in connection with the challenges arising from new technologies;
(b) seek, receive and respond to information, while avoiding duplication, from States, the United Nations and its agencies, programmes and funds, regional human rights mechanisms, national human rights institutions, civil society organizations, the private sector, including business enterprises, and any other relevant stakeholders or parties;
(c) identify possible obstacles to the promotion and protection of the right to privacy, identify, exchange and promote principles and best practices at the national, regional and international levels;
(d) raise awareness concerning the importance of promoting and protecting the right to privacy, with a focus on particular challenges arising in the digital age, consistent with international human rights obligations;
(e) integrate a gender perspective throughout the work of the mandate;
(f) report on alleged violations of the right to privacy, set out in article 12 of the Universal Declaration of Human Rights and article 17 of the International Covenant on Civil and Political Rights; and
(g) submit an annual report to the Human Rights Council and to the General Assembly.[4]

However, there appears to be little reference to harmonization or convergence of principles, law, policy and rules for personal data protection. Moreover, Kinfe Micheal Yilma highlights significant limitations to the current UN framework in rela-

---

[3] United Nations, Resolution adopted by the General Assembly on 18 December 2013, 68/167. *The right to privacy in the digital age*, http://www.un.org/ga/search/view_doc.asp?symbol=A/RES/68/167, accessed 20 March 2018.

[4] Human Rights Council Resolution 28/16.

tion to data privacy. The two major limitations are normative[5] and institutional.[6] That is, the current data privacy norms are ad hoc and scattered across various UN binding and nonbinding instruments. Moreover, it is argued that there is no single dedicated institution within the UN that has sole responsibility for data protection or privacy.

16.4 Organization for Economic Development [OECD]

Arguably, the OECD has been the most active when it comes to data protection. The OECD had developed the *Guidelines on the Protection of Privacy and Transborder Flows of Personal Data* ("the Personal Data Guidelines") as far back as 1980. These were recently revised in 2013. This is further underpinned by the OECD Principles for Internet Policy Making 2011, which recognizes that supporting the free flow of data needs to be achieved in the context of these other goals, stating that, "while promoting the free flow of information, it is also essential for government to work towards better protection of personal data, children online, intellectual property rights, and to address cyber-security".[7]

Currently, there are only 35 members of the OECD. In the context of this book, Australia and most member states of the EU are members of the OECD. However, there are no countries from Asia or Central Asia who are members. Apart from anything else, it could be time as the economic landscape evolves from industrial to digital that ASEAN and central Asian countries to consider membership to the OECD. Even though the OECD has a limited membership, this would not stop nation states in adopting the principles set out in the Personal Data Guidelines. It is argued that even though Singapore, Japan and Malaysia are not members, the principles and concepts found in the OECD Personal Data Guidelines have found their way into their respective data protection laws, to vary degrees. Even though Indonesia and India are in the process of developing their specific data protection or privacy laws, and to a lesser extent their current sectorial law, they are both considering, and do have elements of the OECD principles and concepts within their current legal frameworks.

Importantly, the Personal Data Guidelines provide a number of principles that recognize more extensive and innovative uses of personal data bring greater eco-

[5] Micheal Yilma K *The United Nations data privacy system and its limits*, International Review of Law, Computers & Technology, (2018). Normative limitations include the 'problem of normative dispersion' in that existing data privacy norms are a mere patchwork of rules dispersed across various instruments that lessened their accessibility and hence effectiveness. Secondly the existing data privacy norms are embodied in an exceeding set of soft law instruments. Thirdly, the existing UN data privacy rules are normatively inferior to their regional counterparts.

[6] Ibid, Institutional limitations include the lack of clear institutional arrangements. Owing to the dispersed nature of existing norms, the monitoring responsibility falls on various bodies. The institutional arrangement for the monitoring of existing data privacy rules is as fragmented as the rules are. The other institutional weakness is that these various bodies tasked to monitor scattered data privacy rules lack the requisite enforcement powers to oversee the implementation of those rules.

[7] OECD Council Recommendation on Principles for Internet Policy Making, 13 December 2011, p. 4–8.

nomic and social benefits, but also increase privacy risks.[8] The principles, while high level can be found in most national legislation frameworks and include collection, data quality, purpose, limitation, security safeguards, openness, participation, accountability, restrictions and national implementation.

The principles specify that a data controller be established to ensure a central point of contact within an organization has accountability on behalf of the organization for the management and implementation of legislative obligations. One of the key principles is to ensure that the flow of data across international borders is not restricted to a point of diminishing economic activity. National implementation is the key to balancing the needs of business, the economy and protecting individuals, and the Personal Data Guidelines assist national governments to consider and develop robust strategies by adopting relevant laws and ensuring appropriate enforcement is undertaken. The core principles of the Personal Data Guidelines include:

- Collection Limitation;
- Data Quality;
- Purpose Specification;
- Use Limitation;
- Security Safeguards;
- Openness;
- Individual Participation; and
- Accountability.[9]

In addition, the OECD has identified that limitation restrictions and international cooperation are also important to the overall legal framework. However, the OECD Personal Data Guidelines has its problems. Michael Kirby points out that the Personal Data Guidelines have achieved four results including building on predecessor guidelines, *added value to strengthening data protection and privacy, allow for flexible implementation, and survival of the guidelines.*[10] Moreover, and while

---

[8] OECD *Guidelines on the Protection of Privacy and Transborder Flows of Personal Data,* https://www.oecd.org/sti/ieconomy/2013-oecd-privacy-guidelines.pdf, accessed 5 December 2017.

[9] *Guidelines on the Protection of Privacy and Transborder Flows of Personal Data* (2013).

[10] Kirby, M *The history, achievement and future of the 1980 OECD guidelines on privacy International Data Privacy Law*, Volume 1, Issue 1, 1 (2011) p. 6. *Building on predecessors*: by not set out to reinvent the wheel or needlessly to alter sensible approaches that had been adopted by our predecessors. *OECD value added*: There were at least seven features of the Guidelines that constituted the 'value added' that the OECD offered in its project: (1) The Guidelines were expressed in technologically neutral terms. (2) The Guidelines were expressed as non-binding. (3) There was also a broad ambit. (4) The Guidelines acknowledged the value of TBDF in itself. (5) The OECD Guidelines added the 'accountability principle. (6) The Guidelines also called on the OECD member countries to implement the principles and to cooperate with other member countries in such implementation so that gaps would not arise in the operation of the Guidelines between different nation states. (7) Above all, the simple conceptual language of the Guidelines strengthened their influence in the succeeding years. *Flexible implementation*: A key to the success of the OECD Guidelines is the way in which they envisage that national implementation will follow their own regulatory cultures. This had been a large potential obstacle standing in the way of success

Kirby wrote in 2010 when the GDPR and 2013 OECD Personal Data Guidelines had not been implemented, he believes that the future Realism, Protecting privacy, Importance of empiricism, Reconceptualizing issues and New challenges, that will be faced by the community and governments, will be formidable.[11] They will require cross-border collaboration in the same way as international trade and finance, which has provided many benefits to the international community. There has been slow progress to realize and implement these concepts into national and supranational law. However, the world is far from being consistent in their approach, as many nations are no where near establishing these concepts.

Omer Tene believes that the OECD Personal Data Guidelines require further work as they continue to be firmly rooted in principles and laws dating back to the age of punch cards and mainframe computers.[12] Specifically, they fail to address challenges to the definition of personal data and science of de-identification; relying on individuals' consent to legitimize processes far removed from individuals. In addition, they fall short in framing data collection and use as a linear process despite the explosion of user-generated content. Tene argues that the Personal Data Guidelines insist on framing data flows in a geographical context while disregarding the effervescent nature and rapid movement of data across international borders. They lack a coherent model for privacy harms, which would allow policymakers to tailor appropriate responses. These concerns are further heightened when unlike

---

because of the concern in non-European countries about what they saw as the expensive and intrusive bureaucratic tradition of European data protection. *Survival of the Guidelines*: Against this background, the survival of the Guidelines, and their continuing utility 30 years later is remarkable but perhaps understandable. In that 30 years, we have seen the development of the Internet and World Wide Web; of search engines; of technology for location detection; of social networking which challenges the very concept of what is 'private' and what is secret; of biometrics and other technologies.

[11] Ibid, *Realism:* It is important to tackle issues presented to information, computer, and communications policies with realism. Personal data should not be disclosed, made available or otherwise used for purposes with consent or by the authority of law. *Protecting privacy:* having acknowledged the inevitability of some erosion of aspects of personal control over data and individual privacy, it is important not to give up on protection of this value. *Reconceptualizing issues:* to some extent, in the decades since the OECD Personal Data Guidelines were adopted, policy developments have been confined to particular areas of information, computer, and communications policy. *New challenges:* many new challenges face any organization that is addressing computer and communications policy today. Some of these challenges include: (a) the development and implementation of new systems of mass surveillance, including facial recognition, whole body imaging, biometric identifiers, and imbedded RFID tags, which the *Madrid Declaration* suggested should not be implemented at all without 'a full and transparent evaluation by independent authorities and democratic debate'; (b) Privacy protectors must ever be on the lookout for privacy enhancing technology (PET) and the ways in which such technology itself can be invoked to afford more effective and efficient privacy protection for the individual; (c) Cross-border cooperation in drafting, implementing, and enforcing laws for privacy protection is a daily challenge but one that is already attracting responses. (d) End-user education may be necessary to sustain community awareness about the value of privacy.

[12] Tene O (2013) *Privacy Law's Midlife Crisis: A Critical Assessment of the Second Wave of Global Privacy Laws*, 74 OHIOST.L.J.1217, 2013, p. 1222.

many other industries, the rapid change and transition in technology that will capture and use data continues to outpace regulators and policy makers.

The Personal Data Guidelines promote intergovernmental collaboration and cross border co-operation to strengthen information sharing of transborder flows of personal and commercial data. Countries, whether a member of the OECD or not are encouraged to utilize the work undertaken by the OECD for data protection law and policy. They have influenced the development of national data protection legislation and model codes within the OECD member countries. The Personal Data Guidelines have also influenced the development of the APEC Privacy Framework, expanding their reach beyond the OECD membership. They have been influential in most countries in the development of privacy legislation throughout the world, including Australia.[13]

16.5 International Conference of Data Protection and Privacy Commissioners [ICDPPC]

In 2017, there were 192 Member States of the United Nations, while privacy and data protection authorities from approximately 79 states are accredited to the International Conference of Data Protection and Privacy Commissioners (ICDPPC) as at September 2017.[14] Australia and most, if not all member states of the EU, including the Union itself have accredited authorities with the ICDPPC. It is noteworthy that no ASEAN members of other Asian or Central Asian countries have membership, authorities or commissioners that form part of the ICDPPC.

At the 2017 meeting in Hong Kong, the ICDPPC recognized that automation will bring many benefits to society. However, at the same time they noted from the G7 Ministers meeting in June of the same year, the necessity to follow relevant existing guidelines on cyber security and data protection.[15] The ICDPPC called upon standardization bodies, public authorities, vehicle and equipment manufacturers, personal transportation services and car rental providers, providers of data driven services, such as speech recognition, navigation, remote maintenance or motor insurance telematics services, to fully respect the users rights to the protection of their personal data and privacy.[16] This is another example where, personal data and information relating to individuals can be collected, stored and used within the computer systems of private, leased or rental motor vehicles that could fall into the hands of others. The ongoing work undertaken by this group may go some way to legal convergence and harmonization.

---

[13] Greenleaf G *Global Data Privacy Laws: 89 Countries, and Accelerating'* Privacy Laws & Business International Report, Issue 11–175 (2012).

[14] International Conference of Data Protection and Privacy Commissioners, https://icdppc.org/participation-in-the-conference/list-of-accredited-members/ accessed 5 December 2017.

[15] 39 International Conference of Data Protection and Privacy Commissioners Hong Kong, 25–29 September 2017.

[16] Ibid.

16.6 International Law Commission [ICL] – Associations and Organizations

The fragmented approach to data protection and privacy law nationally, regionally and internationally poses many challenges going forward. The ILC has stated that:

> 'the international binding and non-binding instruments, as well as the national legislation adopted by nation states, and judicial decisions reveal a number of core principles' of data protection.[17]

Data protection is an area 'in which state practice is not yet extensive or fully developed', and the Statute of the ILC suggests that codification should take place in fields where there has already been extensive State practice, precedent and doctrine.[18] In addition, work in the area of data protection 'may nevertheless be able to identify emerging trends in legal opinion and practice which are likely to shape any international legal regime which would develop.[19] Therefore, more coordinated work is needed to collaborate on issues that are both in the national, commercial and private interest. Steps have been taken at the regional level to harmonies and develop a consistent framework for the management of data.

There are a number of international organizations that assist and provide government and policy guidance to industry and government on data protection, systems and networks. The importance of understanding the different associations and organizations, also help to formulate future government and industry co-regulation.

The International Organization for Standardization (ISO) systems and standards are encouraged by the GDPR, for organizations to adopt certified schemes. The ISO 27001 provides oversight of three key areas (1) the security regime, (2) the people and (3) the processes and technology. This scheme automatically requires organizations to undertake a comprehensive risk assessment. Secondly, the International Organization of Securities Commissions, provides a policy role for the implementations of standards to prepare capital markets for a larger role in financing economic growth. Committees have been formed for enforcement, data, asset management, bond market liquidity, market conduct, corporate governance, audit quality, long-term financing of small and mid-sized enterprises and infrastructure, and investor protection and education as a means to strengthen investor confidence and create the conditions for sustainable economic growth. This is particularly focused at the banking and finance sector. Since November 2014, the Committee on Payments and Market Infrastructures (CPMI) and International Organization of Securities Commissions (IOSCO) Harmonization Group has worked to develop guidance regarding the definition, format and usage of key OTC derivatives data elements reported to trade repositories (TRs), including the Unique Transaction Identifier (UTI), the Unique Product Identifier (UPI) and other critical data elements.

[17] International Law Commission Report, Annex D, para. 11.

[18] Ibid.

[19] Ibid.

Technical Guidance on the Unique Transaction Identifier (UTI) was published in February 2017 and Technical Guidance on the UPI will be published in Q3 2017.[20]

16.7 World Economic Forum

The World Economic Forum (WEF) argues that the traditional approach to data protection needs to be revised to react recent technological evolutions that have given way to technologies such as Bigdata and blockchain.[21] The traditional approach focuses on the individual's 'consent' at the time of collection and is appropriate when the collected data was used for a specific purpose, and deleted when no longer needed. This approach to data collection fails to account for unforeseen uses of data long after the time of collection, and relies on unrealistic expectations regarding the data subject's ability to protect their privacy. The concept of concept is complex, and in our view, it has not been fully reconciled. While out of scope of this book, there are some in the community arguing that people are becoming consent fatigued. This is a valid position to take. However, others are arguing that consent has not been addressed adequately within national laws. It appears, at this early stage that, consent only really applies to when the data subject provides their first consent. After that, the entity can do whatever they like with the data. This is problematic and needs to be addressed. Even so, the WEF has stated that establishing principles can serve as a global foundation for creating an interoperable, flexible and accountable framework for coordinated multi-stakeholder action. Codes of conduct, technological solutions and contract law can all help translate principles into trustworthy practices that enable sustainable economic growth.[22] This is an important point because, it relies on industry taking responsibility and self or co-regulating with governments to solve important issue. To some extent, most jurisdictions discussed in this book have adopted this roadmap, to varying degrees. The exceptions are India, Indonesia and Thailand.

The WEF went onto to say, that the traditional approaches are no longer fit for the purposes for which they were designed, which was based on the 1970's technology. They fail to account for the possibility that new and beneficial uses for the data will be discovered, long after the time of collection. The current framework does not account for networked data architectures that lower the cost of data collection, transfer and processing to nearly zero, and enable multiuser access to a single piece of data. The torrent of data being generated from and about data subjects imposes an undue cognitive burden on individual data subjects. Many circumstances, as when 'automation' becomes fully operational, is no longer practical or effective to gain the consent of individuals using traditional approaches (see Chap. 17).

---

[20] International Organization of Securities Commissions, Harmonization of critical OTC derivatives data elements (other than UTI and UPI) – third batch.

[21] World Economic Forum White Paper Digital Transformation of Industries: In collaboration with Accenture, Digital Enterprise, (2016), http://reports.weforum.org/digital-transformation/wp-content/blogs.dir/94/mp/files/pages/files/digital-enterprise-narrative-final-january-2016.pdf, accessed 22 November 2016.

[22] Ibid.

16.8 Regional Programs

16.8.1 Asia-Pacific Economic Cooperation [APEC]

APEC is composed of 21 member economies that together represent approximately 55% of the world's GDP, 44% of world trade and 41% of the world's population.[23] Australia along with many Asian countries such as China, Indonesia, Japan, Malaysia, Singapore and Thailand are members. A notable absentee from this club is the emerging economy of India. It is our view that one of the major benefits of the APEC community is the very broad membership that includes Russia, China and the United States, Canada and New Zealand. Having the support and involvement from these major economies provides greater flexibility, and many opportunities for harmonization of laws and rules to occur in regards to data protection and privacy.

APEC has developed several recent data protection initiatives. The three key initiatives include (1) the development of a set of common APEC Privacy Principles; (2) the development of a system for coordinating complaints that involve more than one APEC jurisdiction; and (3) the development of the Cross-Border Privacy Rules system (CBPRs). The APEC CBPR system is an innovative self-regulatory mechanism for allowing the transfer of data between APEC members where a company has voluntarily joined the scheme. While in its infancy, self-regulation with government setting the minimum standards is more efficient for business and government collectively. The APEC privacy framework has set the course for member countries to cinder the following principles, as core elements to their respective legal frameworks. These include:

- Preventing Harm, to prevent the misuse of information;
- Notice, ensuring that individuals are able to know what information is collected about them and for what purpose;
- Collection limitations, of personal information that is relevant to the purposes;
- Uses, should be used only to fulfill the purposes of collection;
- Choice, ensure that individuals are provided with choice in relation to the collection, use, transfer, and disclosure of their personal information;
- Integrity, to ensure personal information is accurate, complete, and kept up to date;
- Security, so as personal information is not used in a way to compromise the individual to who the data applies;
- Access and Correction, so as individuals have the ability to access and correct their personal information; and
- Accountability, to ensure organizations and individuals handling personal data are accountable.[24]

---

[23] Asia-Pacific Economic Cooperation, https://www.apec.org/Groups/Committee-on-Trade-and-Investment/Electronic-Commerce-Steering-Group, accessed 5 December 2017.

[24] Ibid.

APEC member are not obliged to implement domestically the APEC privacy Framework. Thus, there continues to be inconsistencies in approach and adoption of data protection and privacy laws.

More importantly, as APEC is a non-binding forum there is the opportunity for members to discuss current and future data protection issues freely. Doing so will allow greater collaboration with policy makers, the business community, academics to learn about the commercial and privacy issues related to the cross-border flow of personal and commercial data. This will provide the bases for addressing trade and investment issues, and continuing to enhance the global approach to the movement of goods and services.

One of the gaps in the APEC policy framework has been addressing the cross border transfer of data.[25]

Firstly, in 2014, the APEC and the European Union's Article 29 Working Party (on Data Protection) released Binding Corporate Rules (BCR). The BCRs govern international data transfers within companies or groups of companies.[26] They reflect a code of conduct which defines the company policy on data transfers. The EU BCR system and the APEC CBPR system have adopted a similar approach to promote the establishment of codes of conduct for international transfers. They are approved by EU Data Protection Authorities or by APEC recognized accountability agents.[27] Nonetheless, the CBPR also play a role in cyber security and data protection (see Chap. 15).

Secondly, the APEC Cross Border Privacy Enforcement Arrangement (CPEA) 2015, underpins the Policy Framework to establish regional cooperation for enforcing Privacy Laws. Any APEC economy who has a Privacy Enforcement Authority can participate. The CPEA promotes voluntary information sharing and enforcement by facilitating information sharing among privacy enforcement authorities. It also supports effective cross-border cooperation between privacy enforcement authorities through enforcement matter referrals.

However, Bennett and Raab consider that the APEC Privacy Principles are a weaker global standard than the EU, which means that they may serve to slow and even reverse the otherwise halting and meandering walk to higher standards which the EU has inspired.[28] It appears that Graham Greenleaf also shares this view.[29] The most plausible future scenario was 'an incoherent and fragmented patchwork', 'a more chaotic future of periodic and unpredictable victories for the privacy value'.[30]

---

[25] APEC Privacy Framework, paragraphs 46–48.

[26] International data transfers: the WP29 and the APEC developed a practical tool for multi-national organizations 07 March 2014.

[27] Ibid.

[28] Bennett C, Raab C, (2006) *The Governance of Privacy: Policy Instruments in Global Perspective*, MIT Press, 2006, in Greenleaf, Graham, *"Sheherezade and the 101 Data Privacy Laws: Origins, Significance and Global Trajectories"*, JlLawInfoSci 2; 2014, 23(1), Journal of Law, Information and Science 4.

[29] Graham G (2014), *"Sheherezade and the 101 Data Privacy Laws: Origins, Significance and Global Trajectories"*, JlLawInfoSci 2; 23(1), Journal of Law, Information and Science 4.

[30] Ibid.

Thus, reinforcing the position set out in the Chap. 1 of this book that the data protection and privacy laws are not settled, and far from being settled or even harmonized. There is a lot of work to do for the international community if they are serious about regulating technology developers and providers, to ensure greater protection of personal data. More needs to be done to force these entities to take on greater responsibility and develop systems and platforms to strengthen data protection, for future generations. As, highlighted earlier in the book, the June 2019 G20 leaders' meeting in Osaka Japan, went someway to promoting the urgent need for regulation to strike a balance between innovation and protection of personal data and privacy. The June 2019 G20 leaders' declaration is a positive sign that politically more needs to be done to strengthen the regulation of data and other areas of emerging technology.

APEC's Electronic Commerce Steering Group (ECSG) promotes the development and use of electronic commerce by supporting the creation of legal, regulatory and policy environments in the APEC region that are predictable, transparent and consistent. Since 2011, APEC has been undertaking a lot of work in the area of data protection and issues the Cross Border Privacy Rules System. The Rules system balance the flow of data across borders of member countries, and at the same time ensure effective protections are in place for private and personal information. This is a good example where states have come together to develop a framework that has been harmonized for member countries to use. The Rules system enables a third party verifier, to verify and be accountable for the data and information according to the APEC Privacy Framework.

More recently, APEC established the Privacy Recognition for Processors (PRP) System in 2015, which assists controllers in complying with relevant privacy obligations, and helps controllers identify accountable processors. In the same year, APEC released the Privacy Framework that promotes electronic commerce throughout the Asia Pacific region. The framework is consistent with the core values of the OECD's Guidelines on the Protection of Privacy and Trans-Border Flows of Personal Data, and reaffirms the value of privacy of individuals. The privacy Framework goes some way to addressing the gaps in policies and regulatory frameworks on E-Commerce to ensure that the free flow of information and data across borders is balanced with the effective protection of personal information essential to trust and confidence in the online market place. Thus, the continued balancing act of economic activity versus private and commercial protections remains at the forefront of not only this regional framework, but also national and international legal personal data and privacy frameworks.

16.8.1.1 Asia Pacific Privacy Authorities

The Asia Pacific Privacy Authorities (APPA) is the principal forum for privacy and data protection authorities in the Asia Pacific Region. It assist and promotes a partnership approach and exchange of ideas about privacy regulation, new technologies and the management of privacy enquiries and complaints. Members convene twice a year, discussing permanent agenda items like jurisdictional reports from each delegation and an initiative-sharing roundtable.[31] Other topics discussed have been balancing privacy and security, cross-jurisdictional law enforcement in the Asia Pacific, cryptography, social media, international data transfer and the de-identification of data. Australia, Singapore and Japan are members, however, there are no other countries from the ASEAN group, South East or Central Asia that are members of this forum. To become a member they need to be an accredited member of the International Conference of Data Protection and Privacy Commissioners (ICDPPC). In addition, a member needs to be a participant in the APEC Cross-border Privacy Enforcement Arrangement (CPEA) or a member of the Global Privacy Enforcement Network (GPEN).[32]

16.9 Association of South East Nations [ASEAN]

ASEAN is made up of countries from South East Asia.[33] Dialogue partners of ASEAN include Australia, China, Japan, New Zealand and the United States. ASEAN is considered another supranational polity, however the major difference is that it has not have the institutions or legal framework the EU has. Importantly, the ASEAN Declaration on Human Rights sets the scene for the application of human rights across its member states. Article 21 states the:

> Every person has the right to be free from arbitrary interference with his or her privacy, family, home or correspondence including personal data, or to attacks upon that person's honour and reputation. Every person has the right to the protection of the law against such interference or attacks.[34]

Arguably, this is an important step in the recognition, understand, management and protection of individual's personal data over the Internet. It serves to provide a level of privacy protection over the Internet. Although the Deceleration does not require member states to fully implement binding legal principles to protect personal data and privacy. It is our view that the Declaration provides the basis for broader consideration of data protection and privacy over the Internet, should privacy take hold and become even more important than the current economic and limited approach taken throughout the ASEAN region. It also enables ASEAN to guide member states to potentially adopt or move closer to EU model for data protection.

Nonetheless, ASEAN has established the Telecommunications and Information Technology Ministers Meeting, endorsed the Framework on Personal Data

---

[31] Asia Pacific Privacy Authorities, http://www.appaforum.org/members, accessed 14 January 2018.

[32] Ibid.

[33] Brunei, Indonesia, Cambodia, Lao, Malaysia, Myanmar, Philippines, Singapore, Thailand, Vietnam.

[34] Association of South East Nations, human Rights Declaration 2012, http://www.refworld.org/docid/50c9fea82.html, accessed 2 November 2018.

Protection.[35] However, and because ASEAN's model of consensus, the framework is non-binding and establishes the following core principles:

- Consent, Notification and Purpose;
- Accuracy of Personal Data;
- Security Safeguards;
- Access and Correction;
- Transfers to Another Country or Territory;
- Retention; and
- Accountability.[36]

These principles replicate the OECD principles for privacy and data protection. The ASEAN Economic Community (AEC) was established and marked an important milestone in ASEAN economic integration.[37] Its aim is to develop a coherent and comprehensive framework for personal data protection. This will require the development of Regional Data Protection and Privacy Principles (Rules System), and identify the responsibilities of businesses in personal data protection between 2016 and 2025.[38] As part of this processes it is also an objective to establish a common ASEAN consumer protection framework through higher levels of consumer protection legislation, improve enforcement and monitoring of consumer protection legislation and make available redress mechanisms including alternative dispute resolution mechanisms. One of the solutions to improving consumer protection is to modernize (taking into account the high level principles) relevant provisions of national consumer protection legislation (particularly in areas of unfair contract terms, ecommerce, product liability-safety, consumer data privacy).

In 2016, the ASEAN cybersecurity strategy was also announced, to ensure funds made available through the Cyber Capacity Programme (ACCP) launched by Singapore to support efforts to deepen cyber capacities across ASEAN. However, since the establishment the ASEAN in 1967, human rights have had a slow road to recognition and implementation. Even though ASEAN issued a Declaration of Human Rights in 2013, across Asia, human rights have been, and continue to be, viewed a Western concept.[39] It was not until the 1980s that ASEAN members including Singapore, Malaysia and Indonesia began to take a greater focus on human rights. However, the ASEAN Declaration on Human Rights has not forced any of the members to ensure the principles, rights and freedoms are implemented national.

---

[35] Telecommunications and Information Technology Ministers Meeting, which endorsed the Framework on Personal Data Protection, http://asean.org/storage/2012/05/10-ASEAN-Framework-on-PDP.pdf, accessed 5 December 2017.

[36] Ibid.

[37] ASEAN Economic Community, http://investasean.asean.org/index.php/page/view/asean-economic-community/view/670/newsid/755/about-aec.html, accessed 2 April 2018.

[38] ASEAN Economic Community 2025 Consolidated Strategic Action Plan, http://asean.org/storage/2017/02/Consolidated-Strategic-Action-Plan.pdf, accessed 2 April 2018.

[39] Thio L (1999) *Implementing Human Rights in ASEAN Countries: "Promises to keep and miles to go before I sleep"*, Yale Human Rights and Development Journal, Vol. 2, Iss. 1, Art. 1, pp. 2–5.

This is because ASEAN itself does not have the legal structure that the EU has developed and cannot mandatorily oblige countries to implement ASEAN policy or law. Unlike Australia, the EU and its member states, it is only recently that some countries across South East Asia have recognised human rights.

In addition, there will be closer cooperation amongst ASEAN member states with a view to enhancing international law enforcement; and an exchange on cyber norms on a regional basis to promote a deeper understanding of the cyber norms and arrive at an ASEAN position. Arguably, as Singapore move ahead with its transformation as a data and technological hub, they will have a significant influence on any future legal framework for the region.

16.10 African Union

The African Union, while outside of the scope of this book is worth mentioning because in 2014, they established the African Union Convention on Cyber-security and Personal Data Protection. The Convention aims to establish regional and national legal frameworks for cyber-security, electronic transactions and personal data protection. The African comprises 55 nation states. Therefore, the Convention has a vast reach. Importantly, the Convention provides the basis for harmonization of data protection laws across the Union, in the same way the EU has developed its legal framework for member states to implement.

16.11 Commonwealth of Nations

Australia, Singapore, India and Malaysia are part of the 52 nation states that make up the Commonwealth of nations, which was born out of the historical connections and territories to the British Empire. In 2017, the Commonwealth Secretariat released the Model Bill (Law) on the Protection of Personal Information.[40] The Model Laws recognizes the privacy of individuals and their personal information that is collected, stored and used by the private sector. The Model Laws set limits on the collection of personal information or data; restricting the use of personal information or data for openly specified purposes; ensuring the right of individual access to personal information relating to that individual. The Model Laws also ensure there is the right to have the data or information corrected and identify the parties who are responsible for compliance. Furthermore, the Model Laws also provide the framework for countries to adopt a process to regulate cross-border disclosure of data and information that requires guarantees of protection. More importantly, and

[40] The Commonwealth Secretariat, http://thecommonwealth.org/sites/default/files/key_reform_pdfs/P15370_6_ROL_Model_Bill_Protection_Personal_Information_2.pdf, accessed 5 December 2017.

consistent with the ICDPPC, the Model Laws encourage the establishment of a Privacy Commissioner so as there is a mechanism of reporting to the relevant Parliament. The establishment of Model Laws is not new and has been applied in international trade for more than two decades. What the Model Law do, is provide a harmonized approach to data protection and privacy laws. The Model Laws have been aligned with the principles established by the OECD and the EU. However, there is no strict adoption required and therefore, countries can choose whether to adopt and apply the Model Laws. What is not clear is the interaction this body has with other regional and international organizations. The lack of collaboration and consultation at the regional level will only continue to result in a patchwork or inconsistent approaches to data protection, both private and commercial.

16.12 European Union

The EU as a supranational polity is a regional Union made up of institutions that harmonize law and policy across most areas of society for its member states, which has been discussed in Chap. 3. The diagram below highlights the steps taken by the European Union, its member states, and Asia Pacific countries to implement international law (human rights, trade and security). Article 288 of the TFEU[41] provides that the legal instruments including Regulations, Directive and Decisions are binding on member states. The GDPR is an EU Regulation, and therefore is binding on all member states to transpose the laws into national laws. In contrast, the countries of the Asia Pacific adopt the right to privacy through a three step process. They do not have the additional layer of a supranational polity. They only need to ratify and implement international human rights law into national law.

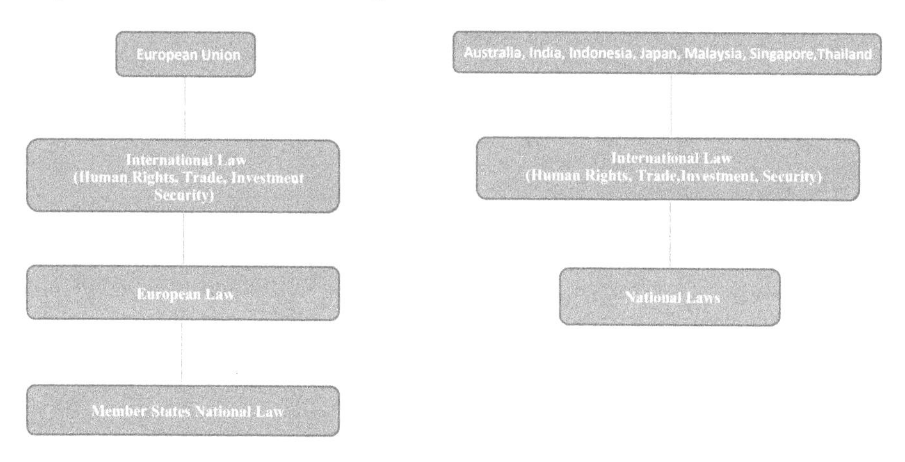

Diagram outlines the hierarchy of legal frameworks in the European Union and Asia Pacific

[41] Treaty on the Function of the European Union, Official Journal of the European Union C 115/47.

16.13 Trade Agreements

Apart from private law playing a predominant role in regulating and governing data protection and privacy, there is also a role for public law. At a country level, Free Trade Agreement, Bilateral Agreements or Multilateral Agreements can be used to manage data flows. At the business to business level, data protection and privacy can be managed and regulated through cross border contracts. The next section provides an example of how the forthcoming Trans-Pacific Partnership Agreement has considered data protection.

The Trans-Pacific Partnership Agreement (TPP) is an example of where data protection is being considered as part of transnational trade.[42] Members to the TTP include Australia, Malaysia, Japan and Singapore amongst others. Even with United States withdrawal, members have recently singed the TTP in February 2018. Thus, when the TPP is fully implemented it will provide another opportunity for harmonization of rules, principles and law regarding private and commercial data. Notably within the TTP, Article 14 pertains to the "Electronic Commerce". Article 14.5 states that in regards to Domestic Electronic Transactions Framework, 'each party shall maintain a legal framework governing electronic transactions consistent with the principles of the UNCITRAL Model Law on Electronic Commerce 1996 or the United Nations Convention on the Use of Electronic Communications in International Contracts.[43] Even though there is no mention of the CISG or PICC, the TPP goes some way to ensuring that states consider aspects of the UNCITRAL framework.

Article 14.8 specifically refers to Personal Information Protection. Article 14.8.1 states that the Parties recognize the economic and social benefits of protecting the personal information of users of electronic commerce and the contribution that this makes to enhancing consumer confidence in electronic commerce. Article 14.8.2 allows each Party to adopt or maintain a legal framework that provides for the protection of the personal information of the users of electronic commerce. In the development of its legal framework for the protection of personal information, each Party should take into account principles and guidelines of relevant international bodies. Article 14.8.3 requires that each Party shall endeavor to adopt non-discriminatory practices in protecting users of electronic commerce from personal information protection violations occurring within its jurisdiction. Each Party should publish information on the personal information protections it provides to users of electronic commerce, including how individuals can pursue remedies; and business can comply with any legal requirements.[44]

---

[42] Greenleaf, G *Asia Data Privacy Laws – Trade and Human Rights Perspectives*, University New South Wales, (2017).

[43] Trans-Pacific Partnership Agreement, Article 14, https://mfat.govt.nz/assets/Trans-Pacific-Partnership/Text/14.-Electronic-Commerce-Chapter.pdf, accessed 5 December 2017.

[44] Ibid, Article 14.8.4.

Article 14.11 relates to Cross-Border Transfer of Information by Electronic Means, and requires that cross-border transfers of personal information be allowed when the transfer relates to the business practices of an organization in a TPP member country. This Article does not prevent a Party from adopting or maintaining measures to achieve a legitimate public policy objective such as the arbitrary or unjustifiable discrimination or a disguised restriction on trade does not impose restrictions on transfers of information greater than are required to achieve the objective.

Article 14.15 recognizes the global nature of electronic commerce is only going to increase and that the Parties shall endeavor to work together to assist business to overcome obstacles to its use and exchange information and share experiences on regulations, policies, enforcement and compliance regarding electronic commerce. The TTP will not only enhance the harmonization processes, but only goes someway to ensuring that data and privacy is managed and appropriately protected.

16.13.1 United States of America (US) and Korean Free Trade Agreement

One of the most notable international trade agreements completed that includes binding rules on cross border data flows, is the 2011 United States of America (US) and Korean Free Trade Agreement.[45] Article 51.8 deals with the Cross Border Information Flows, and recognizes the importance of the free flow of information in facilitating trade, and acknowledging the importance of protecting personal information, the Parties shall endeavor to refrain from imposing or maintaining unnecessary barriers to electronic information flows across borders. While this is rather symbolic and more encouraging provision, it nevertheless obliges both parties not to impose any unnecessary barriers to electronic information flows across international borders.

16.13.2 Proposed Australia and the European Union Free Trade Agreement

Australia and the European Union have begun working on a Free Trade Agreement. Arguably, this provides an opportunity for both jurisdictions to work together to strengthen data protection rules within any future agreement.

[45] United States of America and Korea, Free Trade Agreement, Article 15.8.

16.13.3 Potential Australian and United Kingdom Free Trade Agreement

As the United Kingdom (UK) continues the process of departing from the EU as a formal member, should this be realized, there will be opportunities for countries such as Australia, India, Indonesia, Japan, Malaysia, Singapore and Thailand to establish Free Trade Agreements with the UK. This provides a perfect opportunity for these countries to go beyond the TTP process and strengthen the governance and regulation of data (personal and commercial) and privacy.

16.14 Conclusion

A lot is being done internationally and regionally to better manage data protection and privacy. However, it is adhoc and very fractured. The portable devices like the Iphone, laptop computers and Ipads alone, can access the internet no matter what country they in. In other words, a person who purchases their device in Australia, can travel to Indonesia, Singapore, India, Malaysia, Japan, Thailand and throughout the EU, turn on their device and log into the Internet. The individual can also do this and be connected to and browsing the Internet, whether they are in a plane, boat or motor vehicle as they cross the border from one country to the next. Thus, the Internet and the devices that allow access to the Internet have created an environment of internationalization. Likewise, a person can be located in Switzerland and access national governmental information of Slovenia or Australia via the Internet.

This Chapter briefly looked at APEC, African Union, ASEAN, OECD, ICDPPC, WEF and the Commonwealth of nations to better understand the work being undertaken by these groups. While they are all doing something, or at a minimum identified some of the key concepts and principles established by the OECD, there remains a lot of work to obtain legal convergence and harmonization. It is our view that until there is greater collaboration, this area of the law will remain far from settled, which could have significant economical impacts and slow the growth in data trade. The June 2019 G20 leaders' meeting in Japan has reinforced the need for effective regulation to protect personal data and emerging technologies.

The current international approach has resulted in no single international organization that deals solely with all aspects of data protection and privacy, in the dame way, for example, agriculture under the United National Food and Agriculture program, or The United Nations Commission on International Trade or The International Institute for the Unification of Private Law (UNIDROIT). Steps have been taken internationally and regionally, and as discussed in other Chapters, most jurisdictions have to some degree adopted the OECD principles and concepts pertaining to data protection and privacy. Those that have not are considering them as part of developing their own specific national data protection laws, for instance, Indonesia and India. The increasing proliferation of privacy-threatening technologies and pervasive privacy-invasive practices since the 2000s, and calls for a global data privacy framework have grown exponentially.[46]

However, is it time more is done to encourage nation states to become members of the OECD? With such a low membership base, particularly that no nations from South East or Central Asia being members, further work is needed to get these countries involved, if the international community agree that the OECD continue to be the most active in setting the concepts and principles for future data protection and privacy law. The suggestion that Singapore, Malaysia and Japan join the OECD is however aspirational and it may be argued that by having their membership for the purpose of data protection, adds no value to the overall policy or legal discourse. Nonetheless, it is acknowledged that the OECD may not be the relevant institution to modernizing of regulatory responses to privacy and personal data threats. As highlighted in Chap. 17, it is proposed that a Model Law be considered to enhance legal convergence and harmonization of data protection law for and on behalf of the international community.

Finally, Free Trade Agreement, Bilateral Agreements or Multilateral Agreements have begun to be are being utilized in the transnational flows of personal data. They appear to be an effective mechanism to force other countries to, at least, adopt certain practices related to data transfers. These agreements are international and will continue to form another part of the complex jig saw puzzle that is data protection and privacy law. However, they are rather symbolic, and that poses a continual dilemma for data protection law into the future.

References

Bennett C, Raab C, (2006) *The Governance of Privacy: Policy Instruments in Global Perspective*, MIT Press, 2006, in Greenleaf, Graham, *"Sheherezade and the 101 Data Privacy Laws: Origins, Significance and Global Trajectories"*, JlLawInfoSci 2; 2014, 23(1), Journal of Law, Information and Science 4.

Graham G (2014), *"Sheherezade and the 101 Data Privacy Laws: Origins, Significance and Global Trajectories"*, JlLawInfoSci 2; 23(1), Journal of Law, Information and Science 4

Greenleaf G *Global Data Privacy Laws: 89 Countries, andAccelerating'* Privacy Laws & Business International Report, Issue 11-175 (2012)

Greenleaf, G (2017) *Asia Data Privacy Laws – Trade and Human Rights Perspectives*, University New South Wales

Michael Kirby (2011) *The history, achievement and future of the 1980 OECD guidelines on privacy International Data Privacy Law*, Volume 1, Issue 1, 1, p. 6

Micheal Yilma K (2018) *The United Nations data privacy system and its limits*, International Review of Law, Computers & Technology

Tene O (2013) *Privacy Law's Midlife Crisis: A Critical Assessment of the Second Wave of Global Privacy Laws*, 74 OHIOST.L.J.1217, 2013, p. 1222

Thio L (1999) *Implementing Human Rights in ASEAN Countries: "Promises to keep and miles to go before I sleep"*, Yale Human Rights and Development Journal, Vol. 2, Iss. 1, Art. 1, pp. 2–5

[46] Micheal Yilma K *The United Nations data privacy system and its limits*, International Review of Law, Computers & Technology, (2018).

Chapter 17
What Is at Issue and A Possible Pathway Forward

Abstract This Chapter aims to bring together the research and analysis from each of the jurisdictional Chapters and provides an outline of the key policy and legal gaps facing data protection and privacy. It does not attempt to highlight every gap or issue, due to the many variables that exist. This Chapter will provide a possible pathway to strengthen the fragmented law of personal data protection and privacy. What has emerged are that the legal and policy principles and concepts such as consent, accountability and transparency suggested by the OECD, which now underpin national and supranational data protection law, have been addressed differently by nation states. The current approach taken towards these concepts and principles pose significant issues to address the protection of personal data and privacy, going forward.

It is understood that an international harmonized data protection law in the form of a convention or treaty is currently not attainable. However, what is suggested is that attention should be focused on devising a Model Law on the subject matter. This has become even more difficult to achieve because, currently there is no consensus as to an agreed international model. At best an agreement on a limited number of principles and concepts is detectable in domestic legislation. Arguably, this is an area that needs to be reconciled.

There is no single option or silver bullet available to address the current, ongoing and future personal data protection and privacy issues that will arise. It will take a multifaceted and multilayered approach. Participants will include, but not be limited to, government, all areas of industry that use technology, international and regional bodies, institutions and organizations, the legal and technology professions, along with the wider community. More work needs to be undertaken to look at the systems and legal frameworks to better understand where existing policy and regulatory frameworks and models currently not utilized in this area of law that, can be adapted and developed further. Furthermore, and more pervasively, consideration must be given to regulating the manufacturers of the systems, platforms and infrastructure that collects and used personal data, in the same way as many other industries (food producers, car, airplane, agricultural chemical and pharmaceutical manufacturers). While this was not discussed in the book, the manufacturers are largely producing systems that are not regulated by government [minimum] standards. Finally, this Chapter proposes a possible way forward and calls on

© Springer Nature Singapore Pte Ltd. 2019

R. Walters et al., *Data Protection Law*,
https://doi.org/10.1007/978-981-13-8110-2_17

governments, regulators, technology experts and the legal profession to collaborate to speed up the process of convergence and harmonization of these laws. It poses the question – what is the best model going forward? However, and while some will view this as a formidable task, other areas of the economy, such as those identified above, have successfully achieved this for a number of decades now.

17.1 Introduction

This Chapter begins by highlighting the differences in regulatory approaches taken in the IT sector that provides the infrastructure and platforms for the Internet and other industries that are important to the daily lives of most people. Applying the laws effectively to both the public and private sectors, arguably reinforces key societal principles related to data protection and privacy law that rely on 'trust' and 'certainty'.[1] As highlighted in earlier chapters the principle of trust[2] has evolved as an important issue in, on and over the Internet and the digital economy to ensure that the community and industry are comfortable in using modern technology and hence information gathering techniques. This same approach has been successful now for more than 70 years in the areas of aircraft manufacturing and food production. Without trust and certainty, the technology industry faces greater challenges, from public pressure on governments to increase the level of regulation and for those governments to impose higher level sanctions and standards for the illegal use of personal data that result in privacy breaches.

Apart from the differences in culture, legal families and histories between the jurisdictions discussed in this book, the book has highlighted the many differences, but also similarities in the data protection and privacy laws of Australia, India, Indonesia, Malaysia, Japan, Singapore, Thailand and the European Union. This Chapter does not attempt to identify all the policy and legal issues related to the current legal frameworks of privacy and data protection. It rather aims to highlight some of the important gaps, and where further work is required.

The issue is that globalization of the Internet economy has caused a dramatic increase in personal data being commercialized and now traded. Individuals and entities are profiting enormously from this activity. However, the digitization of this sector of the economy has also created significant complexities in the manner in which personal data is collected, used and processed. For instance, the evolution of technology by enabling the collection and collation of personal data through predictive analysis of people's behavior is changing the way entities, not only capture personal data, but also use the data and hence information, resulting in market dominance. In addition, the full extent and commercial activity, that is, the sale and

[1] Hofman, D., Duranti, L., How, E *Trust in the Balance: Data Protection Laws as Tools for Privacy and Security in the Cloud Algorithms* MDPI (2017).

[2] Ibid.

distribution of personal data over the Internet, is not well understood and is only beginning to emerge. A more pervasive issue is where and what states are collecting, storing and using this data, particularly where nation states are involved, is also only beginning to be understood. The challenge for governments and regulators, including the community is that regulation will not keep up with these advances in data capture and use. The law will continue to lag well behind technology development. Therefore, more than ever, a combination of regulatory and non-regulatory approaches is needed to ensure innovation continues to be developed, while maintaining and strengthening the protection of individual's personal data and subsequently privacy.

17.2 Technology and Regulation

The transformations in technology have seen and will continue to see individuals and entities apply sophisticated techniques through data analytics and new technology such as quantum analytics, to harvest, store and use personal data. Today, some of the most profitable businesses and nation states in the world have adopted this model. These include, but are not limited to Google, Amazon, Apple, Facebook, and Microsoft. It is well understood that these legal entities and many others use people's personal data (both generally and defined by law), within their business model – to make a profit. In some cases, those profits far exceed expectations from the business community and can be millions if not billions of dollars. However, and it is also well understood that, by these and other companies having a substantial economic and market power, does not necessarily mean that they are infringing any competition laws or any other laws *per se*. Moreover, the problem arises with the way in which these companies make use of the data.

What is at issue is the ongoing need for regulators to balance the economic development in personal data, with the protection of that data and privacy. The current regulatory balance is going to be continually challenged and may never be resolved. In other words, the balance between stifling innovation and protecting people's personal data and privacy, walks a thin line along which innovation is becoming increasingly dependent on large scale dissemination of all forms of data. Furthermore, the challenge going forward is to better understand the level of risk or harm from data and privacy breaches and infringements. However, as personal data is increasingly used as a tradeable commodity, it not only heightens the potential for privacy breaches, but can also be used to obtain a market advantage as noted above. Nonetheless, there is no current data regulatory policy in place that can resolve all of these issues and arguably none is foreseeable in the near future. Importantly, technology alone is unlikely to provide a safe and fool-proof system or platform to protect personal data. This is because, for every new system or platform implemented to protect data, is undertaken by either an individual or entity, who can over time penetrate that system or platform.

The technology sector may well argue that new technology (distributed ledger technology), such as blockchain,[3] or quantum is likely to be safe as only registered users are able to gain access. Using blockchain as an example, security begins with the network and the management of the nodes. At the private level it appears to be the security of verification within a blockchain system that is most pertinent. At the public level, an incident highlighted that the code to run on Ethereum is the security issue. However, it was not the Ethereum protocols that were at issue, but rather the codes.[4] Blockchain technology no longer only serves the finance industry; it is also being used in trade, capital raising, clearing settlements, deposits and lending, domestic and global payments, property and insurance, digital identity management and automated compliance.[5] Zetzche, Buckley and Wagner make a very important observation, by highlighting the legal issues that arise from blockchain technology.[6] The authors particularly point to the trust solution. They assert that the trust enhancing function of multiple ('distributed') entities together provide authentication, rather than one 'centralized' ledger, leads to (1) disintermediation of traditional intermediaries and clearing and settlement systems (resulting in greater security and transparency), (2) enhanced efficiency and speed, (3) lower transaction costs, and (4) enhanced market access.[7] They further highlight the potential legal liability for distributed ledger technology, because the distributed ledgers are often put forward as the answer to ever-increasing cybersecurity risks. While distributed ledgers are currently considered more secure than traditional centralized ledgers, recent incidents would prove otherwise.

To illustrate that blockchain is not secure a Luxembourg- and London-based Bitstamp undertook the second largest Bitcoin exchange in terms of volume traded, and suffered from a hot wallet hack leading to the loss of 19,000 Bitcoins, valued at about US$5.1 million. Bitstamp subsequently suspended services for nearly a week during which client deposits were not accessible.[8] A year later, in 2016, US$53 million of the over US$150 million crowd funded assets in DLT-based virtual currency ETHER held in the investor-directed DLT- enabled Decentralized Autonomous Organization (DAO) were channeled to a third-party controlled account after exploiting previously published vulnerabilities in the DAO code. In 2016, a Hong-Kong-based Bitfinex, one of the world's largest bitcoin exchanges, lost 119,756 Bitcoins with a market value at the time of between US$66–72 million in a hack that involved its multi-signature accounts.[9] It was decided to apportion losses from the theft across the company's clients and assets, widening the group of those affected

---

[3] Reijers W., O'Brolcháin F, Haynes P, *Governance in Blockchain Technologies & Social Contract Theories*, 1 LEDGER 134, (2016).

[4] Bourke A *How Safe are Blockchains?* It Depends, *Harvard Business Review*, (2017).

[5] Zetzche., D, Buckley., R, Wagner, D *The Distributed Liability of Distributed Ledgers: Legal Risks of Blockchain,* UNSWLRS 52, (2017), pp. 3–8.

[6] Ibid.

[7] Ibid.

[8] Ibid.

[9] Ibid.

by the losses beyond those holding the multi-signature accounts that were hacked. Accordingly, all Bitfinex clients lost a significant 36% of their holdings.[10]

As noted, no one can be certain whether this technology will create greater safeguards for personal data and subsequently for privacy where there is little to no protection of an individual's personal data. This debate is further complicated when having to balance the proponents for the free market versus proponents for greater rights protections. In other words, the rise of the Internet has seen the rise in business that is driven by data, a development which is unlikely to abate any time soon. The above example, while not directly relating to personal data, highlights how even supposed safe systems and platforms can be infringed to obtain large amounts of money.

17.3 International & Regional Institutions

In Chap. 16, it was highlighted how the high level principles established for data protection and subsequently privacy have largely been set by the OECD. Without repeating what was discussed in Chap. 16, the question is whether the OECD Guidelines, albeit updated in 2013, are current? The broad principles established by the Guidelines include Collection Limitation; Data Quality; Purpose Specification; Use Limitation; Security Safeguards; Openness; Individual Participation and Accountability. It is well understood that within these high level principles, the concept of consent and the definition of personal data has become very important, amongst others. The respective APEC and ASEAN data protection frameworks have also established similar principles and concepts to that of the OECD (see Chap. 16). It is argued that, further work is required to better understand what other, if any, principles and concepts are required to be included and formally recognized by the OECD Guidelines. A further issues from the OECD Guidelines highlights how these concepts and principles when transposed into national law largely deal with the data user and not the manufacturer of the systems, infrastructure or platforms. The problem is that many nations are not members of the OECD and may not be willing to follow that model. Nevertheless, there are other principles and concepts that have found their way into EU law that are beginning to be recognized and adopted internationally. Chapter 3 highlighted how the EU through the recent implementation of the GDPR has strengthened data protection rights and regulation, which is having a significant influence on, and to, other nation states. It is arguable that the GDPR could be the starting point in harmonizing data protection regulations. That is, harmonization is not only driven by world bodies such as UNCITRAL but also by transplantations of legal principles across jurisdictions.

For governments, regulators and policy makers to be effective in the future, further work is required to better understand whether the core principles and concepts underlying data protection and privacy need to become more prevalent at the international level. This is no more important than the trade, dissemination and transfer of data

[10] Ibid.

internationally. The benefit of such internationalization lies in the fact that harmonization and convergence of laws is likely to make them more effective and efficient. However, it may not be politically practicable for countries in the Asia region that are not members of one of these organization to follow transnational efforts of harmonization. This does not mean that countries will not adopt these and other principles, concepts and guidelines enunciated by international and regional organizations, which has in large part been the case. At issue is that these international and regional guidelines and frameworks, are high level developments and arguably, lack the necessary detail needed in the current transformation of personal data protection law.

17.4 Current Data Protection and Privacy Regulation

Arguably, there is no doubt the EU's GDPR has set the benchmark for data protection and privacy around the world. As discussed, and in part, the EU has been very strategic in how it has been able to effectively force other countries into adopting similar laws, or at least, a similar framework that has included similar concepts and principles. While conceptually many of the concepts and principles such as rights, consent, transparency and accountability have influenced, to varying degrees, how legal transplantation is taking place locally and regionally, what is at issue is the lack of convergence and harmonization on a global scale.

This lack of global harmonization has to be one of the most significant issues facing governments, elements of the business sector and the community into the future. However, it is acknowledged that countries are developing their data protection laws to meet their own economic and social needs. It is our view that, while the EU balances the need for privacy and economic outcomes within the GDPR, the EU places the right to privacy at a higher level to most, if not, all other jurisdictions. On the other side, Singapore have placed the business needs and interests at the forefront of their data protection framework; while, it is our view that Australia sits somewhere in the middle – between Singapore and the EU.

The lack of legal harmonization and convergence can create a situation, that has been experienced in international trade for decades. That is, in the same way as companies engage in forum shopping in responding to international trade and investment disputes, the same can and is beginning to develop in data protection. For instance, it is well understood, and a long standing practice in international trade law, when a dispute arises from non-performance of a contract, entities will use the choice of jurisdiction and law, among others, as a tactic to circumvent their obligations.

The technology industry – that is, data protection – is no different, and this calls for greater harmonization and convergence of legal concepts, principles and frameworks. An excellent example of the need to create greater certainty is the recent Facebook issue that resulted in millions of people's personal data being misused. The resulting effect is that Facebook is now looking at how to circumvent the legal framework, particularly in Europe in relation to the obligations imposed by the GDPR, because of one of their headquarters being located in Ireland. It has been reported that they are intending to move an estimated 1.5 billion users out of reach of the GDPR.[11] If successful, the resulting effect will see members in Africa, Asia, Australia and Latin America no longer fall under the European Union's GDPR.[12] Thus Facebook will reduce its exposure to the requirements of the GDPR. The most important issue is that companies that collect and use personal data in the EU must obtain consent from the data subject. This action by Facebook would result in consent not being a requirement for those data subjects located for example in Australia, Asia or Africa.

Where current policy and the law in many regions and countries around the world are different, individuals and multinational entities can easily use or relocate to different regions or countries as a tactic to minimize their exposure to particular laws. This, in itself, will only further complicate the regulation of personal data and subsequently privacy going forward. It will dilute people's (data subjects) ability to control their data and result in less privacy – not more.

17.5 Convergence or Disconnection of Data Protection and Privacy?

The development of data protection and privacy law is fascinating across very different regions and countries. This is no more evident in how the laws have been structured and have influenced the reference to data, information and/or privacy. The EU GDPR, Singapore or Malaysia's respective laws do not mention the word 'privacy' within the formal text of the Regulation. Moreover, and as highlighted in Chap. 16, there are varying degrees of acceptance of privacy as a human right. Singapore, for example, view the concept of privacy as a secondary element to building the economic environment to attract and retain international competitive businesses, and the right to privacy barely exists there. Is it wrong for Singapore to adopt this approach? No, it is our view, that Singapore, like any other country is, and has decided to adopt a model that continues to place that Island state and its people in an economically advantageous position, enabling it to retain its place as a business hub in a highly competitive region of the world. The challenge though is how to balance this with the needs of other nation states and EU's position in the forward regulation of this area of the law.

Arguably, one of the major obstacles to legal convergence and harmonisation is for nation states to come together and agree on what level of right (s) should be afforded consistently to privacy through the Internet and its supporting systems and infrastructure. Likewise, Australia's privacy laws do not mention data protection, but do refer to data more generally. On the other hand, India, Indonesia and Thailand's current laws do not mention either data protection or privacy. However, this is due to the fact that these laws are still under consideration and are currently sectorial.

[11] Facebook is set to move more than 1.5 billion users out of reach of a new European privacy law, which would allow regulators to fine companies for data breaches, http://www.abc.net.au/news/2018-04-20/facebook-to-move-1.5-billion-users-out-of-reach-new-eu-privacy/9678842, accessed 8 August 2018.

[12] Ibid.

Despite the fact that some jurisdictions are yet to develop specific data protection or privacy laws, those jurisdictions that have such laws, either refer specifically to data (personal) protection or privacy. Yet they are, arguably, one and the same thing. The data protection law is protecting and controlling a level of privacy. Therefore, there is some convergence of laws, concepts and principles across data protection and privacy law.

17.6 Case Law

Throughout the book, it has been highlighted that the emerging case law is vast and varied. The current jurisprudence is far from settled. There is no doubt that the EU, through the Court of Justice of the European Union and the European Court of Human Rights has, to a large degree, currently set the tone for jurisprudence in this area of law. The courts established in the EU have certainly lead the way in strengthening the right to privacy and protecting people's personal data, that is defined by the law. However, there has been little jurisprudence in regards to the economic protection of personal data. That is, and as highlighted earlier, the courts have not yet had to fully determine where, when and how the concept of consent is to be applied. Arguably, it is increasingly likely that courts across other jurisdictions will be called upon to make complex decisions to ensure that a meaningful balance is struck between the protection of privacy and personal data, and also with other areas of the law such as intellectual property, transnational contracts and completion law. This is another area of the law that will continue to evolve and become even more complex, if, and possibly when courts are forced to decide on whether the core concepts and principles, such as consent and the definition of personal data or personal information is, in fact adequate. The courts in Singapore, for instance, have been one of the first jurisdictions in Asia to look at these issues. The further complex overlay is how national courts themselves operate, which will continue to be different from one another. Take for example, Australia, Singapore, Indonesia and Malaysia, combined they all have different court systems and legal families that influence court decisions (see jurisdictional Chapters). A more pervasive issue facing the courts will be the development and use of artificial intelligence and quantum technology over the coming decade. No one, to date, can with certainty confirm how this technology will develop and use personal data.

17.7 Data Localization

Data localization has begun to take hold across many countries. In September 2015, Russia implemented what is broadly considered the world's most onerous data localization law applicable to personal data of its citizens. China has also implemented similar laws. As highlighted in Chap. 5, Australia, is a good example of a country that has limited this approach to personal health data. However, this does not mean that countries will continue to retain or strengthen this approach, when considering that people, across the world, may not care in the future where and how their personal data is used. An opposing argument could be, that it is in their interest for these laws to be strengthened to promote and open up more opportunities for another area of the economy to grow and create jobs and prosperity.

Despite the possible economic benefits from the trade in personal data, data localization also creates a framework for countries to have greater sovereign control over their citizen's personal and general data. However, it is arguable that this type of regulatory activity is unlikely to slow, and may actually become the norm in the future. How this development will impact on data protection and privacy and hence influence business decisions is not known, and more work is required to better understand such likely impacts. On the one side, it is likely to result in greater protection of personal data for those citizens to which data localization applies. It will also strengthen governmental approaches to cybercrime and cyber security for its citizens, industries, and public services (law enforcement). On the other side, this approach is showing signs that governments are likely to use their territorial sovereignty as a tool to protect local jobs and local business. One argument is that data localization, may in fact, reduce demand for foreign exchange because local businesses and Internet users would not have to pay foreign companies to host their data offshore.[13] However, John Selby points out that, to date, it is unclear whether (at least in the short term) this would outweigh the increase in imports necessary to build local data centers. He goes onto say that a weakness in this argument is the issue of whether the local country has the technical capability and infrastructure needed to reliably and successfully operate the local data center(s). Countries with inadequate or unreliable power networks, or those that experience hot summer months, bad weather or earthquakes, might face significant hurdles in avoiding significant downtime for their local data centers.[14] Selby concludes that, if local businesses and Internet users have no other options for their data hosting, such infrastructure failures could significantly hurt the local Internet-based economy. Such an approach, could be viewed as stifling innovation and impacting on other areas of the economy and law, such as competition and intellectual property.

Data protection, privacy and competition law all seek to achieve a similar result – they protect the individual (either as a consumer or individual person), albeit in different ways and from different perspective. Although there appears to be significant overlap between the objectives underlying these laws, arguably they can converge and should be harmonized in the future, to provide greater accountability to businesses, no matter in what jurisdiction they are located. This approach will ensure competition in the new digital economy. As highlighted in a earlier chapter the systems and platforms that support the collection of personal data do, in directly create anti competitive behavior. Data localization ca, in part, create the same environment. This is a complex issue for governments to resolve. On the one hand they want to enable economic activity. However, on the other hand, they want to protect the commercial and personal data within their territory.

---

[13] Selby, J *Data localization laws: trade barriers or legitimate responses to cybersecurity risks, or both?* International Journal of Law and Information Technology, 2017, 25, 213–232.

[14] Ibid.

17.8 Storage Limitation

Storage limitations has also become an important obstacle to the overall data protection and privacy framework. It raises more questions than solutions about the globalisation of technology. It also strengthens the sovereign autonomy of nation states over their citizen's personal data. Arguably, this approach is something that citizens of individual countries would expect. In other words, citizens do expect that their governments would protect their personal data. Nothing stops an entity from collecting personal data in one country and storing it in another country for an extended period of time– where there is no storage limitation. This will become even more evident when the laws differ greatly, to the extent that organisations will move their operation to another country where data protection laws are limited, to exploit their business opportunities. It is argued that more work is needed to address these regulatory gaps. Thus, there is a well-founded argument for the need to harmonise data protection and privacy laws.

17.9 Consent

Throughout the book, it has been demonstrated that the concept of consent has arguably emerged as key to strengthening the interrelationship between data protection and privacy law. Consent along with the definition of personal data has arguably become increasingly important where data protection law transcends other areas of law, such as intellectual property, competition, transnational contracts and the right to erasure. It is our view that consent has a dual role. On the one hand, it provides a right to data subjects to control the collection and use of their personal data. Consent, as the underpinning of self-determination, has evolved into a standalone right whose importance is prescribed by law, which also demonstrates how the right to privacy was conceived.[15] On the other hand, consent has an indirect impact on the economic use of personal data, and can either enhance economic activity or restrict it, particularly as economic activity related to personal data increases. In jurisdictions where consent is required by law, the resulting effect is the tradability of that data for economic benefit. However, there are concerns that the concept of consent is not fully functional and in part, is adding to the ambiguity as to what level of control data subjects have over their personal data.

The OECD regards 'consent' as an important part of collecting and processing personal data.[16] However, the level and extent of consent required by the data subject differs from jurisdictions to jurisdiction. Consent can come in three forms

---

[15] Blume P *Data Protection and Privacy – Basic Concepts in a Changing World,* Scandinavian Studies In Law (1999–2015), http://www.scandinavianlaw.se/pdf/56-7.pdf, accessed 16 October 2018.

[16] Organization for the Economic Co-operation and Development, Guidelines on the Protection of Privacy and Transborder Flows of Personal Data 2013. http://www.oecd.org/sti/ieconomy/oecd-guidelinesontheprotectionofprivacyandtransborderflowsofpersonaldata.htm, accessed 20 February 2018.

1). consent from an adult, 2). consent from children and 3). consent by notice. The OECD has left it up to nation states to determine how consent will operate. For instance, states determine whether consent constitutes actual or implied consent, or both. The ASEAN data protection framework has also placed the concept of consent at the forefront of its regulatory framework, to ensure member countries adopt the concept within national laws (see Chap. 16).

The tension in the law between consent and other areas of society and the economy, has yet to be fully realized. That tension will continue to evolve and expose themselves, as other areas of law require that it be resolved. A good example of this is the emergence of personal data being afforded property rights, and that data is being used to create a dominant position in the market. As already highlighted, organizations seeking to avoid the adoption and application of consent, can relocate their operations to other countries where the laws are weaker or even nonexistent. Therefore, based on this issue alone, it can be argued that the harmonization and convergence of data protection and privacy laws, would go a long way to ensuring that the higher level principles outlined earlier, of accountability, transparency, trust, certainty, ownership and control over one's personal data, is not diminished as the digital economy and technology evolves. The converse argument to this position is where states view the concept as restricting innovation, business opportunity and economic activity. Should this be the case, the tension and challenges between the laws of nation states will only increase, while the benefit of legal harmonization will be diminished.

17.10 Definition of Personal Data and Personal Information

The definition of personal data and the concept of consent go hand in hand, as they have to co- exist. There is both general and sensitive personal data, however, as highlighted earlier, some jurisdictions have defined sensitive data, and others have not done so. This, in itself, poses many challenges, not only for individuals but also for organizations that deal with personal data across national borders. This is certainly an area that needs further work. Furthermore, there is little guidance or jurisprudence from the courts on this issue. Scant information or clarification has been provided by the courts confirming what is personal data, how it is defined by law, and what other areas could constitute personal data. Arguably, the expansion in personal data highlights problems arising from the as yet unsettled definition of personal information and data. What could be of greater concern is that the current approach will not meet the needs of future technology or the potential requests for clarification from industry.

For instance in early 2018, Facebook had approached the major banks in the United States[17] to develop a partnership whereby the personal information held by

[17] Glazer E, Seetharaman D, Andri A, *Facebook to Banks: Give Us Your Data,* We'll Give You Our Users Facebook has asked large U.S. banks to share detailed financial information about customers as it seeks to boost user engagement, Wall Street Journal, https://www.wsj.com/articles/facebook-to-banks-give-us-your-data-well-give-you-our-users-1533564049, accessed 12 August 2018. Facebook increasingly wants to be a platform where people buy and sell goods and services,

banks could be integrated with Facebook Messenger platform. The Wall Street Journal believe that, if achieved, it would provide Facebook with access to some of the most important and sensitive personal data and financial information of individuals and entities.[18] While Facebook has tried to sell the proposal as benefiting banks, particularly in regard to fraud and money laundering, the Banks, in turn, have raised serious concerns about privacy.[19] In this context it can be observed, that the personal financial information of individuals may not be fully covered by the definitions adopted by the data protection and privacy laws. Arguably, vigilance and monitoring of data usage is needed, as technology rapidly advances, and of these types of organizations to ensure the current laws provide and protect data subjects' control over this information, and more broadly other data, which could be used to generate high levels of wealth at the expense of those subjects.

Importantly, and despite the varied definitions of "sensitive data", much of this information can be tradable, transported and portable from one jurisdiction to another (see country Chapters). To do so under the current legal framework, requires the consent of the data subject (individual), at least, to the point when the data controller or processor is in full control of that personal data. Therefore, further research is needed to, not only harmonize what different countries define as personal data and information, but also, how regions of the world define such information. The purpose is to promote convergence in the law and to render them far more effective, efficient and fairer.

17.10.1 Ownership

There is some conjecture as to whether personal data is owned by the data subject. That is, does a data subject have a level of ownership of the personal data or information, which has been defined by law? In Chaps. 13, 14 and 15, it has been argued that, today, a level of ownership is being provided to data subjects to their personal data, as defined by the law.

The concept of ownership of data in general and personal information in particular is considered complex and not settled. In 2015, the OECD looked at this issue, questioning whether the concept of ownership would or could exist, given that data typically involves complex assignments of different rights across different stakeholders. The OECD also identified that ownership of personal data affords some stakeholders the ability to access, create, modify, package, derive benefits from, sell or remove data, but also the right to assign these privileges to others.[20] The

besides connecting with friends. The company over the past year asked JPMorgan Chase JPM & Co., Wells Fargo & Co., Citigroup Inc. C and U.S. Bancorp USB to discuss potential offerings it could host for bank customers on Facebook Messenger, people familiar with the matter said.

[18] Ibid.

[19] Ibid.

[20] Organization on the Economic Cooperation and Development *Data driven innovation: Big data for growth and well-being* (2015), OECD Publishing, Paris, available at: http://www.oecd.org/sti/data-driven-innovation-9789264229358-en.htm, accessed 15 October 2018.

OECD further notes that the situation is even more complex in the case of personal data, where certain (non-proprietary) rights of the data subject cannot be waived.

Notwithstanding the OECD's position, it was argued that recent case law discussed in Chaps. 13 and 15, points to far greater control and ownership of personal data by a data subject. Moreover, the GDPR has arguably reinforced a level of ownership provided to data subjects, of the personal data that has been defined by the law. For instance, throughout this book, there are a number of rights provided by law that strengthen the ownership and control of personal data. These include, but are not limited to, the right to be forgotten (deletion of personal information); the right to access one's personal data; and the right to correct one's personal data. Furthermore, the ability for an individual to provide (actual or implied) and withdraw their consent to an organization for the collection, use, storage and dissemination of personal data, reinforces the earlier point that personal data has a level of property right. This proposition has been further advanced with the recent introduction of rules around data portability, arguably, which strengthen the concept of ownership of personal data for data subjects-.

To a lesser extent, it is our view that the ownership of personal data has been further strengthened with the introduction and need for organizations to provide dedicated points of contact. This, to varying degrees, can be seen with the establishment of Data Controllers, Processors or Business Operators. It is these individuals within an organization that are responsible for managing consent, and the deletion or removal of personal data. They are also responsible for the transfer of personal data between organizations. The ownership of personal data can also be ascertained, indirectly, in those jurisdictions that have established a complaints mechanism, allowing the data subject to inform an authority, such as Commission or Commissioner or both, of an alleged breach, or mishandling of personal data.

Finally, what can be determined, at this stage, is that jurisdictions continue to develop law, to varying degrees, that enable data subjects to decide on how and where their personal data can be used. Therefore, it appears that the baseline has been set, whereby it is a matter of time before ownership of personal data, by a data subject will become more secure. Furthermore, it is our view this will only be strengthened as courts around the world take a similar approach to that which has been developed in the United Kingdom (see Chaps. 13 and 15). However, this does not guarantee that the courts will reinforce, let alone strengthen the idea that people have, or will continue to have, some ownership of their data.

17.11 Adequacy

Arguably, one of the most influential elements of the EU GDPR is Article 45, which allows the EU to determine whether the data protection or privacy laws of third countries (countries outside of the EU) provide adequate protection (see Chap. 3). It is a form of equivalence, because of the process by which the EU expects other countries to implement similar laws. To date, there is a limited number of third countries that have received the adequacy protection tick from the EU, as highlighted in Chap. 3. These countries include Andorra, Argentina, Canada, Faroe Islands, Guernsey, Israel, Isle of Man, Japan, Jersey, New Zealand, Switzerland, Uruguay and the United States. Notably, only one of the countries (Japan) discussed in this book has obtained the adequacy tick of approval. Other countries have also begun to adopt a similar or reciprocal approach through their legal frameworks. However, this does not mean that their laws are harmonized. The extent of reciprocity only seeks to achieve that level of comfort that can be used to allay the concerns of individual data subjects that similar controls are in place across these jurisdictions. This is certainly a step in the right direction, even though it is limited to particular personal data and not data used in areas of law enforcement.[21] It is our view, that to a limited extent, this reciprocal practice is forcing at least some minimal level of legal convergence. More work can be undertaken in this area, to promote and obtain harmonization of data protection and privacy law. Nonetheless, the question arises, do countries actually want to obtain this adequacy tick/approval? Some countries may decide that it is against their national (economic and social) interest to do so.

17.12 Measuring the Harm in Data Breaches

Another complex area that has surfaced from comparing the data protection and privacy laws from different jurisdiction is: what methodology has been used to measure harm from a breach of the law? In other words, how has, or can, the level of harm be measured when an individual's personal data and information, which is defined by the law, has been used unlawfully. The book contends that this probably constitutes one of the most pressing and complex questions throughout the lifecycle of data protection and privacy.

17.12.1 What Is a Privacy Harm?

Therefore, the particular question: What is a privacy harm, and what is a data breach? The conceptualization of a privacy harm can be best described as "cognizable," "actual," "specific," "material," "fundamental," or "special" harm before a court will consider awarding compensation.[22] Ryan argues that the subjective and objective categories of privacy and data protection harm are distinct but not entirely separate. As an example, assault and battery are two distinct and well known torts in common law jurisdictions.[23] That is, each can occur without the other, and have very different

[21] European Commission, Adequacy of protection of personal data in non-EU countries, https://ec.europa.eu/info/law/law-topic/data-protection/data-transfers-outside-eu/adequacy-protection-personal-data-non-eu-countries_en, accessed 14 August 2018.

[22] Ryan, C *eBoundaries of Privacy Harm,* Indiana Law Journal: Vol. 86: Iss. 3, (2011) Article 8.

[23] Ibid.

elements. Both do involve the physical person. Arguably, they are also linked because one is the apprehension of the other. Ryan further argues that the harm of assault is an internal or subjective state, specifically, the apprehension of unwanted touching.[24] On the other hand, the harm of battery is the unwanted physical contact itself.

Moreover, the two components of privacy and personal data harm are interrelated. The objective privacy harm is the actual adverse consequence of the theft of identity (personal data defined by the law) itself or the formation of a negative opinion which flows from the loss of control over information or sensory access to it.[25] The subjective privacy harm is the perception of loss of control that results in fear or discomfort.[26] The two categories are distinct but somewhat related. They constitute and result in the same harm – that is, the loss of control over personal information. Yet, this does not provide a clear view of what is harm, the level of that harm, and how that harm has impacted the data subject – when related to the Internet. These uncertainties are accentuated because understanding the level of harm relates to persons who, firstly, would not know their personal data has or is being used illegally, and secondly, to persons who are even unaware that a breach has occurred until well after the fact. It is understood today that, in general, measuring the level of harm from the result of a physical or mental impairment can be undertaken by a suitably qualified practitioner today. It is also true that the harm arising from data or privacy breaches can be detected on Internet platforms, including by professionals who are able to detect both the nature of the breach and its likely impact upon, and to, data subjects. Such detection is aided, for example in judicial decisions in Australia, by the codification of criteria by which to identify and measure such breaches.

However, understanding and measuring the harm of a data or privacy breach over the Internet is arguably far more subjective; and, the traditional statutory and judicial elucidation of personal harm and damage. Privacy breaches to a person's property also not only limited in general in most of the jurisdictions studied: they are even more limited in relation to such breaches over the Internet.

Furthermore, subjective privacy harm (s) over the Internet can also be identified where human beings are not physically able to review personal information in order to determine its adverse impact on them, and how to remedy it. There does not have to be a human observer who gathers and identifies the misuse of personal data and engages in ameliorating action. Danielle Citron explains that:

> In the past, computer systems helped humans apply rules to individual cases. Now, automated systems have become the primary decision makers. These systems often take human decision making out of the process of terminating individuals' Medicaid, food stamp, and other welfare benefits. Computer programs identify parents believed to owe child support and instruct state agencies to file collection proceedings against those individuals. Voters are purged from the rolls without notice, and small businesses are deemed ineligible for federal contracts.[27]

[24] Ibid.

[25] Ibid.

[26] Ibid.

[27] Citron, D *Technological Due Process*, 85 WASH. U. L. REV. 1249, (2008) p. 1252–1254.

Citron explores the harm of automated decision making from the perspective of due process.[28] However, such automated decisions can also constitute privacy harms where, as often, they involve the unanticipated or coerced use of sensitive[29] personal data and information that has been defined by the law.

Richard Posner believes that no privacy harm occurs unless and until a human sees the data or information at issue.[30] In many respects Posner is correct, because as already highlighted, data subjects will only be aware of the harm when (a) there has been a breach of the law and (b) they themselves are informed that their personal data has been illegally obtained or used or (c) those data subjects come across information which confirms their personal data is being misused. To minimize the harm to data subjects, a number of current day laws have implemented controls over the data, such as consent, the definition of personal data, and controls over the transfer of personal data by controllers who are responsible for that data. However, as highlighted throughout the book, the concepts, principles and the laws are far from being consistent or settled across the countries under study.

Nevertheless, it is our view that those jurisdictions which have established specific data protection and privacy laws are indirectly codifying the potential level of harm, by imposing penalties for the illegal use of data. Furthermore, the EU, UK and Canada have begun to identify key concepts and principles both by the courts and statute to help guide the measure of harm from the misuse of personal data.[31] However, it is out of scope of this chapter to comprehensively discuss this area of the law. That is, the level of penalties imposed by jurisdictions are significant when compared to many other areas of the law.

17.12.2 Penalties & Enforcement

Enforcement and fines (penalties) are intertwined, making it useful to identify variations in fines in different countries, in order to determine, for example, whether fines are perceived as being too small or alternatively, adequate. The approach to enforcement is vast and varied, depending on the region and country being examined. There are signs that governments are taking the illegal use of data more seriously in recent times, such as in the EU, Australia and Singapore. Even in jurisdictions that have raised the level of fines that can be imposed on large organizations by their respective laws, it is questionable whether these go far enough, given the size of profits these companies make. What has developed is how the fines are imposed by a regulator rather than the court. In most, if not all situations over the past 2 years it has been the regulators that have imposed the fine on an organization for the misuse of personal data. Furthermore, not only are

[28] Ibid.

[29] Ibid.

[30] Posner, R *Privacy, Surveillance, and Law*, 75 U. CHI. L. REV. 245, 251 (2008).

[31] Leon Trakman, Robert Walters, Bruno Zeller, *Tort in Data Protection Law - are there any lessons to be learnt?* forthcoming.

corporate profit margins often in the millions, if not, billions, such data users usually have the financial resources even after a huge fine is imposed, to appeal the fine through the applicable judicial system. The question going forward is whether countries or possibly regions of the world will adopt comparable approaches to the severity and level of penalty to be imposed on organizations that are in breach of the law.

As highlighted in previous Chapters, in Australia, the *Privacy Act 1988* provides that a corporation can be fined up to AU$420,000, for a serious or repeated interference with privacy (see Chap. 5). In addition, a corporate entity can be penalized up to AU$2.1 million, where a privacy interference had been proved. Singapore, similarly provides that regulatory bodies and courts can impose fines of up to SG$1 million for non-compliance with any the data protection laws (see Chap. 4). On the other hand, the EU appears to have set the harshest level of penalties when compared with the Asia-Pacific Region. There, a company can be fined up to €10 million, or 2% of the worldwide annual revenue of its prior financial year. A higher level of penalty can be imposed of up to €20 million, or 4% of its worldwide annual revenue in the prior financial year, whichever is higher. The criminal sanctions imposed in Japan are both imprisonment and a modest fine of YEN 500,000, equivalent to US$6200. In Malaysia, a fine of up to RM 20,000 can be imposed for breach of the PDPA.

Arguably, these penalties, other than in the EU, do not fully reflect the revenue and profit that can be made by entities, even though the penalties can be criminal or civil, or a combination of both and can be imposed by courts and sometimes, regulatory authority. Even though the level of penalties varies significantly, the EU is leading the push for significant fines to be imposed on individuals and particularly entities. It is our view that, by jurisdictions imposing such significant high level fines, the EU recognizes that there is an implied level of harm arising from the illegal use of personal data. It also highlights that the EU, in particular, understand that this sector has a high level of profitability.

Singapore's privacy watchdog has fined 22 organizations – one of them twice – a total of SG$216,500 over the past 2 years for security breaches that have exposed the personal details of Singaporeans. Another 19 organizations have been censured for their data breaches and shortcomings.[32] The numbers compiled by The Straits Times give the clearest indication yet of how deep the problem of securing personal data runs. It has also raised concerns among experts that organizations are still not taking this issue seriously. Of particular worry is the fact that nearly every fine issued by the PDPC centered around the same type of offence, namely, inadequate security measures for personal data. Experts say this points to a lack of understanding over how data laws apply to daily operations, even more than 3 years after the Personal Data Protection Act was fully enforced in July 2014.

In 2018, the European Union regulators imposed on Google a record $6.85 billion antitrust fine for using its Android mobile operating system to squeeze out rivals.[33] This fine arose from Google's anti-competitive practices relating to its

[32] The Strait Times, Privacy watchdog fines 22 in past 2 years over security breaches http://www.straitstimes.com/tech/privacy-watchdog-fines-22-in-past-two-years-over-breaches

[33] Google fined a record US $6.8 billion over Android mobile systems, http://www.abc.net.au/news/2018-07-19/eu-fines-google-a-record-6.8-billion-over-android-mobile-system/10010510, accessed 5 August 2018.

online shopping search services.[34] The penalty is nearly double the previous record of $3.7 billion which the United States tech company was ordered to pay last year over these services.[35] An important observation in all of this is that the fine only represents just over 2 weeks of revenue for Google parent Alphabet Inc. and would scarcely dent its cash reserves of almost $140 billion.[36] Indeed, the tech giant reportedly has enough cash reserves of almost $140 billion to appeal fines imposed on it.[37] It is most likely to appeal this fine.

More recently, the scandal has exposed Facebook's disregard of data protection laws is likely to lead to many different and varied responses taken by individuals, entities and government, depending on the country and region of the world. For instance, in Australia companies and individuals are gearing up to take class actions against Facebook over Cambridge Analytica.[38] In Europe and America, governments have been actively involved in undertaking their own investigations to better understand the privacy and data protection breaches.

The broader public policy issue for governments and these multinational organizations is not so much the fine or enforcement action that is taken, but the indirect impact that the public will impose on them. That is, these and other unreported breaches erode public trust in government policy and regulation as not being able to stop this corporate misuse and abuse of personal data. It also erodes confidence and trust in the technology industry and organizations that collect and store such data.

It is our view that the level of harm, penalties and enforcement are yet another piece of the data protection and privacy jigsaw puzzle that is unlikely to be settled any time soon. Moreover, the fluid nature of data protection and privacy law, along with developments and changes in technology are only going to challenge and further complicate how the level of harm, penalties and enforcement (a) can and will be measured, (b) developed into concrete policy objectives, and (c) implemented – when this is a global issue. Thus, this proposition reinforces and strengthens our argument for the need to obtain greater legal convergence and harmonization of data protection and privacy law.

17.13 Pathway Forward

What is the pathway forward for data protection and privacy law? Data protection and privacy law is not limited to any single nation state or region of the world. The legal framework encompasses a multilayered approach, which makes it very complex to develop a single pathway forward. This section does not attempt to provide

[34] Ibid.

[35] Ibid.

[36] Ibid.

[37] Ibid.

[38] Facebook staring at Australian Class Action, https://www.itnews.com.au/news/facebook-staring-at-australian-class-action-497592, accessed 8 August 2018.

all the answers, as there is no single solution or silver bullet that will address all the diverse issues associated with this area of law. Arguably, there are significant global problems in policy, regulatory and legal issues that will require a global, or at least a regional response.

One of the major obstacles is the varied levels of data protection and privacy laws that exist, because of the cultural, religious and economic differences across regions and countries. In addition, government policy also plays a part in deciding the level of data protection accorded. What can be said is that most countries have recognized that there is a problem related to personal data and privacy over the Internet. The proposal put forward in this Chapter may be viewed by some as not being compatible with existing data protection and privacy law. Nevertheless, it is argued that there needs to be a starting point that goes beyond the current framework which the OECD has established to protect personal data and privacy. The proposal below will go some way to assist in dealing with the operational issues, and will add a layer of protection based on existing legal concepts, principles and guidelines. At a minimum, the objective is to stimulate thought and further discussion between countries and the global community.

Firstly, as already highlighted above, the core principles that have been set by the OECD need to be reviewed to ensure they are still adequate. The starting point is to understand what other principles and concepts have emerged from national laws that need to be placed into the international arena. Arguably, to overcome the hurdles posed by the lack of harmonization, it could be left to UNIDROIT or UNCITRAL to develop Model laws. Two reasons are persuasive. Firstly, the two organizations are devising instruments though diplomatic conferences. Secondly, they are practiced in devising conventions and model laws for transnational commercial law. These two observations are elucidated below.

The UNCITRAL or UNIDROIT model and framework have been very successful and continue to pave the way in many area of commercial law since WWII.[39] Many countries across Europe, North America and the Asia Pacific have adopted this framework into their national laws.[40] The most recent addition is the Model Law on Electronic Transferable Records (MLETR).[41] Today, personal data is a tradable commodity (see Chaps. 13, 14, 15). This aims to enable the legal use of electronic transferable records both domestically and across borders, and applies to electronic transferable records that are functionally equivalent to transferable documents or instruments.[42] The connection is that data used in these transmissions is likely to include personal data and information on individuals, even though that data may not be covered under the definition of the current day data protection and privacy laws. However, it does not exclude that this personal data could not be included.

[39] United Nations Commission on International Trade Law, http://www.uncitral.org/uncitral/en/uncitral_texts/electronic_commerce/2017model.html, accessed 9 August 2018.

[40] Ibid.

[41] Ibid.

[42] Ibid.

A Model Law would not only provide the bases for setting a standardized approach to defining personal data.[43] It could also extend to other areas such as consent, including withdrawing consent, transparency, accountability, data portability, transfer of data to third countries, data retention (particularly forcing entities to delete specific data once a person has died, within a specific timeframe) and data localization. In addition, it would provide greater clarity and certainty for the business community, providing a basis for managing personal data in relation to anti-competitive behavior, transnational contracts and intellectual property.

Chapter 1 of the book has highlighted the significant differences in the way regions of the world operate and develop their respective laws, including but not limited to data protection and privacy. It is well understood that the EU imposes its laws upon all member states. Thus, nation states can continue to rely on the EU to set the entire policy and legal direction. The advantage is that most, if not all EU states, are represented in the UNCITRAL and hence that organisation can profit form the knowledge of the EU in devising a model law.[44]

The alternative approach is for greater collaboration across nation states in general. As Graham Greenleaf highlights, countries that have adopted data protection laws are mostly based on European standards.[45] However, ASEAN operates in a very different way, by seeking to progress the values, economic and social development of member countries under a consensus approach. Arguably, ASEAN as a supranational polity is not as formally advanced as the EU because ASEAN member states have not transferred any of their sovereignty to ASEAN itself. However, there continues to be work undertaken through ASEAN and its affiliate organizations to, not only obtain consensus on many issues, but also to harmonize areas of law that mutually benefit the member states and their citizens. Thus, there is the opportunity for countries, including the EU, to build on their current programs, and possibly to follow the model of ASEAN to harmonize this area of law. One possible drawback from the consensus approach is that it could be perceived as being too slow. A response to this drawback is for reform to include influential organizations such as the ASEAN Law Association or Asian Business Law Institute, and with country partners like Australia, to combine their collective research on how best to achieve greater legal convergence and harmonization.

The level of fines available in cases in which there have been extensive breaches of the law, if indeed, such breaches are detected at all or fully, are simply inadequate. The reality is that fines often do act as a wholly effective deterrent because the profit incentives for breach far exceed the level of penalty imposed. Even though there will be both corporate and some regulatory opposition to increasing the level of fines, jurisdictions will need to take this into consideration. While the GDPR has raised the amount an entity can be fined, a concerted and harmonized approach would act as a greater deterrence. However, this is a difficult area to harmonize as a

---

[43] Ibid.

[44] Ibid.

[45] Greenleaf G *Global Analysis of Data Privacy Laws and Bills* Privacy Law and Business International Report 145: (2017) pp. 14–26.

fine in Thailand might be exorbitant, whereas the same fine in Australia might be too lenient. However, a model law will overcome this issue as the fines can be leveled on a national, not a supranational basis.

Trade Agreements offer another source of regulation, operating outside the private sphere. They, too, should be more focused on strengthening the governance of data protection and privacy. Chapter 16 highlighted some of the work the United States and other countries have begun to include in their trade agreements in protecting data and privacy across signatory states.

Programs such as the Adequacy test established by the EU, also goes someway to harmonizing the data protection and privacy laws. However, it does not go far enough to harmonise the law, but rather, only recognises that inadequacies of controls. Further consideration could be given to expanding such a program, not necessarily to force other countries to follow or even adopt the EU model, but to seek greater convergence and harmonization. The drawback is that states may not favor the reforming their regulatory frameworks, or may not seek to gain adequacy recognition from the EU, let alone participate in a broader program that can synchronize this area of law. They may also be wary of trying to balance the need not to stifle innovation with the virtue of protecting peoples' personal data and privacy. Engaging in that balance, is complex and often very difficult because innovation is becoming dependent on large scale trade in personal data.

Another option available to resolve these ongoing gaps, variables and tensions, is by having recourse to the courts. The courts within the jurisdictions discussed throughout the book are all starting to consider the broader issues related to data protection and privacy. However, this approach will not resolve the need for legal harmonisation and convergence in the respective laws because most national courts will hand down decisions based on local needs, rather than regional or international needs. Exceptions to this have been the Court of Justice of the European Commission and the European Court of Human Rights that have developed jurisprudence, which has not only seen EU member states adopt those decisions, but other nation states have looked of the EU for guidance and direction. Another drawback is the fact that harmonization through the courts takes a long time to come to fruition.

The problem is that nation states continue to be reactive to, not only the law, but also to societal needs that require intervention. This is an issue that is common across jurisdictions and without regard to the area of law and policy in issue. In other words, states can continue to sit back and rely on, or, wait for the EU to set the entire policy and legal direction. Or, other approaches and mechanisms can be developed to improve upon the law. What those approaches and mechanisms might be, are vast and varied as is highlighted in this chapter. However, a Model Law has to be one option. Graham Greenleaf reinforces this important point, and argues that approximately 120 countries have adopted, or are considering adopting, some form of data protection or privacy legislation, which are mostly based on European standards.[46] Therefore, the legal convergence and harmonization of these laws is well

---

[46] Greenleaf G *Global Analysis of Data Privacy Laws and Bills* Privacy Law and Business International Report 145: (2017) pp. 14–24.

underway. However, such legal innovation is far from consistent and often falls well short in satisfying measures of equivalency.

17.14 Conclusion

Australia, Indonesia, India, Malaysia, Japan, Singapore and Thailand are currently in a new phase of investigation into the further regulation that is needed in this area, notably following the introduction of the GDPR in 2018. Some countries are far more advanced than others in regard to such regulation. Arguably, Indonesia, India and Thailand being the least advanced in this area of law and face the greatest challenge as to their respective pathway forwards. There are promising signs from these countries that are not only looking inwards, but also looking outward to the EU and other countries like Australia and Singapore, for guidance. However, there are many questions related to the future needs of data protection and privacy law. That is, how the next generation perceived their personal data being used and traded? Whether the community disregard the use of the internet in exchange for a loss of privacy and misuse of their personal data? Thus, more needs to be also done to better understand community perceptions of data protection across all countries and not just a few.

However, it remains to be seen whether the countries modelled in this book are willing to harmonize their respective approaches in the law – over the longer term. The more pressing issue is the need to close gap between the policy objectives of competition and privacy. More importantly, is how nation states balance the needs of innovation (technology), economic activity from the trade in personal data and the community expectation that, their personal data will be protected – to some level. This alone is going to challenge governments, regulators and policy makers. There are many unanswered questions because the area continues to evolve, change and overlap. More transparency and vigilance is needed by regulators and the community as to what is actually happening behind the veil of the computer screen, software, platforms, hardware and systems, to name a few, to better understand the gaps in these systems. There is also greater need to expose personal data and privacy beyond the current regulatory dilemma in which only a few control the activity or the right.

Arguably, a consensus approach is needed to harmonize key definitions, concepts and principles with the law so as to strengthen the governance of data protection and privacy. It was highlighted earlier that one approach to overcome some of the hurdles to harmonization is for countries and the EU to work together and develop a Model Law in conjunction with UNCITRAL or some other institution. The convergence of these laws is more likely than not to improve the balance between market forces, innovation and protecting people's privacy. Converging and harmonizing laws will also assist in measuring the harm and risks to the community from data breaches. It will strengthen trust in the Internet economy, which is vital for future trade. Legal harmonization will also provide a consistent approach to penalties and enforcement action to deter the illegal use of personal data. The question arises, whether the international community see these as issues that require redress? The alternative approach for the international community is to maintain the status quo, and wait for jurisdictions such as the EU to force the hand of others to adopt similar laws.

In summary, the jury is out regarding the direction of where privacy, personal data and the legal concepts and principles that form part of these laws is at, and at what stage they will be settled. The questions are to determine the appropriate body or forum for this to take place, and how such action is to be conducted. While answers to these questions would not be the silver bullet that would resolve all of the vexing issues, they would go some way to standardizing the law and dealing with a growing international problem. What can be confirmed is the continued ad hoc and fragmented approach will only benefit those governments and business entities that operate in jurisdiction (s) where there is no, or immature, laws. The lack of knowledge by data subjects of what is occurring behind the computer screen, is a formidable concern. However, a telling response is whether this lack of knowledge is due to people's ignorance, in not wanting to know, or more to a lack of transparency across the sector. Public awareness of the Internet and data protection and privacy is being understood across all the jurisdictions discussed. However, further work is needed to inform the broader community of what the risks and impacts might be of deficient and inefficient regulatory patterns.

A final thought is that the internationalization of technology has resulted in nation states developing, adopting and applying broad conceptions of data protection and privacy laws to address local economic and social needs. What model will meet the needs of the international community in the future? Currently, many countries have followed the EU's direction, which tells us they are focusing on data protection as a tool to protect privacy online, grounded in a distinctly human rights approach. If that is the case, and people around the world are increasingly viewing privacy over the Internet as an issue, the EU model is likely to prevail. This is particularly so when these legal innovations are coupled with other areas of the law, such as competition, intellectual property, transnational contract law and cybercrime-security law. Moreover, other challenges that are also emerging, but not limited to, data protection in Artificial Intelligence and Quantum technologies. However, should data subjects become accepting and possibly indifferent to the misuse of their personal data, then another model may well prevail. For instance, Singapore's business friendly model, or even Australia's balanced model that sits somewhere between the two could emerge as future benchmarks. Should the direction change, policy makers, government and regulators will be further challenged.

References

Bourke A *How Safe are Blockchains?* It Depends, *Harvard Business Review*, (2017).
Blume P *Data Protection and Privacy – Basic Concepts in a Changing World,* Scandinavian Studies In Law (1999–2015), http://www.scandinavianlaw.se/pdf/56-7.pdf, accessed 16 October 2018.
Greenleaf G *Global Analysis of Data Privacy Laws and Bills* Privacy Law and Business International Report 145: (2017) pp. 14–26
Hofman, D., Duranti, L., How, E *Trust in the Balance: Data Protection Laws as Tools for Privacy and Security in the Cloud Algorithms* MDPI (2017)
Posner, R *Privacy, Surveillance, and Law*, 75 U. CHI. L. REV. 245, 251 (2008)
Reijers W., O'Brolcháin F, Haynes P, *Governance in Blockchain Technologies & Social Contract Theories*, 1 LEDGER 134, (2016)
Ryan, C *eBoundaries of Privacy Harm,* Indiana Law Journal: Vol. 86: Iss. 3, (2011)
Selby, J *Data localization laws: trade barriers or legitimate responses to cybersecurity risks, or both?* International Journal of Law and Information Technology, 2017, 25, 213–232
Zetzche., D, Buckley., R, Wagner, D *The Distributed Liability of Distributed Ledgers: Legal Risks of Blockchain,* UNSWLRS 52, (2017), pp. 3–8

CPSIA information can be obtained
at www.ICGtesting.com
Printed in the USA
LVHW081552271220
675120LV00009B/343

Data Protection Law